Advances in Experimental Medicine and Biology

Volume 947

More information about this series at http://www.springer.com/series/5584

Lang Tran • Miguel A. Bañares • Robert Rallo
Editors

Modelling the Toxicity of Nanoparticles

Springer

Editors
Lang Tran
Institute of Occupational Medicine
Edinburgh
United Kingdom

Miguel A. Bañares
Institute of Catalysis & Petrochemistry, CSIC
Madrid
Spain

Robert Rallo
Universitat Rovira i Virgili
Departament d'Enginyeria Informatica i Matematiques
Tarragona
Spain

Frontispiece: Uschi M. Graham and Günter Oberdörster, 2016

ISSN 0065-2598   ISSN 2214-8019 (electronic)
Advances in Experimental Medicine and Biology
ISBN 978-3-319-47752-7   ISBN 978-3-319-47754-1 (eBook)
DOI 10.1007/978-3-319-47754-1

Library of Congress Control Number: 2016963324

Printed on acid-free paper

This Springer imprint is published by Springer Nature
The registered company is Springer International Publishing AG
The registered company address is: Gewerbestrasse 11, 6330 Cham, Switzerland

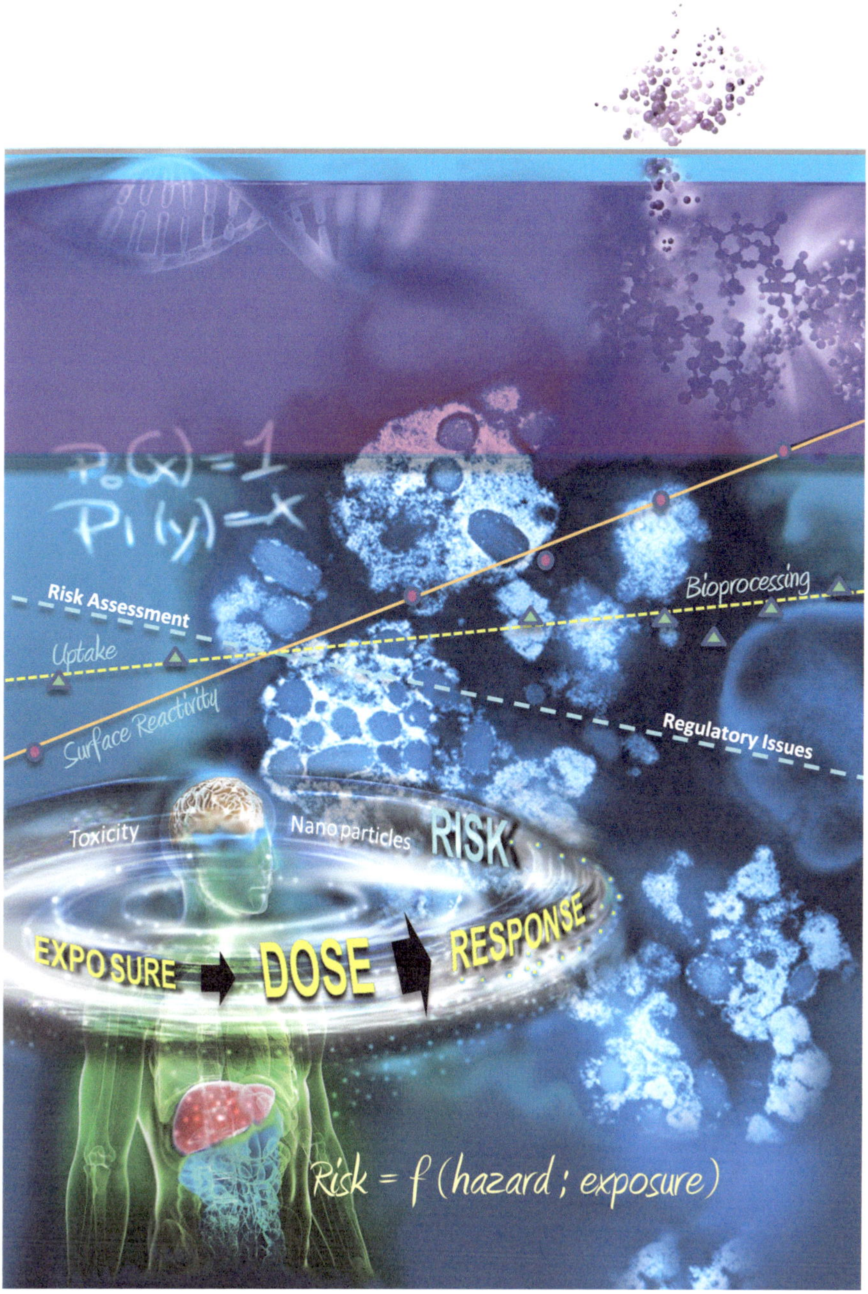
Risk Assessment
Uptake
Surface Reactivity
Bioprocessing
Regulatory Issues
Toxicity
Nano particles
RISK
EXPOSURE
DOSE
RESPONSE
Risk = f (hazard ; exposure)

Introduction

Nanotechnology is recognised as one of the most important new technologies of the twenty-first century. The global investment in nanotechnology from public sources for 2008 exceeded $7 billion. The market size for nanotechnology is expected to grow to over $64.2 billion by 2019 with over commercial 50,000 products containing engineered nanomaterials (ENMs).

Nanotechnology promises new materials for industrial applications by having new or enhanced physico-chemical properties that are different in comparison to their micron-sized counterparts. However, as in all industrial applications, the potential exposure of humans and the environment to these materials is inevitable. As these materials go through their life-cycle – from development, to manufacture, to consumer usage, to final disposal – different human groups (workers, bystanders, consumers), animal species (e.g. worm, fish or humans through secondary exposure) and environmental compartments (air, soil, sediment, ground and surface water) will be exposed to them. A growing body of evidence has shown a range of toxic effects from ENM suggesting that even their low mass exposure will result in a risk to human health or the environment. Furthermore, the toxicity of ENM can be attributed to some of their physico-chemical properties such as surface area, charge or reactivity. There is thus a clear need for a better understanding of the relationship between ENM structure and properties and the adverse responses they evoke in living organisms. Clearly, understanding this relationship will greatly help in designing future ENM by using the 'safe by design' approach.

The European Union (EU) has set ambitious plans for the future of nanotechnology. Accordingly, the regulatory agencies are urgently seeking ways of assessing the potential health risks caused by exposure to ENM. Available data are so far insufficient to meet this need. This is because research to determine impacts of ENM on diverse biological systems, although essential for assessing their hazard, is time-consuming and expensive, and has ethical implications when animals are used. *In silico* methods for predicting biological effects of ENM play an important, complementary role to that of experimental research. Due to the complexity of interactions of ENM with living organisms, and the increasing use of high-throughput screening methods to generate large in vitro datasets, statistical modelling (e.g. quantitative structure-activity relationships, QSAR and machine learning methods (e.g. neural networks) are becoming methods of choice. They have been applied successfully to the development of pharmaceuticals and crop protection agents over the past several decades.

QSAR methods are increasingly being used by regulatory agencies for chemical risk assessment. They have also been applied more recently to modelling the properties of materials, including ENM. Although QSAR techniques have only started to be used to predict biological effects of ENM, they have shown encouraging initial results. However, ENMs present significantly different obstacles to modelling compared to drugs and industrial chemicals, and specific models based on the structures and properties of ENMs are therefore needed. Issues pertinent to the development of computational methods for modelling ENM properties and their biological effects are central to this book, together with development in research that are required if the regulation of nanomaterials is to be assisted by computational tools within the next decade. Preliminary work demonstrates that these models show considerable promise for modelling ENM toxic effects. They are also useful for predicting toxic effects of new ENM based on their material properties and for classifying ENM according to common properties or common biological endpoints. Ultimately, the predictive power of these models will lead to considerable reduction in the use of animal experimentation.

In the frame of the FP7 program, the EC funded a total of six research projects and the COST Action TD1204 Modena, specifically devoted to various aspects of nanotoxicity modelling. All the projects worked in close coordination to develop new methodologies and tools suitable for transforming raw nanosafety data into new insight on the mechanisms governing nano-bio interactions. The most relevant outcomes of some of these modelling projects are presented either as chapters on modelling or as integrated case studies in the safety/hazard assessment of ENMs.

This book reflects on the co-operation between international scientists; the contributors cover a truly multidisciplinary spectrum of scientific disciplines including material sciences, metrology, exposure sciences, mammalian toxicology and ecotoxicology. In addition, transversal approaches entangle these disciplines providing the breadth and depth necessary to foster understanding the molecular rationale of toxicity of nanoparticles; these include computer science (databases, data mining) mathematics (data analysis ad modelling) and systems biology so that toxicity can be understood and tackled at any possible level, ranging from molecular to living organism.

Edinburgh, UK Lang Tran
Madrid, Spain Miguel A. Bañares
Tarragona, Spain Robert Rallo

Contents

About The Editors

Miguel A. Bañares (PhD in Chemistry, University of Salamanca, 1992) is Full Research Professor at ICP Institute for Catalysis, CSIC (Spanish National Research Council), where he joined in 1996 and Associate Editor for *Catalysis Today*. He was postdoctoral fellow at Lehigh University, PA, and at the University of Notre Dame, IN. Bañares has recently been distinguished by the "Otto Mønsted" Visiting Professorship at DTU University, Denmark (2014) and the FEDER-Region Base de Normandie, "chaire d'excellence" at CNRS Caen, France, during 2013–2015.

Bañares has been Deputy Vice-President of CSIC, Spanish National Research Council (2014–2015), chairman of the Management Committee of COST Action D36 (Molecular structure-performance relationships at the surface of functional materials) (2006–2011), and is Vice-Chairman of the Management Committee of COST Action TD1404 MODENA (Modelling Nanomaterial Toxicity), on toxicity of nanoparticles (2012–2016). He is co-founder of Advanced Dispersed Particles, a CSIC spin-off company, in 2011

Bañares' research focuses on understanding structure-property relationships at a molecular scale at the surface of functional materials for catalysis and toxicity by combining in situ and operando analyses with computational description of the structure, spectra and reactivity of functional materials. He has more than 190 publications in international journals and an "h" factor or 37 with more than 27 plenary/keynote lectures. Bañares' research stands on the use of real-time Raman spectroscopy during catalysis combining in situ spectroscopic characterisation of functional materials with simultaneous performance measurement. He coined the term "operando" to identify this advanced in situ spectroscopic methodology.

Dr. Lang Tran is the Principal Toxicologist at the Institute of Occupational Medicine. Lang has contributed considerably on the toxicology of inhaled particles and fibres. His early work was in the mathematical modelling of the retention and clearance of man-made mineral fibres and mineral dusts. He has led the investigation and the risk assessment of inhaled poorly soluble particles which helped NIOSH in establishing the control limit for fine and ultrafine TiO2 aerosol.

From his experience with particle toxicology, Lang has contributed also in many European Research Projects through the Framework 6 and 7 Programmes. Most notable are the project FP7 ENPRA (Risk Assessment of Engineered Nanoparticles) and FP7 MARINA (Managing Risks of Nanoparticles). Lang is currently the chairman of the COST Action TD1204 MODENA on modelling the toxicity of nanoparticles. Lang is an editor of the *Nanotoxicology* and *Particle and Fibre Toxicology* journals. He also contributes in the organisation of the Nanotoxicology Symposium series.

Dr. Robert Rallo is Associate Professor of Computer Science and Artificial Intelligence in the Departament d'Enginyeria Informàtica i Matemàtiques at Universitat Rovira i Virgili, Tarragona, Catalunya. He directs the Research Group on Bioinformatics & Computational Engineering (BIOCENIT, www.biocenit.cat) and is also the Director of the Advanced Technology Innovation Center (ATIC) at the Universitat Rovira i Virgili. In 2007 and 2009, he was a visiting professor at the University of California Los Angeles where he is also faculty member of the Center for Environmental Implications of Nanotechnology. Since 2013, he serves as Chair for the Modelling WG in the EU NanoSafety Cluster. From 2013 to 2015, he was the EU co-chair of the US-EU Nano-Dialogue Community of Research on Predictive Modelling and Human Health. His research focuses on multi-scale modelling of complex systems of industrial or environmental relevance. Dr. Rallo serves regularly as reviewer for high-ranked scientific journals and international research organisations.

Contributors

Villem Aruoja Laboratory of Environmental Toxicology, National Institute of Chemical Physics and Biophysics, Tallinn, Estonia

Rambabu Atluri National Research Centre for the Working Environment (NRCWE), Copenhagen, Denmark

Aggelos Avramopoulos Institute of Biology, Pharmaceutical Chemistry and Biotechnology, National Hellenic Research Foundation, Athens, Greece

Nazanin Golbamaki Bakhtyari Laboratory of Environmental Chemistry and Toxicology, Istituto di Ricerche Farmacologiche Mario Negri, Milan, Italy

Markus Bamler UFZ Department of Ecological Chemistry, Helmholtz Centre for Environmental Research, Leipzig, Germany

Institute for Organic Chemistry, Technical University Bergakademie Freiberg, Freiberg, Germany

Miguel A. Bañares Institute of Catalysis & Petrochemistry, CSIC, Madrid, Spain

Emilio Benfenati Laboratory of Environmental Chemistry and Toxicology, Istituto di Ricerche Farmacologiche Mario Negri, Milan, Italy

M. Eileen Birch CDC/NIOSH DART, Cincinnati, OH, USA

Erik G. Brandt Department of Materials and Environmental Chemistry, Stockholm University, Stockholm, Sweden

Martin Brehm UFZ Department of Ecological Chemistry, Helmholtz Centre for Environmental Research, Leipzig, Germany

Jaanus Burk Institute of Chemistry, University of Tartu, Tartu, Estonia

Peeter Burk Institute of Chemistry, University of Tartu, Tartu, Estonia

Ran Chen Institute of Computational Comparative Medicine, Nanotechnology Innovation Center of Kansas State, Kansas State University, Manhattan, KS, USA

Mark T.D. Cronin School of Pharmacy and Biomolecular Sciences, Liverpool John Moores University, Liverpool, England

Burtron H. Davis Center for Applied Energy Research, University of Kentucky, Lexington, KY, USA

Lisa DeLouise University of Rochester, Rochester, NY, USA

Alan K. Dozier CDC/NIOSH DART, Cincinnati, OH, USA

Alison Elder University of Rochester, Rochester, NY, USA

Laura Escorihuela Departament d'Enginyeria Química, Universitat Rovira i Virgili, Tarragona, Spain

Alberto Fernández Departament d'Enginyeria Química, Universitat Rovira i Virgili, Tarragona, Spain

Agnieszka Gajewicz Laboratory of Environmental Chemometrics, Institute for Environmental and Human Health Protection, Faculty of Chemistry, University of Gdańsk, Gdańsk, Poland

Francesc Giralt Departament d'Enginyeria Química, Universitat Rovira i Virgili, Tarragona, Spain

David González-Gálvez LEITAT Technological Center, Terrassa (Barcelona), Spain

Uschi M. Graham Center for Applied Energy Research, University of Kentucky, Lexington, KY, USA

Andrea Haase German Federal Institute for Risk Assessment, Department of Chemical and Product Safety, Berlin, Germany

Peter Hoet Center for Environment and Health, KU Leuven, Leuven, Belgium

Gary Jacobs Center for Applied Energy Research, University of Kentucky, Lexington, KY, USA

Gemma Janer LEITAT Technological Center, Terrassa (Barcelona), Spain

Keld Alstrup Jensen National Research Centre for the Working Environment (NRCWE), Copenhagen, Denmark

Araceli Sánchez Jiménez Centre for Human Exposure Science (CHES), Institute of Occupational Medicine, Edinburgh, UK

Harald Jungnickel German Federal Institute for Risk Assessment, Department of Chemical and Product Safety, Berlin, Germany

Alexander Kafka UFZ Department of Ecological Chemistry, Helmholtz Centre for Environmental Research, Leipzig, Germany

Faculty for Chemistry and Mineralogy, University of Leipzig, Leipzig, Germany

Anne Kahru Laboratory of Environmental Toxicology, National Institute of Chemical Physics and Biophysics, Tallinn, Estonia

Ralph Kühne UFZ Department of Ecological Chemistry, Helmholtz Centre for Environmental Research, Leipzig, Germany

Georgios Leonis Institute of Biology, Pharmaceutical Chemistry and Biotechnology, National Hellenic Research Foundation, Athens, Greece

Pinpin Lin National Health Research Institutes, Zhunan Town, Taiwan

Jing J. Liu Institute of Particle Science and Engineering, School of Chemical and Process Engineering, University of Leeds, Leeds, UK

School of Chemistry and Chemical Engineering, South China University of Technology, Guangzhou, China

Vladimir Lobaskin School of Physics, Complex and Adaptive Systems Lab, University College Dublin, Dublin, Ireland

Jean-Pierre Locquet Solid State Physics and Magnetism Section, KU Leuven, Leuven, Belgium

Hender Lopez School of Physics, Complex and Adaptive Systems Lab, University College Dublin, Dublin, Ireland

Andreas Luch German Federal Institute for Risk Assessment, Department of Chemical and Product Safety, Berlin, Germany

Alexander Lyubartsev Department of Materials and Environmental Chemistry, Stockholm University, Stockholm, Sweden

Cai Y. Ma Institute of Particle Science and Engineering, School of Chemical and Process Engineering, University of Leeds, Leeds, UK

Lutz Mädler Foundation Institute of Materials Science (IWT), Departmet of Production Engineering, University of Bremen, Bremen, Germany

Dominik Mertens Center for Environment and Health, KU Leuven, Leuven, Belgium

Alexander Mirzoev Department of Materials and Environmental Chemistry, Stockholm University, Stockholm, Sweden

Günter Oberdörster University of Rochester, Rochester, NY, USA

Ceyda Oksel Institute of Particle Science and Engineering, School of Chemical and Process Engineering, University of Leeds, Leeds, UK

Manthos G. Papadopoulos Institute of Biology, Pharmaceutical Chemistry and Biotechnology, National Hellenic Research Foundation, Athens, Greece

Suman Pokhrel Foundation Institute of Materials Science (IWT), Departmet of Production Engineering, University of Bremen, Bremen, Germany

Tomasz Puzyn Laboratory of Environmental Chemometrics, Institute for Environmental and Human Health Protection, Faculty of Chemistry, University of Gdańsk, Gdańsk, Poland

Robert Rallo Departament d'Enginyeria Informatica i Matematiques, Universitat Rovira i Virgili, Tarragona, Spain

Renaud Regret Rhenovia Pharma, Mulhouse, France

Andrea-Nicole Richarz School of Pharmacy and Biomolecular Sciences, Liverpool John Moores University, Liverpool, England

Christian Riebeling German Federal Institute for Risk Assessment, Department of Chemical and Product Safety, Berlin, Germany

Jim E. Riviere Institute of Computational Comparative Medicine, Department of Anatomy and Physiology, Kansas State University, Manhattan, KS, USA

Richard L. Marchese Robinson School of Pharmacy and Biomolecular Sciences, Liverpool John Moores University, Liverpool, England

Carlos P. Roca Departament d'Enginyeria Química, Universitat Rovira i Virgili, Tarragona, Spain

Gerrit Schüürmann UFZ Department of Ecological Chemistry, Helmholtz Centre for Environmental Research, Leipzig, Germany

Institute for Organic Chemistry, Technical University Bergakademie Freiberg, Freiberg, Germany

Janeck Scott-Fordsmand Department of Bioscience, Aarhus Universit, Silkeborg, Denmark

Mariliis Sihtmäe Laboratory of Environmental Toxicology, National Institute of Chemical Physics and Biophysics, Tallinn, Estonia

Lauri Sikk Institute of Chemistry, University of Tartu, Tartu, Estonia

Institut de Chimie de Nice (UMR CNRS 7272), Université Nice Sophia Antipolis, Nice, France

Peter B. Sorensen Department of Bioscience, Aarhus Universit, Silkeborg, Denmark

Tarmo Tamm Institute of Technology, University of Tartu, Tartu, Estonia

Kaido Tämm Institute of Chemistry, University of Tartu, Tartu, Estonia

Lang Tran Institute of Occupational Medicine, Edinburgh, UK

Michael T. Tseng University of Louisville, Louisville, KY, USA

Martie van Tongeren Centre for Human Exposure Science (CHES), Institute of Occupational Medicine, Edinburgh, UK

Socorro Vázquez-Campos LEITAT Technological Center, Terrassa (Barcelona), Spain

Gemma Vilar LEITAT Technological Center, Terrassa (Barcelona), Spain

Alejandro Vílchez LEITAT Technological Center, Terrassa (Barcelona), Spain

Hanne Vriens Center for Environment and Health, KU Leuven, Leuven, Belgium

Xue Z. Wang Institute of Particle Science and Engineering, School of Chemical and Process Engineering, University of Leeds, Leeds, UK

School of Chemistry and Chemical Engineering, South China University of Technology, Guangzhou, China

Terry Wilkins Institute of Particle Science and Engineering, School of Chemical and Process Engineering, University of Leeds, Leeds, UK

Dmitry Zhurkin Department of Materials and Environmental Chemistry, Stockholm University, Stockholm, Sweden

Part I
Characterisation

Chapter 1
Engineered Nanomaterials: Their Physicochemical Characteristics and How to Measure Them

Rambabu Atluri and Keld Alstrup Jensen

Abstract Numerous types of engineered nanomaterials (ENMs) are commercially available and developments move towards producing more advanced nanomaterials with tailored properties. Such advanced nanomaterials may include chemically doped or modified derivatives with specific surface chemistries; also called higher generation or multiconstituent nanomaterials. To fully enjoy the benefits of nanomaterials, appropriate characterisation of ENMs is necessary for many aspects of their production, use, testing and reporting to regulatory bodies. This chapter introduces both structural and textural properties of nanomaterials with a focus on demonstrating the information that can be achieved by analysis of primary physicochemical characteristics and how such information is critical to understand or assess the possible toxicity of engineered nanomaterials. Many of characterization methods are very specific to obtain particular characteristics and therefore the most widely used techniques are explained and demonstrated.

Keywords Nanomaterials • Nanoparticles • Nanostructures • Physico-Chemical Characterization • Properties • Microscopy • Spectroscopy • Specific Surface Area • Functionalization

1.1 Introduction

In today's world, there has been a change in the comfort of human life in many ways ranging from smart and light-weight materials, technologically advanced buildings with self- or easy-to-clean coatings, small communication systems, functional foods, advanced medication system and light-weight and/or high-speed transport systems. This progress is not only linked to the development of specialty materials

R. Atluri (✉) • K.A. Jensen
National Research Centre for the Working Environment (NRCWE),
Lerso Parkallé 105, 2100 Copenhagen, Denmark
e-mail: rba@nrcwe.dk

L. Tran et al. (eds.), *Modelling the Toxicity of Nanoparticles*, Advances in Experimental Medicine and Biology, Vol. 947, DOI 10.1007/978-3-319-47754-1_1

but also the methods for their characterization and analysis. Invention of new materials leads to categorize materials research into different branches such as nanoscience and nanotechnology i.e. fundamental principles of molecules, structures at 1–100 nm and their application of these structures into useful nanoscale devices and functional products.

Looking into the history of material development, nanomaterials as such are not new. They have been in use since B.C and A. D. An iconic example of it is Lycurgus cup, which exhibits the most remarkable characteristic of color diffraction between green to red under different lighting conditions [16]. This is due to the suspension of tiny gold (nano) particles within the glass matrix, whose diameters are comparable to the wavelength of visible light. The quantity of these particles was quite low and even the glass makers did not know these features. If the glass manufacturers had known the explanations for the color changes and a way to investigate the properties, nano-products we might have reached beyond the nanotechnology stage by the twenty-first Century [7]. Lack of instrumentation and methods to study the internal, external and structural properties at nanoscale levels have led a little progress until late 1930s, where commercial electron microscopes were established. At the same time spectroscopic methods became available beginning with the discovery of X-ray diffraction by Max von Laue and his colleagues in 1912 [14], which enabled determination of nano- to atomic scale structures in materials. In later years development of quantum science and computing has added speed by which material innovation may occur.

Today a large number of nanomaterials exist and are still developed as needs and ideas arise for new applications and new properties. Developments especially occur in the development of advanced nanomaterials with tailored properties. Such nanomaterials include nanoparticles which have been chemically doped or modified to have a specific tailored surface chemistry. Therefore particulate nanomaterials (nano-objects in ISO terminology) are grouped into mono and multi-constituent nanomaterials (ISO/TR 11360:2010(E)). Mono-constituent nanomaterials are also referred to as being first order nanomaterials whereas multiconstituent nanomaterials are second, third generation materials etc., which describes the presence of one or two to the n'th additional compounds enclosed or coating the nano-object considered the core or skeleton of the nanomaterial. Clearly this plurality of possible structural and compositional combinations can require the combination of several techniques to characterize a nanomaterial.

In this chapter, we will focus on demonstrating the information that can be achieved by analysis of primary physicochemical characteristics of nanomaterials and how such information can be used to identify and understand the nanostructures within nanomaterials. A range of methods used in the analysis of nanomaterial characteristics is quite large. Many of these methods are very specific for particular characteristics and therefore most widely used techniques are explained. Compared to bulk materials, nanomaterials need one or more analysis, sometime a specific technique to clearly understand the properties and applications associated with it. Moreover, it is described how such information is key to understand or assess the possible toxicity of engineered nanomaterials.

1.2 Types of NM and Physicochemical Properties to Measure

In principle nanomaterials can be produced for all non-gaseous elements in the entire periodic system. Some nanomaterials are produced in very large amounts (e.g., carbon black, SiO_2, TiO_2, ZnO, Fe-oxide, carbon nanotubes, and nanoclays) whereas others, but still technologically important materials (e.g., Ag, Ce-oxides, quantum dots such as CdSe, and nanoporous materials) are produced in much smaller quantities. Nanomaterials, such as nano-cellulose and graphene, also have high industrial potential and currently have very high focus [10, 24–26, 47]. A recent analysis of nanomaterial producers in the world showed 22 materials produced by up to 300 manufacturers (Table 1.1; [3]). However, a world list of nanomaterials in production covers a much wider range without including considerations of chemical derivates achieved by e.g., surface chemical modifications and also considering variations in dimensions and structural variations, which is important for e.g., carbon nanotubes. Future Markets estimates the 2010 worldwide production of nanomaterials was 21,713 tons, a tenfold increase from 2002 and is estimated to more than double to 44,267 tons by 2016, driven by demand from applications in electronics, energy, medicine, chemicals, coatings and catalysts.

As for all other particulate materials, there are a number of physico-chemical characterization end-points define the different nanomaterials. The characteristics and properties to be assessed depend strongly on chemical type and chemical-structural complexity of the nanomaterials and on the purpose. One may discriminate between the end-points required to identify the material versus the end-points required to characterize its dispersibility into a given matrix versus characterization required for chemicals registration and finally for full risk assessment [22, 49]. The characterization can be divided into characterization of primary physicochemical properties and secondary properties, which describe the state, reactivity and fate of the nanomaterial during and after release and/or exposure. The primary character-

Table 1.1 List of nanomaterials in production according to Future Markets (2014)

Carbon based	Metal oxides and Metalloid oxides	Metals, Salts	organics	Others
Single Wall CNTs Double wall CNTs Multi Wall CNTs Fullerenes Graphene Nano-carbon black	Silicon dioxide Titanium dioxide Aluminium oxide Antimony tin oxide Bismuth oxide Cerium oxide Cobalt oxide Copper oxide Iron oxide Magnesium oxide Manganese oxide Yttrium oxide Zinc oxide Zirconium oxide	Gold Nickel Silver Palladium Quantum dots	Dendimers Nano-cellulose Organic dyes Organic pigments Polymers	Nanoclays Tungsten carbide

ization end-points including average size/size-distribution including agglomeration/aggregation state, shape/morphology, surface area and porosity, atomic structure, chemical composition, surface chemistry and a long list of available analytical techniques were listed as candidates to enable such data. We refer to the chapters by Zuin et al. [49] and Jensen et al. [22] and additional references [1, 2] for more detailed information on these end-points and characterization techniques.

1.3 Physicochemical Characterization to Identify Nanomaterials

1.3.1 Chemical and Structural Properties

The chemical composition and structure of nanomaterials, similar to regular materials, is the key to identify and group the nanomaterial into its material class. A combination of chemical and structural information is particularly important if different generations of nanomaterials should be identified as proposed in Atluri and Jensen [5].

1.3.1.1 Composition

The chemical composition is the key to classifying various nanomaterials. Groups could be ceramic, metallic, semi-metallic/semi-conducting, polymers, carbon-based, and organic/inorganic as proposed in ISO/TR 11360 (2010), but more detailed classification is most likely needed for both material and regulatory purposes and already proposed for carbon allotropes (e.g., [18]) and in more detailed as exemplified for fullerenes [12].

The chemical composition of a nanomaterial refers to entities of which the material is composed. The function of nanomaterials is influenced by the chemical composition and hence different physical, chemical, mechanical and biological properties. The observed toxicity of nanomaterials often linked to their composition in the form of coating, and impurities. For example, it has been shown that CNTs show considerable toxicity especially because of catalyst metal contaminants such as the metal oxides, introduced during production and purification process. The toxicity of CNTs is due to the release of metal contaminants and their ability to cross the cell membrane [35]. Others demonstrated that CNTs containing different surface charge and modification, size, and length, could influence the potential toxicity. Especially the presence of individual separated stiff fibrils or fibers are considered one of the key characteristics leading to severe pulmonary toxicological effects [13, 23, 33]. Understanding the effects of CNTs on the biocompatibility, toxicity, and risk assessment may sometimes lead to conflicting results and hard to predict where the toxicity comes from [30]. In addition, there is no consistency of constituent species and amounts of metal impurities in CNTs made from the same process but of different batches and are quite different in materials obtained from different vendors, which makes it more complex to generalize the toxicity of CNTs. Table 1.2 shows chemical compositions of different Multiwall-CNTs as derived by Wave-Dispersive X-Ray fluorescence spectroscopy (WDXRF). All the materials were purchased from different vendors in the form of pristine and surface functionalized CNT. Irrespective of their group, the purity of MWCTNs as pure carbon ranges between 86 and 97 % and a wide range of metal oxides as impurities.

The chemical composition can be analyzed using a range of methods. Depending on the method used, the elemental analysis will range from qualitative to quantitative analysis. The most common characterisation methods for analyzing the chemical composition of nanomaterials are X-Ray fluorescence spectroscopy (XRF), Energy Dispersive X-Ray Spectroscopy (EDS/EDS), Electron Energy Loss Spectroscopy (EELS), X-Ray photoelectron spectroscopy (XPS), Inductively Coupled Plasma-Mass Spectrometer (ICP-MS), Nuclear Magnetic Resonance (NMR), Raman Spectroscopy, and Static secondary Ion Mass Spectrometry (SIMS). However, all the methods are not equally suitable for quantification of the material chemical composition. For example variation depending on extraction procedures may be important for indirect methods such as ICP-MS.

1.3.1.2 Crystalline Phases

Identification of the atomic structure of materials has been essential for their identification for several decades and is the whole foundation for modern mineralogy and materials science. It is also evident from the current toxicological literature that the toxicological effects of engineered nanomaterials can vary considerably depending on the structural properties [27]. In particular, the biological responses of many nanoparticles largely depend on the crystal phases but of similar in their composition. For example, the composition of silica is ***stoichiometrically*** similar; various forms of silica differ in their physicochemical and toxicological properties. It exists in crystalline and amorphous state with long and short range order, respectively. It is well known that inhalation of crystalline silica shown to be serious adverse effects among workers in the form of increased lung cancer and has been classified as a human lung carcinogen with important differences between quartz [trigonal (α-quartz) or hexagonal (β-quartz)] and the less abundant tridymite [orthorhombic (α-tridymite) or hexagonal (β-tridymite)] polymorphs. However, synthetic amorphous silica (SAS) has so far not shown any adverse effects because of their amorphous state [29]. Similar distinctions between rutile and anatase phases of titanium oxide can be made [42].

The powder X-ray diffraction technique is a fundamental technique for the identification of crystalline phase of nanomaterials. When an incident beam of X-rays (a form of electromagnetic radiation with a wavelength of 1Å) interacts with a target sample, the waves are scattered from lattice planes separated by an interplanar distance d. The scattered waves interfere constructively, and the path difference between two waves undergoing constructive interference is given by $2d\sin\theta$, where

Table 1.2 Chemical composition of various carbon nanotubes measured by WDXRF from [19]

MWCNT group	MWCNT code	Type	C	Al_2O_3	CaO	CoO	Fe_2O_3	La_2O_3	MgO	MnO	NiO	P_2O_5	SO_3
Standard materials	NRCWE-026	Pristine	84.4	14.97	–	0.11	0.29	–	–	–	–	–	–
	NM-401	Pristine	99.7	–	–	–	–	–	–	–	–	0.14	–
	NM-402	Pristine	96.1	2.43	–	–	1.31	–	–	–	–	0.16	–
	NM-403	Pristine	99.1	0.24	–	0.12	–	–	0.19	0.16	–	0.14	–
	NRCWE-006	Pristine	99.6	–	–	–	–	–	–	–	–	0.14	–
Group I	NRCWE-040	Pristine	98.6	–	–	–	0.20	0.32	–	–	0.56	0.15	–
	NRCWE-041	OH	99.2	–	0.13	–	0.13	–	–	–	0.31	0.15	–
	NRCWE-042	COOH	99.2	–	0.25	–	–	–	–	–	0.21	0.14	–
Group II	NRCWE-043	Pristine	98.5	–	–	–	–	–	–	–	1.20	0.15	–
	NRCWE-044	OH	98.6	–	–	–	–	–	–	–	1.04	0.14	–
	NRCWE-045	COOH	96.3	0.52	–	0.25	1.17	–	–	–	1.34	0.16	–
Group III	NRCWE-046	Pristine	98.7	0.29	–	0.25	–	–	0.22	0.30	–	0.14	–
	NRCWE-047	OH	98.7	0.27	–	0.25	–	–	0.22	0.30	–	0.15	–
	NRCWE-048	COOH	98.8	0.26	–	0.24	–	–	0.19	0.28	–	0.14	–
	NRCWE-049	NH_2	98.8	0.26	–	0.25	–	–	0.19	0.29	–	0.15	–
Reference	Printex 90	–	99.3	–	–	–	–	–	–	–	–	–	0.66

To ease interpretation, all data were reported in wt % of the oxides of the elements determined and their content below 0.1 % is not reported

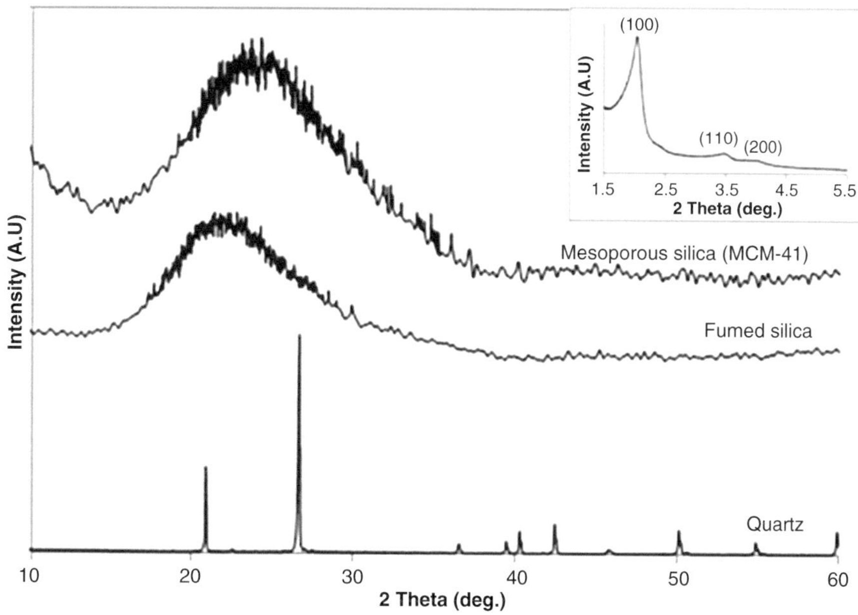

Fig. 1.1 Powder X-Ray diffraction (XRD) graphs of quartz, fumed silica, and mesoporous silica (Inset: shows low angle peaks of mesoporous silica)

θ is the scattering angle. The intensity of the scattering wave as a function of scattering angle gives a diffraction pattern. Both the positions and the relative intensities of the diffraction peaks are indicative of a particular structure, such as cubic, hexagonal and help to determine crystal symmetry to structure determination. One possible limitation of XRD while using for nanomaterials is their peak-broadening at lower scale where nanomaterials become amorphous structure, but may still be crystalline [34]. Though, the size limitation is more of a material specific and indeed depends on the accuracy of diffractometer, users must be aware of the problem while measuring nanomaterials.

Figure 1.1 shows the X-Ray diffraction patterns of different silica materials such as quartz (Sigma-Aldrich), Mesoporous Silica (MCM-41, [8]) and fumed silica (Sigma-Aldrich). Between 20 and 30^0 2θ, the amorphous nature of silica is indicated by a single broad peak, for both MCM-41 and fumed silica. Due to ordered pore structure, mesoporous silica shows (Fig. 1.1 inset) three peaks (100, 110 and 200) between 2theta angles of 1.5–6°, which are consistent with 2 dimensional (2D) hexagonal cylindrical porous network with space group symmetry, *p6mm* [6]. On the other hand, the crystallinity of quartz results in a range of diffraction peaks, consistent with hexagonal crystal system with a space group, $P3_121$.

Compared to laboratory X-Rays, the synchrotron radiation sources give more intense and higher energy radiation with very shorter angles of scattering <0.10. In addition, an in-situ Small-angle X-ray scattering (SAXS) pattern can be recorded in a short time while the XRD pattern needs at least a couple of minutes for acceptable data. The structural growth information from the nucleation stage during the nanoparticle synthesis can thus be accessible with the SAXS technique. SAXS is also an X-ray diffraction-based technique, where synchrotron radiation is used as a source, which occurs when charged particles are accelerated in a curved path or orbit. Any charged particle which moves in a curved path or is accelerated along a linear path will emit electromagnetic radiation. When the wavelength of the electromagnetic radiation is of the same order as the length of a sample particle, the particle will scatter the radiation. Detection and analysis of this scattering pattern can yield valuable information about the size, shape, and internal structure of the particle.

Both XRD and SAXS only give a diffraction pattern for further interpretation of the crystal phases of nanoparticles. However, microcopy methods such as AFM, SEM, and TEM give 2-dimensional images of the nanoparticles at the atomic scale. In particular, direct imaging by TEM images gives not only the amplitude but also the phase information of the structure factors of the crystal. The indexed Fast Fourier Transform (FFTs) of the TEM images can be further used to refine the lattice parameters and thereby the crystal symmetry. TEM information is obtained from a single crystal (~108 times smaller than the size of a specimen for XRD) whereas XRD data is from the bulk material. The importance of TEM imaging for structural investigation is best explained, for instance, in refining the symmetry of mesoporous silicas and their pore connectivity in the amorphous silica network [4, 6]. Typical TEM images recorded along the [1 0 0], [1 1 0] and [1 1 1] direction and corresponding FFT-diffractograms of all the samples are shown in Fig. 1.2. Crystallographic reconstruction procedure and image processing of the images gives 3D-electrostatic potential density model as shown in Fig. 1.2d. The reflection conditions derived from the TEM images and their corresponding FFT-diffractograms, confirms the space group symmetry of mesoporous silica as P*m-3n*. The pore structures of a sample may be visualized from the 3D-electrostatic potential maps reconstructed from the structure factors obtained from electron crystallography. The boundary between pore and the pore-wall is determined by the so-called threshold value (related to the pore volume fraction), derived from the mesopores volume and the silica wall density (2.2 g/cm^3). Overall, HRTEM combined with electron crystallography gives not only the 2D images of nanomaterials at atomic resolution but also their tomographic view at macro-scale.

This type of analysis can be used to describe the nanostructures as well as location of doped or nature of nanomaterial cores, which is important for identification of nanomaterial class and nanomaterial generations if not known beforehand.

1.3.1.3 Surface Modification/Functionalization

Surface treatment or surface modification or functionalization or doping of nanomaterials induces distinct chemical and physical properties compared to their pristine form. For example, the surface treatment of silica with methyl groups is an effective way to disperse the silica nanoparticles in a wide range of organic

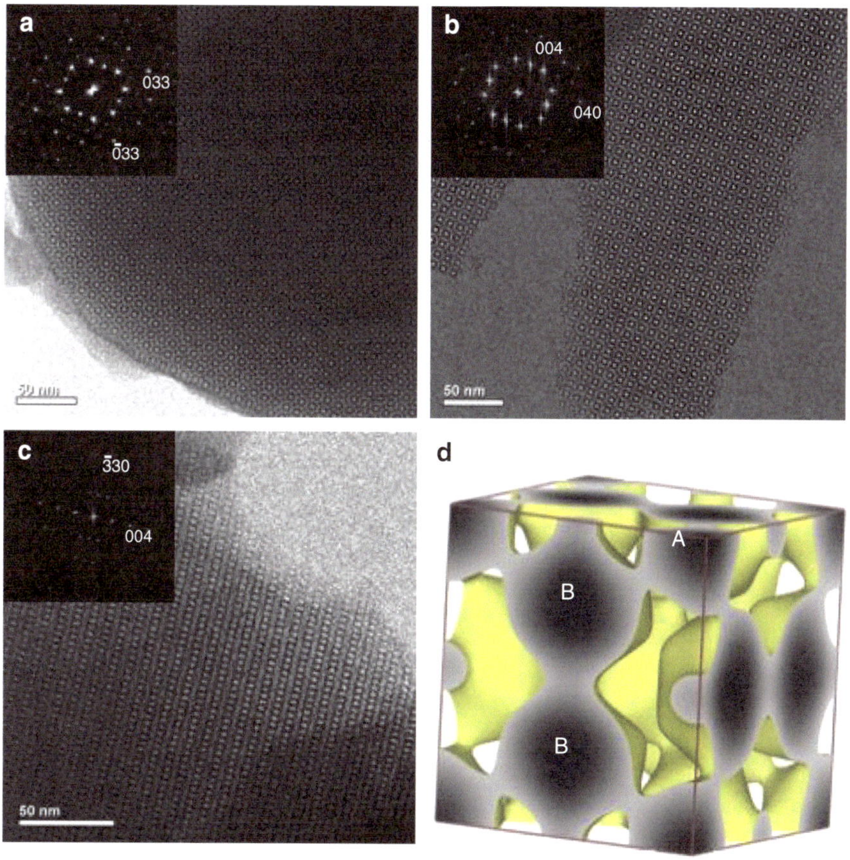

Fig. 1.2 HRTEM images of mesoporous material SBA-1 type, viewed parallel to the [100] (**a**), [111] (**b**) and [110] (**c**) directions. Inset shows FT diffractograms recorded from regions of the particles shown. (**d**) The 3D-electrostaic potential map of mesoporous silica showing two types of cavities (*A*, *B*) with open cavity-connecting windows to the neighboring cavities [6][1]

solvents. Functionalization has been used to conjugate drug molecules, polymers and organic groups to NPs. It has been demonstrated that non-covalent attachment of polyethyleneimine (PEI) polymers to the silica surface not only increases cellular uptake but also generates a cationic surface to which DNA and siRNA constructs could be attached [43]. In another case, functionalization has also been shown to protect NPs against agglomeration and render them compatible in other phases. Silica coating on semiconductor materials such as CdS nanoparticles

[1]Reprinted from Microporous and Mesoporous Materials, 133 /1–3, Rambabu Atluri, Zoltán Bacsik, Niklas Hedin, Alfonso E. Garcia-Bennett, Structural variations in mesoporous materials with cubic Pm3¯n symmetry, 27–35., Copyright (2016), with permission from Elsevier.

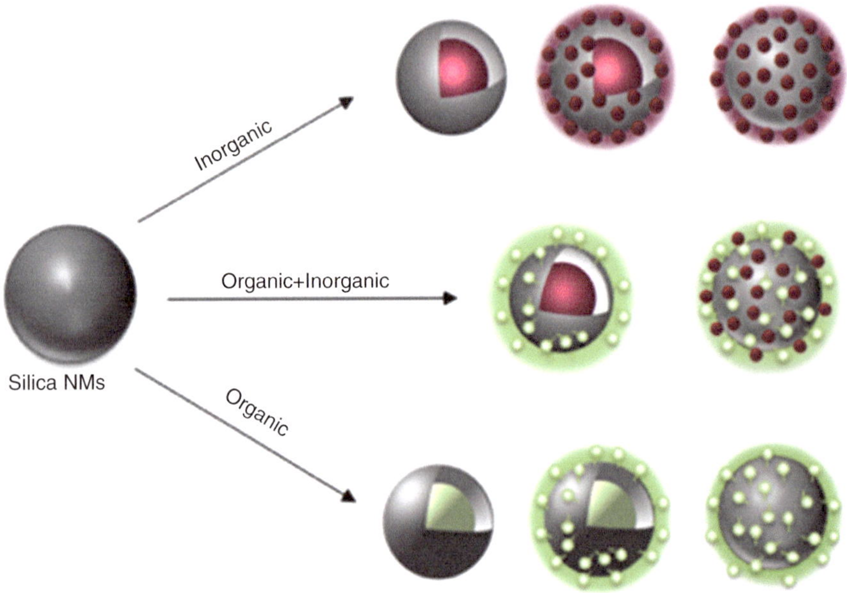

Fig. 1.3 Schematic illustration of several derivations after surface modification/functionalization of silica nanoparticles. *Green*: organic, *Plum*: Inorganic, *Light Black*: Silica

improves the stability of the particles as well as prevents coagulation during the chemical or electronic processing [20, 38, 40, 44, 45]. Therefore, surface functionalization of NPs is rather necessary to render specific functionality over the core NPs. Similar surface-chemical modification of carbon-based materials such as fullerenes and CNTs are usually considered essential for dispersion and any application in matrix nanocomposites [18].

In general, surface modification of nanoparticles can be achieved by chemical and physical modifications. Most often a combination of both methods being used for higher generation of engineered nanomaterials. The most frequently used and industrially relevant method for nanocomposites based on polymers/nanoparticles includes solution belding or melt blending [48]. Depending on the function and application, various surface treating agents such as organic and/or inorganic in the form of elements, compounds and materials can be used to modify the surfaces of nanoparticles. The complexity of the nanoparticles derived by their chemical doping and surface modifications, which depends on the extent and location of the surface treating agents. Complexity is the location of the compounds either external (E) or as a core (C) of nanoparticle matrix or combination of both, thanks to the chemistry of nanomaterials for merging many phases into a single but multifunctional system. Figure 1.3 shows possible derivations of silica nanoparticles after modification; mostly rely on chemical and physical or combinations of both the methods. Apart from organic compounds for surface treatment, various inorganic nanoparticles used to construct hybrid architectures of silica including nanoparticles assembled at the surface of a silica particle, encapsulation of nanoparticles in a silica shell etc. Several physico-chemical characterization methods can be used to determine the chemical complexity of doping and surface chemical modification of nanomaterials both quantitatively or qualitatively, including electron microscopy, XRF, XPS, molecular spectroscopy such as FT-IR and X-Ray, Raman, and mass spectrometry (GC/LC/ICP). In addition, Thermal Gravimetric Analysis (TGA) is the most widely used and relatively easy method for quantitative analysis of surface modifications.

1.3.2 Textural Properties

1.3.2.1 Size and Size Distribution

Materials exist in different forms and size depending on the source and manufacturing process. Size and size-distribution is the key characteristic of nanomaterials and key for their identification for regulatory purposes [17, 21]. Nature has a large number of nano-sized materials with properties that are distinct and relatively comparable to man-made nanomaterials. The carbon based nanoparticles such as Buckminsterfullerene and graphene are typical examples of it.

As nanomaterials (NMs) are generally defined by having small dimensions in the nanoscale, i.e. between ca. 1 and 100 nm and have large surface to volume ratios, their physicochemical properties are different from the properties known for bulk materials [36]. As the size of a particle decreases, the proportion of atoms on the surface of the particle increase and, consequently, the physicochemical properties will be different from the properties known for non-nanomaterials.

Therefore, significant efforts have been put forward by different government bodies and policy makers to define the size limits of a nano-object. According to the EU recommendation, "*Nanomaterial' means a natural, incidental or manufactured material containing particles, in an unbound state or as an aggregate or as an agglomerate and where, for 50 % or more of the particles in the number size distribution, one or more external dimensions is in the size range 1–100 nm*".

Particle size measurements have a direct impact on the reliability of products containing nanoparticles and provide a basis for toxicological studies of nanoparticles. However, selecting right particle size methods have a great impact on the reliability of the size and size distribution. Different methods (Table 1.3) often have different bases for the particle size and size distribution, for example dry samples vs. wet sample dispersions. In addition, the principles of methods such as light diffraction or electron microscopy should also be taken into account for a correct interpretation of the measurements. The quality of measurements also depends on the different screening criteria such as size measurements for spheroidal and non-spheroidal particles, agglomerates, and aggregate particles and size measurements for environmental, health and safety evaluations. Currently, methods are rather advanced for quantitative size-distribution analysis of granular nanomaterials (see e.g., [11, 31]).

Table 1.3 Table shows different particle size methods and typical size ranges

Method	Typical measurement range	Type of size distribution	Available standards
SEM	1 nm–10 μm[a]	Number based	ISO 16700: 2004
TEM	0.5 nm–1 μm[a]	Number based	ISO 13322–1: 2004 ISO 13322–2: 2006
SAXS	>5 nm	Scattering intensity based	ISO/TS 13762:2001
AFM	>1 nm	Number based	ASTM E2859–11
XRD	<100 nm	Scattering intensity based	ISO 20203: 2005
Centrifugal liquid sedimentation	>20 nm	Extension Intensity based	ISO 13318–3: 2004 ISO 13318–2: 2007
DLS	1 nm – several micro meters[b]	Number and volume based	ISO 22412:2008
Particle tracking analysis	>25 nm	Number based	ISO/DIS 19430
Field flow fractionation	1 nm–200 nm	Intensity based	–

[a]Some instruments have lower detection limits
[b]Ranges may be wider and vary considerably with instruments and software

1.3.2.2 Aggregation/Agglomeration

Due to high surface energies, nanoparticles have tendency to form agglomerates/aggregates. As summarized in Schneider and Jensen (2009) [50], the agglomeration of nano-objects in nanomaterials can be due to different phenomena, which can be grouped into: electrical properties (e.g., van der waal forces), magnetic properties (e.g., ferromagnetism), physical interlock (e.g., entanglement), and bridging (liquid film or greasy coatings). Some of these agglomeration phenomena have weak forces and require moderate energy for dispersion, whereas others have large effect or binding forces and require dedicated efforts to disperse the nano-objects. Aggregates consist of nano-objects bound together with significant interface contacts and high binding energies. Aggregates can therefore not be separated without breaking the material.

Agglomerates or aggregates make it especially difficult to explore the properties and applications of nanomaterials. In reality the nano-objects in nanomaterials rarely exist as separate units unless this is a particular target in the manufacturing process and in many cases the particles are not in uniform shape, particularly when the materials are scaled up. Particles do agglomerate and aggregate and have different size ranges depending on the use and the environment. In many synthetic processes for nanoparticles, especially surfactant-free chemical reactions, aggregation or agglomeration occurs immediately as particles are generated. The terms agglomeration and aggregation are still rather confusing for defining the particle state.

If the properties associated with the aggregation or agglomeration is not understood, the analysis may sometimes give misleading results on data such as particle size and dispersion level. In another context, increasing use of manufactured

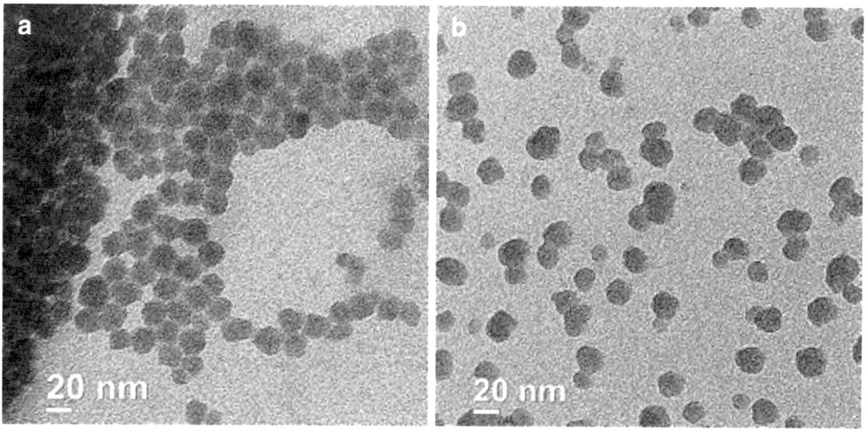

Fig. 1.4 Transmission electron microscopy images of silica nanoparticles showing their agglomeration state depending on the sample preparation methods such as (**a**) drop-on-grid and (**b**) grid-on-drop

nanoparticles ensures these materials will make their way into the environment during their lifecycle. The state of these nanoparticles in the environment and biological systems remains a question, which stress the importance of understanding the mechanisms of aggregation and agglomeration.

Methods for determining the state of aggregation and agglomeration are mainly the same techniques as used for the nanoparticle size determinations. It is highly recommended to follow a special preparation protocol before doing the measurements, to avoid misleading information on the state of particles in the dry state vs. dispersion state. In this aspect, preparation of TEM grids shows (Fig. 1.4) a nice attempt on the importance of methods such as drop-on-grid and grid-on-drop showing the dispersion of particles on the grid.

Strategies for preventing aggregation mainly come from conventional colloid science in which particles are coated with foreign capping agents and/or the surface charges are tailored to separate them via electrostatic repulsions [28]. High energy mechanical methods such as grinding, sonication have also been used as a post treatment for making the dispersions and fine particles more homogenous. ISO TR 13097 2013–06 on dispersion reports the methods that can be used to monitor the state of nanoparticles in solutions, acceleration procedures and data evaluation. The OECD is now looking to prepare a new test guideline, which can refer to existing ISO standards.

1.3.2.3 Shape

The beauty of engineered nanoparticles is their hierarchal shapes and associated physical – chemical properties. Unlike bulk materials, the thermodynamic and surface energy considerations at nanoscale are more complicated by the high surface area to volume ratio. For a material with a perfect symmetric sphere, the total surface energy is lowered by decreasing the amount of surface area corresponding to a given volume. On the other hand, faceted nanoparticles show high number of reactive and high atom density facets and may influence the properties such as dissolution, aggregation, and reactivity. Bottom-up methods are used for producing a large variety of nanoparticles. Depending on the synthesis conditions and composition, variety of shapes such as spheres, wires, rods, plates, spheroids, tubes etc. have been developed.

The shape of nanoparticles plays an important role in understanding the properties associated at nanoscale. Particularly for nanomedicine applications, the shape of nanoparticles has recently been identified as a key factor influencing circulation time, bio-distribution, cellular uptake, as well as targeted drug delivery. The realization of shape factor came from the non-spheroidal shapes of various and bacteria, but still improving their ability to evade an immune response [41]. However, shape effect studies show considerable toxicity to human cells and question the health and environmental fate of nanoparticles. It was shown that wire- shaped silver nanoparticles induced a strong cytotoxicity to human cells (A549) than spherical silver nanoparticles. It was argued that small diameter of nanowires shown to induce cell membranes but the large length of silver wires does not allow a complete entry as compared to the spherical nanoparticles [39]. In another example, ZnO nanoparticles of different shapes also show toxicity to marine algae [32]. Therefore, shape is an important parameter when considering the fate of nanoparticles.

For shape determination of nanoparticles, electron microscopes such as SEM, TEM and SPM (Scanning Probe Microscopy) are used. However, for intrinsic properties such as crystal structure, symmetries and surface morphology, electron microscopy combined with tomography are used. Figure 1.5 shows SEM images of different zinc oxide particles, synthesized under various reaction conditions and concentrations. The morphology of zinc oxide formed by the reaction of zinc salts and hydroxide ions is very dependent not only on pH, temperature, concentration, and reaction time but also on the stirring time and water addition sequence of the reactions. By controlling these factors, flower-like, needle-like, star-like and spherical morphologies were obtained. Likewise, different metal, non-metal, and metal oxide nanoparticles including silica, TiO_2, iron oxide, gold, and, silver has been developed with various shapes and aspect ratios.

1.3.2.4 Surface Area and Porosity

A distinct characteristic of nanomaterials as compared to their bulk form is the area of accessible surface, as described by the specific surface area. Nanomaterials possess high surface area per unit mass due to a high portion of atoms at the surface relative to the atoms in the interior of the particles. Due to their large surface area to volume ratio, nanomaterials are highly attractive, and therefore, lead to a lot of new properties stemming from quantum effect and surface/interface effect. For applications requiring a large surface area per unit weight such as hydrogen storage for

Fig. 1.5 Different shapes of ZnO particles, under various synthesis conditions such as aging time, concentration, pH (=10, 8, 7), and temperature gives (**a**) Flower-like, (**b**) Prism-like, (**c**) Needle-like, (**d**) Star-like grains, (**e**) Aggregated spheres, and (**f**) Sheet-like flakes, respectively. Scale bar is 1 μm.

vehicles, chemical sensing, light harvesting, and for catalysts, nanomaterials or nanostructured materials be promising candidates. A typical example and so far holds the highest experimental surface areas of any porous materials reported to date is metal-organic frameworks (MOFs), displaying ~7000 m^2/g of surface area [15]. To put it another way, just a few grams of nanoparticles offers a surface area equivalent to the size of a football stadium. On the flip side, a greater toxicity was found from nanomaterials than from their larger counterparts associated by high absorption capacity. It was shown that ultrafine titanium oxide nanoparticles are more inflammogenic and cytotoxic than when compared to the fine sized particles of lower surface area [37].

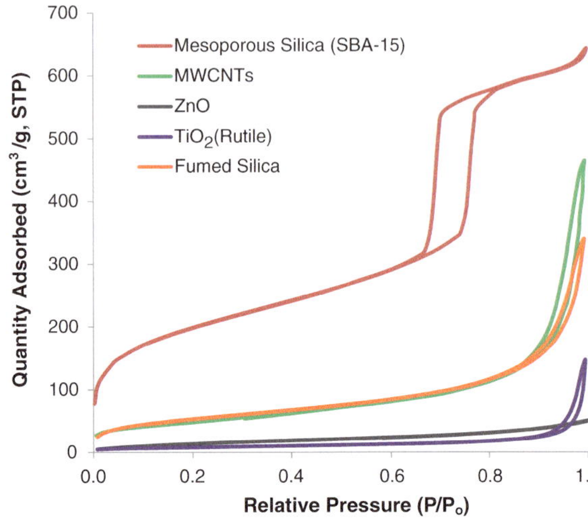

Fig. 1.6 Nitrogen adsorption-desorption isotherms of various nanoparticles

Having high surface and interface areas of nanoparticles induce a strong interaction with the surrounding liquid or gaseous adsorbents. Depending on the strength of the interaction and surface charge, the adsorption can be described as being either physical or chemisorption. Chemisorption is characterized mainly by a strong interaction between adsorbate-adsorbent while physical adsorption is due to mainly dispersion forces, i.e. weak intermolecular forces between non-polar molecules. The latter is more favored for surface area and porosity measurements because of its non-destructive nature and the ease of quantification of the adsorbate.

For powders, the specific surface area is usually determined by the nitrogen adsorption technique using the BET (Brunauer, Emmett and Teller) method [9], by which one may also measure the surface area given by nanoporosity. Though, other techniques such as SAXS (Small Angle X-Ray Scattering) developed for solutions, but the methods are not routinely usable.

Gas adsorption is a prominent technique for the comprehensive characterization of surface area and porosity measurements. The adsorption of gasses such as N_2, He, Ar (adsorbate) at various relative pressures on a solid (adsorbent) gives information of textural properties including surface area, pore volume, and pore size. The measurement is performed volumetrically; calibrated volumes of gas are added to a sample tube that is immersed in liquid nitrogen with a known amount of sample. The amount of gas adsorbed can be calculated from the measured pressure difference in the sample tube after the addition of a known volume. If the amount of adsorbed gas is plotted against the pressure, an isotherm is obtained. From this plot, surface area, pore volume, and pore size can be derived. Unfortunately, the gas adsorption method for surface area and porosity measurements is not applicable for liquid based materials.

Figure 1.6 shows nitrogen adsorption-desorption isotherms of various porous and nonporous nanoparticles. The shapes of the isotherms are different, and the adsorption volumes of nitrogen differ extensively between the nanoparticles. Mesoporous silica,

SBA-15 [46] shows high absorption capacity than the other nanoparticles, indicative of high surface area material. This is due to the ordered pore structure at mesoscale and possesses a narrow pore size distribution as evidenced from the reversible hysteresis loop (Type IV). MWCNTs and Fumed silica exhibit a moderate amount of adsorption at low relative pressures and resembles the inter-particle pore characteristics. The other nanoparticles exhibit nonporous characteristics and most of the adoption comes from the surface of the particles and hence exhibit low surface areas. The shape of the isotherm sometimes distinguishes the porous vs nonporous materials and even on the types of various porous materials. Most of the materials exhibit six types of adsorption isotherms. For microporous materials, the pore filling occurs in a continuous way (Type I) and the majority of micropore filling occurs at relative pressures below 0.1. The Type II isotherm is typical for macroporous and non-porous solids where monolayer coverage followed by multilayers at high relative pressures. Type III and Type V isotherms are characteristic of weak adsorbate-adsorbent interactions and are most commonly associated respectively with non-porous and microporous adsorbents and microporous or mesoporous solids. Type IV isotherms are typical for mesopores where pore-filling occurs by pore condensation.

The BET equation was used to calculate the surface area (S_{BET}) from the adsorption data obtained in the relative pressure (P/Po) range from 0.05 to 0.3. The total pore volume (V_{tot}), which is an important parameter for porous nanoparticles calculated from the amount of gas adsorbed at the highest relative pressure (P/Po), implies how open the pore structure is. The pore size distributions (PSD) of the nanoparticles are calculated from the adsorption branch of the isotherm and fitting various pore shapes such as cylindrical, bottleneck, slit type pores.

Most widely accepted PSD models are the Barrett-Joyner-Halenda (BJH), Horvath-Kawazoe (HK), and the Non-local density functional theory (NLDFT) models. The Surface area and porosity properties derived from the nitrogen isotherms on selected nanoparticles are shown in Table 1.4. From the table, it is clear that porous nanoparticles show high porosity properties than nonporous particles and mesoporous silica (SBA-15) being the high in pore characteristics.

Finally, BET surface area has also been considered as an important parameter to identify nanomaterials. As per the EU recommendation, volume specific surface area (VSSA) proposed as a complementary definition to distinguish nanomaterials from non-nanomaterials. The recommended VSSA (i.e. greater than 60 m^2/cm^3) corresponds to a 100 nm sphere.

1.4 Conclusions and Closing Remarks

Engineered nanomaterials represent one of the most fascinating developments in recent years. They are increasingly used to construct products with attractive features and uses. More than 20 nanomaterials of type are already used in consumer products and estimates to demand 50,000 tons/year in near future. With industrial production specific reporting requirements also arise for chemical registration and

Table 1.4 Surface area and porosity of selected nanoparticles

Nanoparticles	BET surface area, m^2/g	Pore volume, cm^3/g	Type of porosity
Mesoporous Silica (SBA-15)	670	0.95	Porous (meso)
MWCNTs	170	0.67	Porous (meso)
TiO_2 (Rutile)	28	0.20	Nonporous
ZnO	51	0.09	Nonporous
Fumed Silica	190	0.44	Porous (meso)

toxicological testing. It is evident that the many different nanomaterial and derivatives result in different properties of the specific nanomaterials. Therefore, there are increasing demands for proper methods and standard operational procedures for physicochemical characterization of nanomaterials and updated systems for chemical classification. As a first step it is important to be able to identify a nanomaterial, which is a key topic in ongoing research projects.

In this chapter, we gave examples on the key physicochemical characteristics and the type of data one can achieve. A variety of techniques can be used to obtain both structural and textural properties, which are essential to identify and report uniqueness of specific nanomaterials. Electron microscopy is the most widely used instrument for chemical and structural investigation of nanomaterials at relatively high resolution and even close to atomic levels. In principle, other non-destructive methods are also applicable, such as X-ray diffraction and gas absorption isotherms. It is anticipated that this type of data will be important in the next generation of material grouping and categorization systems as well as to understand what drives particle toxicity in greater depth. Even-though, significant progress have been made over the last decade, there is still no full understanding on how the different physicochemical parameters relate to nanomaterials hazards. In fact the parameters are still only used to some extent in this types of exploration. The challenge will be even greater acknowledging that complex nanomaterials have entered industrial scale production and needs to be included in such analysis.

Acknowledgements We gratefully acknowledge that this chapter was written with financial support from the EU FP7 project NANoREG (Grant 310584) and the Danish Centre for Nano-Safety funded by the Danish Work Environment Fund (Grant 49803).

References

1. Appendix R7-1 recommendations for nanomaterials applicable to chapter R7a endpoint specific guidance (2012)
2. Appendix R7-1 recommendations for nanomaterials applicable to chapter R7b endpoint specific guidance (2014a)
3. Nanomaterials producers directory 2014–2015 (2014b) Future Markets Inc
4. Atluri R, Bacsik Z, Hedin N, Garcia-Bennett AE (2010) Structural variations in mesoporous materials with cubic Pm(3)over-barn symmetry. Micropor Mesopor Mat 133(1–3):27–35

5. Atluri R, Keld Alstrup J (2014) Classification and reporting of nanostructured silica materials. Manuscript
6. Atluri R, Sakamoto Y, Garcia-Bennettt AE (2009) Co-structure directing agent induced phase transformation of mesoporous materials. Langmuir 25(5):3189–3195
7. Barber DJ, Freestone IC (1990) An investigation of the origin of the color of the lycurgus cup by analytical transmission electron-microscopy. Archaeometry 32:33–45
8. Beck JS, Vartuli JC, Roth WJ, Leonowicz ME, Kresge CT, Schmitt KD, Chu CTW, Olson DH, Sheppard EW, McCullen SB, Higgins JB, Schlenker JL (1992) A new family of mesoporous molecular-sieves prepared with liquid-crystal templates. J Am Chem Soc 114(27):10834–10843
9. Brunauer S, Emmett PH, Teller E (1938) Adsorption of gases in multimolecular layers. J Am Chem Soc 60:309–319
10. Castro Neto AH, Guinea F, Peres NMR, Novoselov KS, Geim AK (2009) The electronic properties of graphene. Rev Mod Phys 81(1):109–162
11. De Temmerman PJ, Verleysen E, Lammertyn J, Mast J (2014) Semi-automatic size measurement of primary particles in aggregated nanomaterials by transmission electron microscopy. Powder Technol 261:191–200
12. Delgado JL, Filippone S, Giacalone F, Herranz MA, Illescas B, Perez EM, Martin N (2014) Buckyballs. Top Curr Chem 350:1–64
13. Donaldson K, Murphy FA, Duffin R, Poland CA (2010) Asbestos, carbon nanotubes and the pleural mesothelium: a review of the hypothesis regarding the role of long fibre retention in the parietal pleura, inflammation and mesothelioma. Part Fibre Toxicol 7:5
14. Eckert M (2012) Max von Laue and the discovery of X-ray diffraction in 1912. Ann Phys 524(5):A83–A85
15. Farha OK, Eryazici I, Jeong NC, Hauser BG, Wilmer CE, Sarjeant AA, Snurr RQ, Nguyen ST, Yazaydin AO, Hupp JT (2012) Metal-organic framework materials with ultrahigh surface areas: is the sky the limit? J Am Chem Soc 134(36):15016–15021
16. Freestone I, Meeks N, Sax M, Higgitt C (2007) The lycurgus cup – a Roman nanotechnology. Gold Bulletin 40(4):270–277
17. Gaffet E (2011) Nanomaterials: a review of the definitions, applications, health effects. How to implement secure development. C R Phys 12(7):648–658
18. Georgakilas V, Perman JA, Tucek J, Zboril R (2015) Broad family of carbon nanoallotropes: classification, chemistry, and applications of fullerenes, carbon dots, nanotubes, graphene, nanodiamonds, and combined superstructures. Chem Rev 115(11):4744–4822
19. Jackson P, Kling K, Jensen KA, Clausen PA, Madsen AM, Wallin H, Vogel U (2014) Characterization of genotoxic response to 15 multiwalled carbon nanotubes with variable physicochemical properties including surface functionalizations in the FE1-Muta(TM) mouse lung epithelial cell line. Environ Mol Mutagen 56(2):183–203
20. Jana NR, Earhart C, Ying JY (2007) Synthesis of water-soluble and functionalized nanoparticles by silica coating. Chem Mater 19(21):5074–5082
21. Janez P (2010) European Commission recommendation on the definition of the term "Nanomaterial". O J E U 275 (L)(L):38–40
22. Jensen KA, Pojana G, Bilanicora D (2014) Characterization of manufactured nanomaterials, dispersion and exposure characterization for toxicological testing. In: Monterio NA, Tran CL (eds) Nanotoxicology: progress towareds nanomedicine. Taylor & Francis, Boca Raton, pp 45–73
23. Johnston HJ, Hutchison GR, Christensen FM, Peters S, Hankin S, Aschberger K, Stone V (2010) A critical review of the biological mechanisms underlying the in vivo and in vitro toxicity of carbon nanotubes: the contribution of physico-chemical characteristics. Nanotoxicology 4(2):207–246
24. Kim JH, Shim BS, Kim HS, Lee YJ, Min SK, Jang D, Abas Z, Kim J (2015) Review of nanocellulose for sustainable future materials. Int J Precis Eng Manuf-Green Tech 2(2):197–213
25. Klemm D, Kramer F, Moritz S, Lindstrom T, Ankerfors M, Gray D, Dorris A (2011) Nanocelluloses: a new family of nature-based materials. Angew Chem Int Ed 50(24):5438–5466

26. Klemm D, Schumann D, Kramer F, Hessler N, Hornung M, Schmauder HP, Marsch S (2006) Nanocelluloses as innovative polymers in research and application. Polysaccharides II 205:49–96
27. Kunzmann A, Andersson B, Thurnherr T, Krug H, Scheynius A, Fadeel B (2011) Toxicology of engineered nanomaterials: focus on biocompatibility, biodistribution and biodegradation. BBA-Gen Subjects 1810(3):361–373
28. Li D, Kaner RB (2006) Shape and aggregation control of nanoparticles: not shaken, not stirred. J Am Chem Soc 128(3):968–975
29. Lin W, Huang Y w, Zhou XD, Ma Y (2006) In vitro toxicity of silica nanoparticles in human lung cancer cells. Toxicol Appl Pharmacol 217(3):252–259
30. Liu Y, Zhao YL, Sun BY, Chen CY (2013) Understanding the toxicity of carbon nanotubes. Acc Chem Res 46(3):702–713
31. Murthy CR, Gao B, Tao AR, Arya G (2015) Automated quantitative image analysis of nanoparticle assembly. Nanoscale 7(21):9793–9805
32. Peng XH, Palma S, Fisher NS, Wong SS (2011) Effect of morphology of ZnO nanostructures on their toxicity to marine algae. Aquat Toxicol 102(3–4):186–196
33. Poulsen SS, Saber AT, Williams A, Andersen O, Kobler C, Atluri R, Pozzebon ME, Mucelli SP, Simion M, Rickerby D, Mortensen A, Jackson P, Kyjovska ZO, Molhave K, Jacobsen NR, Jensen KA, Yauk CL, Wallin H, Halappanavar S, Vogel U (2015) MWCNTs of different physicochemical properties cause similar inflammatory responses, but differences in transcriptional and histological markers of fibrosis in mouse lungs. Toxicol Appl Pharmacol 284(1):16–32
34. Prymak O, Ristig S, Meyer-Zaika V, Rostek A, Ruiz L, Gonzalez-Calbet J, Vallet-Regi M, Epple M (2014) X-ray powder diffraction as a tool to investigate the ultrastructure of nanoparticles. Russ Phys J 56(10):1111–1115
35. Pulskamp K, Diabate S, Krug HF (2007) Carbon nanotubes show no sign of acute toxicity but induce intracellular reactive oxygen species in dependence on contaminants. Toxicol Lett 168(1):58–74
36. Roduner E (2006) Size matters: why nanomaterials are different. Chem Soc Rev 35(7):583–592
37. Sager TM, Kommineni C, Castranova V (2008) Pulmonary response to intratracheal instillation of ultrafine versus fine titanium dioxide: role of particle surface area. Part Fibre Toxicol 5:17
38. Sivakumar S, Diamente PR, van Veggel FC (2006) Silica-coated Ln(3+)-doped LaF3 nanoparticles as robust down- and upconverting biolabels. Chem Eur J 12(22):5878–5884
39. Stoehr LC, Gonzalez E, Stampfl A, Casals E, Duschl A, Puntes V, Oostingh GJ (2011) Shape matters: effects of silver nanospheres and wires on human alveolar epithelial cells. Part Fibre Toxicol 8:36
40. Tallury P, Payton K, Santra S (2008) Silica-based multimodal/multifunctional nanoparticles for bioimaging and biosensing applications. Nanomedicine 3(4):579–592
41. Truong NP, Whittaker MR, Mak CW, Davis TP (2015) The importance of nanoparticle shape in cancer drug delivery. Expert Opin Drug Deliv 12(1):129–142
42. Warheit DB, Webb TR, Reed KL, Frerichs S, Sayes CM (2007) Pulmonary toxicity study in rats with three forms of ultrafine-TiO2 particles: differential responses related to surface properties. Toxicology 230(1):90–104
43. Xia T, Kovochich M, Liong M, Meng H, Kabehie S, George S, Zink JI, Nel AE (2009) Polyethyleneimine coating enhances the cellular uptake of mesoporous silica nanoparticles and allows safe delivery of siRNA and DNA constructs. ACS Nano 3(10):3273–3286
44. Yang H, Zhuang Y, Hu H, Du X, Zhang C, Shi X, Wu H, Yang S (2010) Silica-coated manganese oxide nanoparticles as a platform for targeted magnetic resonance and fluorescence imaging of cancer cells. Adv Funct Mater 20(11):1733–1741
45. Yi DK, Selvan ST, Lee SS, Papaefthymiou GC, Kundaliya D, Ying JY (2005) Silica-coated nanocomposites of magnetic nanoparticles and quantum dots. J Am Chem Soc 127(14):4990–4991

46. Zhao DY, Feng JL, Huo QS, Melosh N, Fredrickson GH, Chmelka BF, Stucky GD (1998) Triblock copolymer syntheses of mesoporous silica with periodic 50 to 300 angstrom pores. Science 279(5350):548–552
47. Zhu YW, Murali S, Cai WW, Li XS, Suk JW, Potts JR, Ruoff RS (2010) Graphene and graphene oxide: synthesis, properties, and applications. Adv Mater 22(35):3906–3924
48. Zou H, Wu SS, Shen J (2008) Polymer/silica nanocomposites: preparation, characterization, properties, and applications. Chem Rev 108(9):3893–3957
49. Zuin S, Pojana G, Marcomini A (2007) Effect-oriented characterization of nanomaterials. Nanotoxicology: characterization, dosing, and health effects. Taylor & Francis, New York, pp 19–57
50. Schneider T, Jensen KA (2009) Relevance of aerosol dynamics and dustiness for personal exposure to manufactured nanoparticles. J Nanopart Res 11 (7):1637–1650

Part II
Exposure

Chapter 2
Assessment of Human Exposure to ENMs

Araceli Sánchez Jiménez and Martie van Tongeren

Abstract Human exposure assessment of engineered nanomaterials (ENMs) is hampered, among other factors, by the difficulty to differentiate ENM from other nanomaterials (incidental to processes or naturally occurring) and the lack of a single metric that can be used for health risk assessment. It is important that the exposure assessment is carried out throughout the entire life-cycle as releases can occur at the different stages of the product life-cycle, from the synthesis, manufacture of the nano-enable product (occupational exposure) to the professional and consumer use of nano-enabled product (consumer exposure) and at the end of life.

Occupational exposure surveys should follow a tiered approach, increasing in complexity in terms of instruments used and sampling strategy applied with higher tiers in order tailor the exposure assessment to the specific materials used and workplace exposure scenarios and to reduce uncertainty in assessment of exposure. Assessment of consumer exposure and of releases from end-of-life processes currently relies on release testing of nano-enabled products in laboratory settings.

Keywords Engineered nanomaterials • Occupational exposure • Consumer exposure • Tiered approach

2.1 Occupational Inhalation Exposure Assessment of Engineered Nanomaterials

2.1.1 Why Carry Out an Exposure Assessment?

Exposure assessments can be carried out for a variety of reasons. For example, for (i) regulatory risk assessment, i.e. to provide evidence that exposures are below Derived No Effect Levels (DNELs) under the EU REACH (Registration, Evaluation, Authorization and Restriction of Chemicals) regulations; (ii) for risk management,

A.S. Jiménez • M. van Tongeren (✉)
Centre for Human Exposure Science (CHES), Institute of Occupational Medicine, Research Avenue North, Riccarton, Edinburgh EH14 4AP, UK
e-mail: Martie.VanTongeren@iom-world.org

L. Tran et al. (eds.), *Modelling the Toxicity of Nanoparticles*, Advances in Experimental Medicine and Biology, Vol. 947, DOI 10.1007/978-3-319-47754-1_2

i.e. by checking whether exposure is below Occupational Exposure Limits (OELs); (iii) to check the effectiveness of exposure control measures; (iv) to contribute to epidemiological studies or health surveillance.

Strategies for conventional chemicals are reasonably well established and in general, relatively easy to measure mass-based metrics are used for exposure assessment. In contrast, for engineered nanomaterials (ENMs) there are no established standard measurement protocols and no consensus on the most appropriate single exposure metric. In fact, the REACH Implementation Projects on Nanomaterials (RIP-oNs) recommended using more than a single metric [1]. The design of the measurement strategy will depend on the purpose of the study. ENMs are manufactured in many variations of size, shape, structure, and surface modifications. Exposure to ENM can occur as primary particles, aggregates or agglomerates (usually referred as nanostructured particles), as well as ENMs embedded in a matrix. The exact physical-chemical composition of the ENM can change across its lifecycle and can also change over time following release into environmental media to such an extent that exposure measurement is a challenging process.

There is strong evidence of a particle size-related and morphology-related health risk following inhalation of some aerosols. Fibre-shaped nanomaterials such as carbon nanotubes (CNT) and other high aspect ratio nanomaterials (HARN) such as graphene, nanoclay or silver nanowires, have been shown to pose particularly high risk to the respiratory system after inhalation exposure [17, 23, 47, 50, 54, 56]. Relevant information about exposure to nanomaterials can be gained from number and surface area concentration measurements. In addition, further characterisation of chemical and physical properties of airborne particles collected on filter samples is recommended [40].

2.1.2 Occupational Exposure Limits for Nanomaterials

There are very few OELs specifically for ENMs. For carbon nanotubes (CNT) (all types) and carbon nanofibers (CNF), NIOSH advocates a Recommended Exposure Limit (REL) of 1 $\mu g\ m^{-3}$ elemental carbon in the respirable fraction as an 8-h time-weighted average (TWA). This REL was established based on a review of animal studies and other toxicological data relevant to assessing the potential non-malignant adverse respiratory effects of CNT and CNF [41]. NIOSH recognizes that the REL level may not fully protect workers' health but will help to minimize the risk of developing lung disease. In order to test compliance with this mass-based limit, NIOSH recommends measurement of airborne elemental carbon as a proxy for CNT/CNF, according to the NIOSH method 5040 [37].

NIOSH also recommends a REL of 0.3 mg/m^3 for ultrafine (including nanoscale) TiO_2 as a TWA concentration for up to 10 h/day during a 40-h work week [40]. When the particulate exposure consists predominantly of TiO_2, then the NIOSH method 0600 [38] for measuring fine and ultrafine TiO_2 is recommended. If there is also exposure to other airborne particulates or when the size distribution of TiO_2

(fine vs. ultrafine) is unknown, other measurements and/or analytical techniques may be needed to characterize ultrafine TiO_2. Either NIOSH Method 7300 based on ICP-AES [39] can be used to analyse TiO_2, or electron microscopy, equipped with X-ray energy dispersive spectroscopy (EDS), to identify TiO_2 particles.

It is argued that for particles smaller than 100 nm due to their low mass compared to their larger particle number and surface area, mass may not be the most appropriate metric for health risk assessment [16, 44] and surface area is the preferred probably most relevant exposure metric [25]. NIOSH (like ISO) also acknowledges that surface area may be a more appropriate metric than mass; however, since there are currently no established analytical methods to assess specific particle surface area for TiO_2, mass-based measurements are accepted as a surrogate metric.

The British Standard Institute (BSI) has proposed bench-mark values for four types of ENMs [12]. For fibrous materials, the bench-mark value (0.01 fibres/ml) is based on the clearance limit in the UK for asbestos removal operations. For other ENMs, the bench-mark values are derived from the OEL of the corresponding micro-sized bulk material. For insoluble ENM this is 0.066 × OEL, for soluble is 0.5 × OEL and for carcinogenic, mutagenic, asthmagenic or reproductive toxin in bulk form ENM this is 0.1 × OEL. It should be noted these bench-mark values are for guidance only and should not be considered to representative of safe workplace exposure levels, as they have not been linked to toxicological end-points. They have been developed under the assumption that the hazard potential of the nanoparticle form is greater than the micron-sized particle. Van Broekhuizen et al. [52] introduced the concept of bench-mark values and proposed so called non-substance specific nano reference values. When the exposure exceeds an 'action level' more specific measurements or exposure controls are required.

In Germany, the Institute for Occupational Safety and Health of the German Social Accident Insurance (IFA) has established limit values for airborne particles between 1 and 100 nm based on the particle number concentration. For metals, metal oxides and other biopersistent granular ENMs with a density over 6,000 kg/m^3, concentrations should not exceed 20,000 particles/cm^3. For ENMs with densities below 6,000 kg/m^3 the concentration (1–100 nm) should not exceed 40,000 particles/cm^3 [24].

2.1.3 Measurement Devices

There are a number of different techniques to measure real-time particle number, mass, size distribution and surface area of airborne particles covering the particle size range from 3 nm to 20 μm. However, the principle of operation of particle size instruments limits the particle range that a single instrument can measure, and therefore in order to acquire the full size distribution a range of different instruments is employed. It should be noted that despite the fact that the definition of ENM refers to materials with one dimension <100 nm, their agglomerates and aggregates can reach micron sizes and therefore measurements should – cover both nano- and micron-size ranges [45]. There is currently no agreement on the upper size limit that should be assessed. It is likely that agglomerates/aggregates of nanosized particles will be largely in the respirable fraction ($D_{50} < 4$ μm). As well as being present as agglomerates/aggregates, ENMs can be scavenged by large background particles and ENM can also be released as part of relatively large particles consisting mainly of the matrix in which the ENM is embedded. High aspect ratio nanomaterials such and CNT and graphene may also be characterised by one dimension in the nano-size range but have large (>4 μm) physical sizes and lengths in the micron-size range [41, 47]. Hence, in addition to measurement of the respirable fraction, it may also be appropriate to include the inhalable fraction (D_{50} up to 100 μm).

Condensation particle counters (CPC) are the most common instruments used to measure the total (i.e. not size resolved) particle number concentration. They are available as portable and hand held devices making them suitable for screening assessment. The size range can be from up to 2.5 nm to 10 μm depending on the model. CPCs can be used in combination with a differential mobility analyser (DMA) to measure size – resolved particle number concentrations.

CPCs generally have two counting modes: a single particle count mode (up to 10^4–10^6 particles cm^{-3}) where each particle is counted individually and the photometric mode for concentrations above 10^6 particles cm^{-3} where the light scattered by all particles is measured and compared with calibration levels. Time resolutions are often down to 1 s measurement intervals. The accuracy of the single particle count mode is usually ±10, and the photometric mode is less accurate (±20). Accuracy of the CPC may also depend on the condensation fluid and particle. For example, the accuracy in the photometric mode differs between hygroscopic and hydrophobic particles when water is used as the condensation fluid. In recent years portable electrical diffusion chargers (e.g. DISCmini, Partector, NanoTracer) have been developed which can be used for personal monitoring. The electrical current, stemming from unipolar diffusion charged particles, is coincidentally proportional to the lung deposited surface area (as long as the particle size range is within 20 nm $\leq d_p \leq$ 400 nm) [4]. In addition these instruments also provide estimates of the total number concentration (usually 10–300 nm and or up to 10 μm in the case of the Partector), along with the mean particle diameter. Size-resolved instruments use electrical mobility analysers or differential mobility analysers. The most frequently used instruments are the FMPS (Fast Mobility Particle Sizer), and the SMPS (Scanning Mobility Particle Sizer). The Electrical Low Pressure Impactor (ELPI) can also be used to estimate the mass if the charge and the density of the particle are known and has the additional advantage that airborne particles are collected to allow for off-line analysis.

Most of these instruments are calibrated with spherical, compact, non-porous particles of a specific density. However, nano-sized particles and their agglomerated/aggregated forms tend to have a fractal-like structure and this can affect the accuracy of the measurements taken by these instruments. This type of instrumentation will also have limitations for the assessment of releases of fibre or platelet-shaped nanomaterials. Instrument-specific effects such as counting efficiency (e.g. CPC) and multiple charging (e.g. SMPS, FMPS) can also affect the measurement accuracy.

Since the instruments are calibrated with specific particles they only provide an equivalent diameter: electrical equivalent mobility diameter when sizing is by an electric field, diffusive (or thermodynamic) equivalent diameter, thermophoretic equivalent diameter or aerodynamic equivalent diameter for separation by impaction. For inhalation exposure the aerodynamic diameter is the most relevant equivalent diameter in the size range above approximately 100 nm. However, for particles below 100 nm diffusion due to Brownian motion is a more dominant deposition mechanism in the respiratory system and therefore the mobility-equivalent diameter is more relevant [27]. However, it is still unclear how the shape and density affect the electrical mobility diameter (and any other equivalent diameter) and therefore for nanofibres and nanoplates further studies are required to understand how they behave following inhalation [25].

Several studies have compared the performance of these devices [2, 22, 29, 53] using different aerosols morphologies and concentrations and found differences for the total number and the sizing of up to 30 %.

As stated above, the range of instruments deployed in nanoparticle detection should not be restricted to instrumentation covering the ENM primary size since airborne ENMs easily agglomerate into particles larger than several hundred nanometers. Aerodynamic Particle Sizers (size range 0.5–20 μm) using a time-of-flight light-scattering technique that measure aerodynamic diameter in real time can be usefully applied in exposure assessment. Aerodynamic diameter is a significant aerosol size parameter as it determines the particles' behaviour while airborne. Particle classification results from differences in the mobility of particles based on their size, density and charge as they travel through an optical detector. Results are presented as aerodynamic equivalent diameter, defined as the physical diameter of a unity density sphere that settles through the air with a velocity equal to that of the particle in question. Particles that have the same aerodynamic diameter will exhibit the same airborne behaviour and knowledge of the aerodynamic diameter subsequently allows determination of where the particle will be deposited in the human respiratory tract [13] and whether the particle will penetrate a filter, cyclone or other particle-removing device.

Measurements taken with direct reading instruments are useful to study variations in the metric assessed and size distribution during nano-related activities for comparison with background values. However, results of direct reading instruments alone should be interpreted with extreme caution [11, 15] in particular when used to derive mass related values for assessment of the exposure dose.

Another aspect that should be taken into consideration when using these instruments for exposure assessment is that they do not discriminate between the ENM and any other nano-sized particles present in the environment. Therefore to confirm the presence of the ENMs, characterization according to structure, size and morphology (Scanning/Transmission electron microscopy (SEM/TEM) and chemical identification (e.g. Energy Dispersive X-Ray (EDX); X-ray Photoelectron Spectroscopy (XPS), X-ray fluorescence (XRF)) of the particles collected on a filter is required. SEM or TEM are the most common methods used for particle characterization. However, SEM/TEM only provides information on the surface of a sample and therefore will not register ENM embedded in a matrix (e.g. polymer fragments with CNT). For nanocomposite materials atomic force microscopy (AFM) has been proved useful to characterize the ENM below the surface of composites [28, 59]. There are very few samplers specifically designed to collect the nano size fraction (<100 nm). Two samplers allow particle collection directly on the TEM grid: The Aspiration Electron Microscopy Sampler designed by VTT Technical Research Centre of Finland (from where it is commercially available) and the Mini Particle Sampler (MPS) developed by INERIS and distributed by EcoMesure.

2.1.4 Exposure Assessment Approaches of Engineered Nanomaterials

In contrast to conventional chemicals, where there are international standards for measuring, analysing and reporting of occupational exposure, for ENMs no established standard methods are available.

The International Standard Organization issued some guidelines in 2007 [25] and 2011 [26]. The reports provide very useful information on the available characterization methods but do not include details on how to analyse and interpret the measurement results and how to differentiate ENM from other nano-sized particles present in the workplace. In recent years several approaches have been published [3, 5, 6, 8, 35, 36, 40, 41, 57] and a number of initiatives have emerged across the nanosafety community to harmonize and standardize measurement strategies for ENMs. A series of international workshops "Global Harmonization of Measurement Strategies for Exposure to Manufactured Nano-Objects" have been organised since 2012 [9]. The European partnership of Occupational Health and Safety research (PEROSH) group has created a Nano Exposure and Contextual Information Database (NECID) to collect exposure measurement in a harmonized way.

These publications and workshop discussions formed the basis for the development of harmonized tiered approach published by the Organisation for Economic Co-operation and Development Working Party on Manufacture Nanomaterials (OECD WPMN SG8) [43]. The European Committee for Standardization is also preparing a document: 'Workplace Atmosphere- Assessment of inhalation exposure to manufactured nano-objects and their agglomerates (NOAA)' (CEN TC 137).

Most of the measurements strategies suggest a tiered approach:

- In the first tier contextual information on the materials, activities and exposure factors (e.g. amount of material used, ventilation, protective equipment, number of workers, frequency of exposure) is gathered to confirm that work with ENM is being carried out and exposure is possible.
- In the second tier, the concentration of airborne nanomaterials in the workplace is measured using a non-size selective real time portable particle number concentration instrument (e.g. CPC, NanoTracer, DISCmini, NanoCheck). The concentration during the activity is compared to background concentrations to assess

any potential increase in the particle number during the handling of the ENMs. In addition, some of the approaches recommend the collection of filter samples for off-line analysis using SEM or TEM coupled with a chemical identification technique. This helps to discriminate between the ENMs and nano-sized background particles. If activity concentrations are significantly increased over the background, the assessor may choose to evaluate the risk management measures and repeat a tier two assessment or to move directly to a tier three assessment.

- During a tier three assessment a more detailed survey is carried out which may include the measurement of personal exposure and/or the use of more complex equipment that provide real-time data on size-resolved particle number concentrations (e.g. FMPS, SMPS, ELPI), particle mass and/or surface area.

The main challenges highlighted in these approaches are (1) to distinguish the ENM from the background NMs (natural or incidental materials generated during the process, e.g. polymer particles release during extrusion); (2) to decide when to move from a basic survey to an in-depth campaign; and (3) to estimate quantitative exposure concentrations that can be used in health risk assessment.

Background particle concentration (especially when measured as particle number concentration) usually has a high spatial and temporal variability as they are affected by multiple emission sources (e.g. passing vehicles or nearby processes [34].

The approaches indicate three main strategies to assess the background:

- To measure during the activity under the same conditions but without using the ENM under investigation. This type of background allows determining the contribution of process-generated nano particles and therefore is the preferred method. However, this approach is not often feasible (e.g. for bagging activities).
- Far-field background: measurements collected at the same time as the activity in a place where no contribution of ENM is expected. This background concentration does not allow differentiation of process-generated nano particles and ENMs.
- Before & after the activity: again background measurements collected in this way do not allow differentiation of process-generated and other nano particles (e.g. from vehicles) and the ENMs of interest.

The nanoGEM approach [3] proposed to subtract the arithmetic mean (AM) of the background particle number concentration (measured for at least 45 min) from the AM measured during the activity. If the difference of the activity minus the background is larger than three times the SD of the background then release (if measurements are collected near the source) or exposure (if measurements are collected near the breathing zone of the operator) can be confirmed. The assumption is that that the background concentration will remain stable. However in practice this is may not be the case and this approach may not be applicable. However careful study design and interpretation can enable discrimination from the background in some circumstances. Considering the multiple sources of NMs in the workplace and the challenges to identify the ENM of concern, it is important to gather contextual information on other sources that could generate airborne nanoparticles.

Another important issue when carrying out a measurement survey is where to locate the measuring equipment in relation to the processes being monitored. Most studies target measurements areas in close proximity to the ENM source (<30 cm). While this is informative to determine whether there is emission of nanoaerosols into the workplace atmosphere, it is usually not representative of workers' exposure. Aerosol concentrations change over time due to deposition, diffusion and the effects of ventilation. This is particularly important for nano-sized aerosols due to their high diffusion rates and the effects of agglomeration and scavenging by background particles resulting in lower concentration and a shift of the size distribution. Consequently, measurements near the source or at any fixed point may not represent accurately the exposure of workers. Considering that only a few personal monitors are currently commercially available, the assessment of personal exposure is quite challenging. In this aspect modelling can be a very useful tool for exposure assessment and some advances have been made in relation to modelling of airborne nanoparticles in the work environment (both in field and theoretical; [19, 33, 48, 49]). In addition, if nanoaerosols are released into the workplace environment, they may deposit on work surfaces and act as a secondary exposure source to workers [55].

Regarding the decision criteria to move from a basic assessement to an in-depth monitoring survey the different approaches highlight different considerations. Witschger et al. [57] argues that the decision to carry out an in-depth monitoring campaign has to be taken considering the knowledge and experience of measurement of nanoaerosols, availability of instruments and methods, reachability of working location, compatibility of the instruments with the working environment and existence of previous measurements from the same place. Their basic level also includes chemical analysis of collected filter samples (e.g. TEM + EDX) in addition to CPC measurements. They also suggest that if the measurement campaign is likely to be challenging due to the working environment, release studies in a laboratory (e.g. dustiness measurement) can be carried out instead.

Brouwer et al. [8] proposed to divide the particle number concentration into those <100 nm and >100 nm. For both size fractions the decision to move to tier three should be based on a Student t-test on the concentrations during the activity and background; the ratio of those concentrations, the results from the TEM and EDX analyses and observations of the activities/processes taken place at the work place (Table 2.1).

Brouwer et al. [10] highlighted that data from direct reading instruments with short measurement intervals are autocorrelated and therefore the use of parametric methods, which are designed for independent data, is not appropriate. ARIMA (Autoregressive Integrated Moving Average) models could be used to estimate whether the activity has an effect on the level of particles compared with background levels [31, 32].

As part of the 7th Framework Project MARINA (Managing the Risk of Nanomaterials, http://www.marina-fp7.eu/) an exposure assessment strategy specifically developed for human risk assessment was developed. In this approach, also consisting of three tiers, the decision to move to a higher tier is not based solely on

Table 2.1 Decision criteria to move to tier three in the exposure assessment Brouwer et al. [8]

p-value (t-test)	Ratio AM-nanoactivity/ AM background	TEM	EDX	Observations	Overall likehood
<0.05	≥2.0	<100 nm + agglomerates	Yes	Absence other sources	Likely
<0.05	1.05–2.0	Agglomerates, few particles		Intrusion outdoor air	Possibly
>0.05	<1.05	Large particle agglomerates	Not	Other sources	Not likely

AM arithmetic mean, *TEM* transmission electron microscopy, *EDX* energy dispersive X-Ray

the exposure but on an assessment of risk (i.e. the combination of the exposure and hazard). In tier one all the available information is used to make a decision about the likelihood of emission into the workplace atmosphere. The approach points to several tools that could be used for such assessment (e.g. MARINA exposure library, control banding tools). If the health risk is not considered to be negligible, the user moves to tier two, where the emission of ENM is confirmed through measurements (off-line particle characterization and chemical identification) [47].

The strategies described above do not provide a consistent framework for – reporting the measurements. Some recommendations and guidelines are given; e.g. Brouwer et al. [8] recommends that the results should be summarised into size bins of ≤100 nm and >100 nm. Other issues such as the number of measurements required to obtain a representative concentration, assessment of personal exposure and transport processes from the source to the receptor are not fully addressed in any of the proposed strategies.

2.2 Consumers Exposure

The development of nanotechnology has unleashed the manufacture of consumer products containing ENMs. The Woodrow Wilson inventory (http://www.nanotechproject.org/cpi/) currently lists 1,600 products in the market claiming to contain ENMs. The Nanowerk database (http://www.nanowerk.com/nanomaterial-database.php) has information on the characteristics and suppliers of 3,000 different types of ENMs. The types of products spread across a wide range of categories, from building materials, sport equipment, electronics and automotive materials, nanomedicine, to personal use products that are used/applied directly on the skin such as clothes, deodorant, cosmetics and sun creams as well as food and food-packing materials [30].

Figure 2.1 shows the estimated maximum volume (metric tons/year) of ENMs used in different product categories in 2010 based on a marker study from Future Markets [18].

While the ENMs are meant to increase the performance of the product, their presence in consumer products has raised concerns over their safety towards human health and the environment. Consumer exposure can occur through direct contact

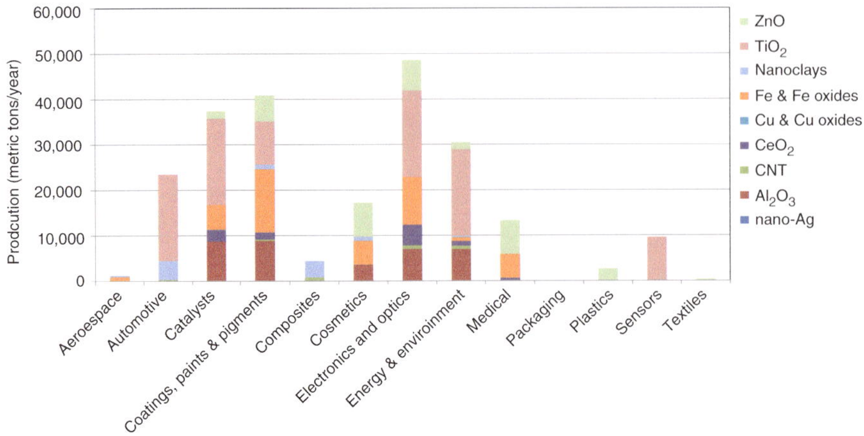

Fig. 2.1 Estimated global volume production of ENM across product sectors (Based on Keller et al. [30])

from the use of products (e.g. sun creams, clothes) or indirectly through the environment (e.g. contact with water, air, soil contaminated with ENMs) or products contaminated with ENMs (e.g. ENMs leached from food packing). Therefore, consumers can be exposed through all exposure pathways (inhalation, dermal, ingestion and eyes). Children exposure through mouthing of materials should also be considered as well as any other susceptible exposed populations.

Several studies have attempted to understand the mechanisms of ENM release from consumers' products (NanoRelease: http://www.ilsi.org/ResearchFoundation/RSIA/Pages/NanoRelease1.aspx: NanoHouse: http://www-nanohouse.cea.fr/; Scaffold: http://www.scaffold.eu-vri.eu/). These studies have provided very useful data on release to evaluate whether consumers can be potentially exposed through the environment; however, in most cases they do not provide quantitative exposure information. Within the NanoRelease project, exposure to consumers from sports equipment and electronics was estimated to be unlikely whilst exposure from tyres and textiles was found to be likely. Therefore, the magnitude of the release from consumer products will largely depend on the type of product, how much energy is applied to it (e.g. tires) and the presence of physical barriers between the ENMs and the consumers (e.g. electronics).The NanoHouse project concluded that for ENM used in paints under hard abrasion and leaching were mainly released embedded within a matrix or in agglomerate form. Very few single nanoparticles were released from paints. Pirela et al. [46] assessed consumers' exposures to particles release during laser printing using inks containing ENMs. The results showed that particles of silica, alumina, titania, iron oxide, zinc oxide, copper oxide, cerium oxide, carbon black among others could be released into the air during printing.

When assessing consumers' exposure it is important to know the conditions of use (specific use and exposure route, frequency, amount used per application), the form and characteristics of the ENM in contact with the human body and how the ENM migrates from the product to the body. ENMs are incorporated in different forms, suspended in liquids, suspended in solids, bound to the surface and embedded in a matrix [20]. It should be noted that the ENM released from the matrix might have different characteristics to the primary ENM incorporated into the manufactured product (e.g. coating may be removed) [42]. In order to understand the potential health effects and to develop acceptable daily intake levels it is important to be able to characterize these changes.

Modelling of consumers' exposure to ENM is less advanced than workers' exposure. Environmental flow dynamic models can predict the concentrations of ENM in different compartment levels providing therefore an insight on the potential indirect exposure to consumers. For example direct exposure to ENM in composites is considered to be unlikely. However, emissions to the environment at the end of life of the ENM are likely. Sun et al. [51] estimated annual emissions in 2014 in the EU to the surface water of 7,610 ton for nano-TiO_2, 1,330 ton for nano-ZnO and 1.27 ton CNT; for natural and urban soil of 2,230 ton for nano-TiO_2, 1,380 ton for nano-ZnO and 26.5 ton CNT; for sludge treated soil emissions were 45,400 for nano-TiO_2, 1.35 ton for nano-ZnO and 8.67 ton CNT and emission in the air of 324 ton for nano-TiO_2, 149 ton for nano-ZnO and 2.80 ton CNT.

Despite these studies and the existing consumer product inventories (e.g. Woodrow Wilson inventory; ANEC-BEUC 2010 inventory of consumer products containing nanomaterials; Wijnhoven et al. [58]), there are few quantitative data available on consumers' exposure.

Using the best estimates available and/or worst-case assumptions, Hansen et al. [21] estimated consumer exposure to be 26, 15, and 44 μg/kg bw/year for a facial lotion, a fluid product, and a spray product containing nanoparticles, respectively. Chen et al. [14] estimated a mass of nano-TiO_2 in the breathing zone of 170 $\mu g/m^3$ during 2.5 min application of a bathroom cleaning propellant spray containing nano TiO_2.

The assessment of consumers' exposure to ENM is more challenging, as the materials contained in the products are not well characterized and are mixed with other chemicals that affect the release and transfer of the ENM and their availability for exposure. Further studies using standard protocols are required to better understand consumers' exposure.

2.3 Exposure During the End of Life

There is little information available on the potential for exposure to nanomaterial-containing products during disposal and recycling stages. Established recycling schemes are available for consumer products such as electronics, packaging as well as large-scale dismantling of appliances, cars, aircraft and structures like wind turbine blades. Releases of ENM from the product matrix and exposure through end-of-life processes are possible as they generally involve high energy, abrasive processes, e.g. incineration, shredding, cutting, bailing and storage in open space environments [7]. Further studies are required to better understand the release scenarios associated with end-of-life processes.

References

1. Aitken RA, Bassan A, Friedrichs S, Hankin SM, Hansen SF, Holmqvist J, Peters SAK, Poland CA, Tran CL (2011) Specific advice on exposure assessment and hazard/risk characterisation for nanomaterials under REACH
2. Asbach C, Kaminski H, Von Barany D, Kuhlbusch TAJ, Monz C, Dziurowitz N, Pelzer J, Berlin K, Dietrich S, Götz U, Kiesling HJ, Schierl R, Dahmann D (2012) Comparability of portable nanoparticle exposure monitors. Ann Occup Hyg 56:606–621
3. Asbach C, Kuhlbusch, TAJ, Kaminski H, Stahlmecke B, Plitzko S, Götz U, Voetz M, Kiesling HJ, Dahmann D (2012) NanoGEM standard operation procedures for assessing exposure to nanomaterials, following a tiered approach
4. Asbach C, Neumann V, Monz C, Dahmann D, van Tongeren M, Alexander C, MacCalman L, Todea AM (2017) On the effect of wearing personal nanoparticle monitors on the comparability of personal exposure measurements. Environ. Sci.: Nano, Advance Article, DOI: 10.1039/c6en00362a
5. BAuA, BG RCI, IFA, IUTA, TUD, VCI (2011) Tiered approach to an exposure measurement and assessment of nanoscale aerosols released from engineered nanomaterials in workplace operations
6. Bekker C, Kuijpers E, Brouwer DH, Vermeulen R, Fransman W (2015) Occupational exposure to nano-objects and their agglomerates and aggregates across various life cycle stages; A broad-scale exposure study. Ann Occup Hyg 59:681–704
7. Boldrin A, Hansen SF, Baun A, Hartmann NIB, Astrup TF (2014) Environmental exposure assessment framework for nanoparticles in solid waste. J Nanopart Res 16:2394
8. Brouwer D, van Duuren-Stuurman B, Berges M, Jankowska E, Delphine B, Mark D (2009) From workplace air measurement results toward estimates of exposure? Development of strategy to assess exposure to manufactured nano-objects. J Nanopart Res 11:1867–1881
9. Brouwer D, Berges M, Virji MA, Fransman W, Bello D, Hodson L, Gabriel S, Tielemans E (2012) Harmonization of measurement strategies for exposure to manufactured nano-objects; report of a workshop. Ann Occup Hyg 56:1–9
10. Brouwer DH (2013) Workplace air measurements and likelihood of exposure to manufactured nano-objects, agglomerates, and aggregates. J Nanopart Res 15:2090
11. Brouwer D, Boessen R, van Duuren-Stuurman B, Bard D, Moehlmann C, Bekker C, Fransman W, Entink RK (2016) Evaluation of Decision Rules in a Tiered Assessment of Inhalation Exposure to Nanomaterials. Ann Occup Hyg 60(8):949–959
12. BSI (2010) Nanotechnologies—part 3: guide to assessing airborne exposure in occupational settings relevant to nanomaterials. British Standards Institution, London (BSI PD 6699-3:2010)
13. European Committee for Standardisation (CEN) (1993) Workplace atmospheres – size fraction definitions for measurement of airborne particles. Standard EN 481, Brussels
14. Chen BT, Afshari A, Stone S, Jackson M, Schwegler-Berry D, Frazer DG, Castranova V, Thomas AT (2010) Nanoparticles-containing spray can aerosol: characterization, exposure assessment, and generator design. Inhal Toxicol 22:1072–1082
15. Dahm MM, Evans DE, Schubauer-Berigan MK, Birch ME, Deddens JA (2013) Occupational exposure assessment in carbon nanotube and nanofiber primary and secondary manufacturers: mobile direct-reading sampling. Ann Occup Hyg 57:328–344
16. Donalson K, Li XY, Macnee W (1998) Ultrafine (nanometer) particle mediated lung injury. Aerosol Science 29:553–560
17. Donaldson K, Murphy FA, Duffin R, Poland CA (2010) Asbestos, carbon nanotubes and the pleural mesothelium: a review of the hypothesis regarding the role of long fibre retention in the parietal pleura, inflammation and mesothelioma. Part Fibre Toxicol 7:5
18. Future Markets (2012) The global market for nanomaterials 2002–2006; production volumes, revenues and end use markets
19. Guichard R, Tanière A, Belut E, Nimbert N (2014) Simulation of nanoparticle coagulation under Brownian motion and turbulence in a differential–algebraic framework: developments and applications. Int J Multiphase Flow 64:73–84
20. Hansen SF, Michelson ES, Kamper A, Borling P, Stuer-Lauridsen F, Baun A (2008) Categorization framework to aid exposure assessment of nanomaterials in consumer products. Ecotoxicology 17:438–447

21. Hansen SF (2010) Regulation and risk assessment of nanomaterials – Too little, too late? PhD Thesis. Technical University of Denmark
22. Hornsbya KE, Pryor SC (2014) A laboratory comparison of real-time measurement methods for 10–100-nm particle size distributions. Aerosol Sci Tech 58:571–582
23. IARC (2015) Carbon nanotubes (CNT), fluoro-edenite and silicon carbide, vol 111. WHO, International Agency for Research on Cancer
24. IFA (2009) Criteria for the assessment of the effectiveness of protective measures. Available at: http://www.dguv.de/ifa/Fachinfos/Nanopartikel-am-Arbeitsplatz/Beurteilung-von-Schutzma%C3%9Fnahmen/index-2.jsp. Last accessed: 25 Aug 2016
25. ISO/TR 27628: 2007 Workplace atmospheres – Ultrafine, nanoparticle and nano-structured aerosols – Inhalation exposure characterization and assessment. http://www.dguv.de/ifa/Fachinfos/Nanopartikel-am-Arbeitsplatz/Beurteilung-von-Schutzma%C3%9Fnahmen/index-2.jsp
26. ISO 28439: 2011 Workplace atmospheres – characterization of ultrafine aerosols/nanoaerosols – determination of the size distribution and number concentration using differential electrical mobility analysing systems
27. ISO/TR 1225: 2012 Nanomaterials-quantification of nano-object release from powders by generation of aerosols
28. Jespersen TS, Nygård J (2007) Mapping of individual carbon nanotubes in polymer/nanotube composites using electrostatic force microscopy. Appl Phys Lett 90:183108
29. Kaminski H, Kuhlbusch TAJ, Rath S, Götz U, Sprenger M, Wels D, Polloczek J, Bachmann V, Dziurowitz N, Kiesling HJ, Schwiegelshohn A, Monz C, Dahmann D, Asbach C (2013) Comparability of mobility particle sizers and diffusion chargers. J Aerosol Sci 57:156–178
30. Keller AA, McFerran S, Lazareva A, Suh S (2013) Global life cycle releases of engineered nanomaterials. J Nanopart Res 15:1692
31. Klein Entik RH, Bekker C, Fransman W, Brouwer D (2015) Analysis of time series of particle size distributions in nano exposure assessment. J Aerosol Sci 81:62–69
32. Klein Entik RH, Fransman W, Brouwer D (2011) How to statistically analyse nano exposure measurement results: using an ARIMA time series approach. J Nanopart Res 123:6991–7004
33. Koivisto AJ, Yu M, Hämeri K, Seipenbusch M (2012) Size resolved particle emission rates from an evolving indoor aerosol system. Aerosol Sci 47:58–69
34. Kuhlbusch TAJ, Asbach C, Fissan H, Göhler D, Stintz M (2011) Nanoparticle exposure at nanotechnology workplaces – a review. Part Fibre Toxicol 8:22
35. Methner M, Hodson L, Geraci C (2010) Nanoparticle emission assessment technique (NEAT) for the identification and measurement of potential inhalation exposure to engineered nanomaterials – part A. J Occup Environ Hyg 7:127–132
36. Methner M, Hodson L, Geraci C (2010c) Nanoparticle emission assessment technique (NEAT) for the identification and measurement of potential inhalation exposure to engineered nanomaterials—Part B: results from 12 field studies. J Occup Environ Hyg 7:163–176
37. NIOSH (1994) Method 5040: elemental carbon (diesel particulate) method. NIOSH manual of analytical methods, 4th edn. NIOSH, Cincinnati, OH. DHHS (NIOSH)
38. NIOSH (1998) Method 0600: particulates not otherwise regulated (respirable) NIOSH manual of analytical methods, 4th edn. NIOSH, Cincinnati, OH. DHHS (NIOSH)
39. NIOSH (2003) Method 7300: elements by ICP. NIOSH manual of analytical methods, 4th edn. NIOSH, Cincinnati, OH. DHHS (NIOSH)
40. NIOSH (2011) Current intelligence bulletin 63: occupational exposure to titanium dioxide. U.S. Department of Health and Human Services, Centers for Disease Control, National Institute for Occupational Safety and Health. DHHS (NIOSH) Publication No. 2011–160, Cincinnati
41. NIOSH (2013) Current intelligence bulletin 65: occupational exposure to carbon nanotubes and nanofibers. Cincinnati, OH: U.S. Department of Health and Human Services, Centers for Disease Control, National Institute for Occupational Safety and Health. DHHS (NIOSH) Publication No. 2013–145
42. Nowack B, Ranville JF, Diamond S, Gallego-Urrea JA, Metcalfe C, Rose J, Horne N, Koelmans AA, Klaine SJ (2012) Potential scenarios for nanomaterial release and subsequent alteration in the environment. Environ Toxicol Chem 31:50–59
43. OECD (2015) Harmonized tiered approach to measure and assess the potential exposure to airborne emissions of engineered nano-objects and their agglomerates and aggregates at work-

places. OECD Health and Safety Publications. Series on the Safety of Manufactured Nanomaterials. No.55. ENV/JM/MONO(2015)19
44. Oberdörster G (1996) Significance of particle parameters in the evaluation of dose-response relationships of inhaled particles. Inhal Toxicol 8 Suppl:73–89
45. Peters TM, Elzey S, Johnson R, Park H, Grassian VH, Maher T, O'Shaughnessy P (2009) Airborne monitoring to distinguish engineered nanomaterials from incidental particles for environmental health and safety. J Occup Environ Hyg 6:73–81
46. Pirela SV, Sotiriou GA, Bello D, Shafer M, Bunker KL, Castranova V, Thomas T, Demokritou P (2014) Consumer exposures to laser printer-emitted engineered nanoparticles: a case study of life-cycle implications from nano-enabled products. Nanotoxicology 11:1–9
47. Schinwald A, Murphy FA, Jones A, MacNee W, Donaldson K (2012) Graphene-based nanoplatelets: a new risk to the respiratory system as a consequence of their unusual aerodynamic properties. ACS Nano 6:736–746
48. Schneider T, Brouwer DH, Koponen IK, Jensen KA, Fransman W, Van Duuren-Stuurman B, Van Tongeren M, Tielemans E (2011) Conceptual model for assessment of inhalation exposure to manufactured nanoparticles. J Expo Sci Environ Epidemiol 21:450–463
49. Seipenbusch M, Binder A, Kasper G (2008) Temporal evolution of nanoparticle aerosols in workplace exposure. Ann Occup Hyg 52:707–716
50. Silva RM, Xu J, Saiki C, Anderson DS, Franzi LM, Vulpe CD, Gilbert B, Van Winkle LS, Pinkerton KE (2014) Short versus long silver nanowires: a comparison of in vivo pulmonary effects post instillation. Part Fibre Toxicol 11:52
51. Sun TY, Bornhöft NA, Hungerbühler K, Nowack B (2016) Dynamic Probabilistic Modeling of Environmental Emissions of Engineered Nanomaterials. Environ. Sci. Technol., 50:4701–4711
52. Van Broekhuizen P, Van Broekhuizen F, Cornelissen R, Reijnders L (2012) Workplace exposure to nanoparticles and the application of provisional nanoreference values in times of uncertain risks. J Nanopart Res 14:770
53. Mills JB, Parka JH, Petersa TM (2014) Comparison of the DiSCmini aerosol monitor to a handheld condensation particle counter and a scanning mobility particle sizer for submicrometer sodium chloride and metal aerosols. J Occup Environ Hyg 10:250–258
54. Murphy FA, Poland CA, Duffin R, Al-Jamal KT, Ali-Boucetta H, Nunes A, Byrne F et al (2011) Length-dependent retention of carbon nanotubes in the pleural space of mice initiates sustained inflammation and progressive fibrosis on the parietal pleura. Am J Pathol 178:2587–2600
55. UK NanoSafety Partnership Group (UKNSPG) 20120 Working Safely with Nanomaterials in Research & Development
56. Verma NK, Moore E, Blau W, Volkov Y, Babu PR (2012) Cytotoxicity evaluation of nanoclays in human epithelial cell line A549 using high content screening and real-time impedance analysis. J Nanopart Res 14:1137
57. Witschger O, Le-Bihan O, Reynier M, Durand C, Charpentier D (2012) Préconisation en matière de caractérisation et d'exposition des potentiels d'emission et d'exposition professionnelle aux aerosols lors d'operations nanomateriaux, INRS – Hygiène et sécurité du travail – 1er trimester 2012, 226:41–55
58. Wijnhoven WEP, Oomen Ir AJ, Sips AJAM, Bourgeois FC, Dorsthorst GJPM te, Kooi MW, Bakker MI (2010) Development of an inventory for consumer products containing nanomaterials. DG Environment within the framework of ENV/D3/SER/2010/0060r
59. Zhao M, Gu X, Lowther SE, Park C, Jean YC, Nguyen T (2010) Subsurface characterization of carbon nanotubes in polymer composites via quantitative electric force microscopy. Nanotechnology, 20;21(33):339801

Chapter 3
The Life Cycle of Engineered Nanoparticles

David González-Gálvez, Gemma Janer, Gemma Vilar, Alejandro Vílchez, and Socorro Vázquez-Campos

Abstract The first years in the twenty-first century have meant the inclusion of nanotechnology in most industrial sectors, from very specific sensors to construction materials. The increasing use of nanomaterials in consumer products has raised concerns about their potential risks for workers, consumers and the environment. In a comprehensive risk assessment or life cycle assessment, a life cycle schema is the starting point necessary to build up the exposure scenarios and study the processes and mechanisms driving to safety concerns. This book chapter describes the processes that usually occur at all the stages of the life cycle of the nano-enabled product, from the nanomaterial synthesis to the end-of-life of the products. Furthermore, release studies reported in literature related to these processes are briefly discussed.

Keywords Life Cycle • Nanomaterials • Nanocomposites • Release • Risk

3.1 Introduction

The increasing use of nanomaterials in industrial and consumer products results in a potential risk for workers, consumers and the environment.

A starting point for any comprehensive risk assessment or life cycle assessment is the identification of all relevant life cycle steps, so that all scenarios with a potential risk can be evaluated. The life cycle is totally product-dependent, as each product has its own manufacturing processes, uses and waste treatment and, so, its own hotspots for nanomaterial release and associated risks. This chapter presents a brief overview of the most common processes that take place at different steps of a product life-cycle (Fig. 3.1) and highlights the potential contribution of each step to the release of nanomaterials and associated risk.

D. González-Gálvez • G. Janer • G. Vilar • A. Vílchez • S. Vázquez-Campos (✉)
LEITAT Technological Center, C/ de la Innovació 2, 08225 Terrassa (Barcelona), Spain
e-mail: svazquez@leitat.org

L. Tran et al. (eds.), *Modelling the Toxicity of Nanoparticles*, Advances in Experimental Medicine and Biology, Vol. 947, DOI 10.1007/978-3-319-47754-1_3

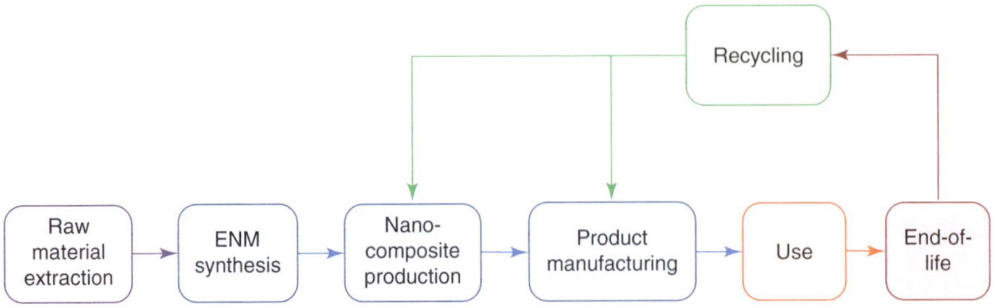

Fig. 3.1 Life-cycle of nano-enabled products

It is also important to clearly identify all the processes implied in each life cycle stage to know which mechanisms drive to release. Apart from release quantification, the form of release (isolated particles, aggregates, embedded in a matrix, surface modified by hydration or oxidation, etcetera) may be determinant for the hazard evaluation.

3.1.1 Production Levels of Engineered Nanomaterials

Some nanomaterials, such as carbon black and silica, have been industrially used for decades. However, during the last decade, new materials and modifications have allowed a dramatic expansion of nanotechnology. Despite the multiple materials that are being investigated at a research scale, at this moment it is estimated that nanomaterials produced at an industrial scale belong to only around 20 chemical classes [1]. At the moment, any attempt to determine nanomaterials production has to be based on estimations as they do not have to be reported. Only France and Denmark have recently regulated nanomaterials in products, so these nano-additivated products have to be registered and labelled in order to inform the consumers [2, 3].

Attempts to estimate the production levels and applications of nanomaterials have been based on information provided by industry through surveys. Sometimes the data collected relates to production capacities and sometimes to actual production amounts and the geographical area under scope also differs [4–6]. One of the most thorough recent surveys is that of Piccino et al., who send a survey to industrial representatives from companies producing or using nanomaterials to estimate the worldwide or Europe-wide production of such materials [7]. A considerable large variability among answers by different industrial representatives reflects the general uncertainties related to the actual worldwide production volumes. However, there was general agreement that silica, titanium dioxide, zinc oxide, carbon nanotubes, iron oxides, aluminium oxides, and cerium oxides are the nanomaterial types with highest production volumes. The median production quantities for each of these nanomaterials ranged between 55 and 5500 tonnes per year worldwide [7], depending on the material. These values are also consistent with estimated annual production volumes in China in 2012, which ranged from 200 to around 1300 tonnes for titanium dioxide, zinc oxide, aluminium oxide, zirconium oxide, and silver [8].

According to the French registry [9], the quantities produced and imported in 2014 are around 274,000 and 122,000 tones, respectively. Carbon black and silicon dioxide are the category of substances with largest produced or imported quantities, both above 100,000 t/year. These are followed by calcium carbonate and titanium dioxide, with volumes ranging 10,000–100,000 t/year. Other materials reported to be produced or imported in amounts above 1000 t/year are aluminium oxide, boehmite (γ-Al(OH)O), calcium 4-[(5-chloro-4-methyl-2-sulphonatophenyl)azo]-3-hydroxy-2-naphthoate, reaction mixture of cerium dioxide and zirconium dioxide, polyvinyl chloride, and magnesium silicate.

3.2 Engineered Nanomaterials Synthesis

The synthesis of ENMs is the step of a nano-enabled product life cycle that has received the highest attention in the literature in relation to the potential risks for human health [10–13]. By contrast, the potential release of ENMs to the environment during this step has received little attention and it is commonly assumed to be low, though this completely depends on the procedures used during the production, cleaning and maintenance [14–17].

Due to the novelty of the field and the continuous research in the development of new nanomaterials, multitude of synthetic methodologies can be found in the literature. Most of these processes are adequate for laboratory scale and even pilot scale synthesis, but completely unworkable at industrial scales. Synthetic methods are often divided in top-down and bottom-up methods [18].

- **Top-down methods**. The successive cutting or slicing of bulk materials into nanomaterials play an important role in industrial synthesis of nanostructures that need specific shapes/sizes such as nanotransistors. Lithography, milling and attrition are the most common top-down processes used at the moment. The principal disadvantages of top-down approaches are the internal stress and the imperfection introduced in the surface structure due to the use of so energetic techniques. Such imperfections may have dramatic effect over surface chemistry and physical properties of such prepared nanomaterials.
- **Bottom-up methods**. Atom by atom chaotic building of nanomaterials comprises most of synthetic methods as these are also the most common procedures in materials science. The main disadvantage of these methods is that usually a distribution of sizes is obtained, but compositions are more homogeneous than in top-down approaches.

Synthetic methods are usually divided in dry or wet synthesis; open or enclosed reactions; and gas-, solid- or liquid-phase reactions. Moreover, depending on the synthetic method and the material, different purification steps may be necessary and, in the case of coated particles, one or more modification steps with extra purification are needed. To facilitate comprehension, this section is organized first by nanomaterial group, and then by synthetic methods.

3.2.1 Carbon Based Nanomaterials Production

Sizable quantities of carbon based nanomaterials can be produced using various methods; among them, plasma based, thermal and hydrothermal syntheses processes are the most used techniques [19–22].

The two most common plasma methods in the literature are Arc Discharge and Laser Ablation. Arc Discharge Method consists in passing current between two graphite electrodes under helium, hydrogen or methane at low pressure in presence of transition metal based catalysts [23–28]. This causes vaporization of graphite that condenses over the cathode (and walls of the reactor). Carbon nanotubes (CNT) can also be produced by Laser Ablation, which is similar to Arc Discharge but the energy is provided by a laser. This laser vaporizes graphite and catalyst, so that nanocatalysts are formed and the carbon nanomaterials grow over them [29–35].

Thermal synthesis methods are also very abundant in the literature for the production of CNT. Carbon Vapour Deposition (CVD) consists in the decomposition of a carbon source (usually a hydrocarbon over a transition metal catalyst). Both type of carbon source and catalyst affect in the CNT growth. Carbon based nanomaterials can also be produced by sono- or hydrothermal methods, which consist on the heating of a hydrocarbon/water mixture under pressure in the presence of a catalyst (usually Ni) [36–38].

These methods usually produce low quantities of carbon based nanomaterials (fullerenes, SWCNT, MWCNT, etcetera) mixed with other allotropic forms of carbon. The conditions used during the synthesis favour one form over the others, but purification steps are always necessary.

Graphene, graphene oxide and derivatives are synthesized very differently [39–43]. The bottom-up approaches used for the rest of carbon based nanomaterials are modified to get 2D carbon layers over a support, which has to avoid 3D growth [44]. Graphene can be also produced by top-down approaches. The purest and most perfect graphene is produced by exfoliation of graphite [45–47], graphite oxide (followed by reduction) [48–50] and carbon nanotubes [51].

Carbon based nanomaterials surface can be modified to improve their dispersibility, their compatibility with a matrix or to functionalize them to add chemical groups that can later react or bond to any other entity, such as antibodies [52, 53], quantum dots [54] or gold nanoparticles [55]. These modifications are usually based on the following approach. In a first step, the nanomaterials are oxidized by a hydrothermal process (i.e.: sonication in presence of diluted nitric acid) that causes defects on the surface. Then, the hydroxyls and carboxylic acids formed are used for subsequent functionalization by traditional chemical reactions. Other strategies include direct arylation, carbene or nitrene addition or Friedel-Craft acylation.

Carbon nanotubes and nanofibers have centred most of the attention of hygienists and toxicologists due to their fast increase in production volume and observed dustiness [16, 56–64]. Most published studies on occupational exposure to CNT show that, as the synthesis processes take place in closed reactors, most of the exposure occurs during material recovery and during cleaning and maintenance operations [57, 65–67].

3.2.2 Metallic Nanomaterials Production

Metallic nanomaterials (MNM) are traditionally synthesized by the reduction of a precursor under controlled conditions and in the presence of a stabilizer [68]. The solvent, conditions and stabilizers used depend on the element of the NP and on the purpose. Wet syntheses have been traditionally considered of lower risk in terms of occupational exposure due to lower aerosol formation compared to the work with powders.

Noble and seminoble metallic nanoparticles (Ag [69], Au [70], Pt [71], Pd [72, 73], Ru [74], Rh [75], Ir [76]) are usually industrially synthesized by reduction of a precursor salt in water (such as $HAuCl_4$ or $RuCl_3$). Once the colloids are synthesized, they can be directly functionalized in situ or phase-transferred to an organic solvent for further surface functionalization when necessary [77]. The conditions and precursors used completely determine the results in terms of size and morphology [78]. Although less common, all these materials can be produced by other methods, such as electrochemical deposition, physical synthesis or sol-gel method.

More reactive metallic nanoparticles are synthesized by similar approaches (which can be also used with noble metals), but under more controlled conditions (air-free atmosphere, organic solvents, ionic liquids, etcetera). The most common strategy consists on direct reduction or decomposition of organometallic compounds or metallic carbonyls, such as $Fe(CO)_5$, $Co_2(CO)_8$ or Ru(cod)(cot). Initially these syntheses used to produce polydisperse NP, but their optimization has improved the control of size and nowadays they are used for the synthesis of zerovalent nanoparticles of several metals: Fe, Co, Ru, Ir, Au, Ni or Rh, and different types of mixtures (core-shell, alloys...) [79, 80]. Surface modification of these materials is usually done *in situ*, and the organic modifier is used as stabilizer.

Though most metallic nanoparticles are produced by wet processes, occupational exposure cannot be neglected [81–86]. Release of metallic nanomaterials to the environment has received some attention, particularly Ag-NPs release, but most studies use assumptions to estimate the release during the MNP synthesis rather than actual measurements [87].

3.2.3 Oxide Based Nanomaterials Production

Several physical and chemical routes for the synthesis of nanometal oxides (NMOx) have been reported. Solution routes are the most widely used at laboratory scale, as they need more easily accessible set-ups and allow having a better control size and shape. In contrast, at the industrial scale, gas phase methods are the most commonly used as they are usually cleaner and better conversions are obtained. Moreover, they can be used for the deposition of thin films or nanostructures over a particular substrate. Both solution and gas-phase methods are widely used to produce different NMOx such as ZnO [88], TiO_2 [89], Fe_xO_y [90], CeO_2 [91, 92], ZrO_2 [93], Cu_xO [94], Al_2O_3 [95], SiO_2 [96–98], Co_xO_y [99], etcetera.

Solution routes are based on the decomposition of a salt or alkoxide precursor in solution, normally by means of a source of energy, to form the nanomaterial. Thereafter, this nanomaterial is separated from the solution by centrifugation, nanofiltration or other nano-appropriate techniques. Some examples of solution routes include solvo-/hydrothermal method, precipitation method, electrochemical synthesis, sonochemical method, sol-gel method and microemulsion.

In gas phase methods, metal vapour is produced by thermal, laser ablation, electron beam, ion beam, molecular beam or by vaporizing and dissociating any metal precursor. This metal vapour reacts with oxygen to produce the metal oxide that is deposited on the bottom and the internal walls of the reactor.

Different surface modification strategies exist. The most commonly reported are: (i) chemical functionalization of the surface by bonding of a silane derivate, which renders a very stable modification; (ii) addition of compounds that have affinity for NMOx surfaces, such as carboxylates or phosphates; and (iii) polymer grafting [100].

Occupational exposure to nanometal oxides in production facilities has not received much attention in the literature. In general, the reported studies show that the exposure is due to specific operations such as reactor opening, material recovery or cleaning and maintenance [86, 101–104].

3.2.4 *Quantum Dots Production*

Quantum dots (QD) can be produced by several methods, both top-down and bottom-up approaches [105–107]. The advantage of top-down processes is that very well-defined QD are produced. This is necessary, for instance, when producing nano-transistors for computing. However, these high energy methods usually result on physical and chemical damage to the particle surface. On the other hand, bottom-up approaches produce smaller and purer particles, as necessary in sensing applications.

Lithography, reactive-ion etching and wet chemical etching are the most commonly used top-down processes. The bottom-up processes are very similar to the ones explained for nanometal oxides synthesis.

The surface of most quantum dots is very easy to modify, as several functional groups have affinity for them (thiols, amines, carboxylic acids, etcetera). Thus, a multitude of papers report QD functionalization with biomolecules (DNA, RNA, proteins...) [108, 109], polymers [110] or other nanomaterials [111].

Despite of the high concern about quantum dots toxicity [106, 112–114] and their potential impact over workers and the environment, there is a lack of experimental data on the exposure to quantum dots during synthesis steps. Moreover, the only study that was found at the moment focused on a labscale synthesis [104].

3.2.5 *Polymeric and Ceramic Nanofibres Production*

Nanofibres are usually produced by the electrospinning method [115–119]. This method produces non-woven fabrics, in which the fibres are randomly oriented and connected by physical entanglements or bonds, without any knitting or stitching. In the electrospinning method, polymers are usually dissolved in a proper solvent (or molten) and the nanofiber is produced by high voltage. Recently, this technique has been extended to ceramic nanofibres synthesis. In this case, polymeric nanofibres loaded with ceramic precursors are prepared by electrospinning and, later, combusted to render the ceramic nanofiber.

Nanofibres production is usually done in closed conditions and neither occupational nor environmental exposure have been reported during this life cycle step. All the publications have focused on secondary manufacturers [120].

3.3 Nanocomposite Production

The incorporation of the nanomaterial in or on a matrix is a key step in nano-enabled products life cycle (except on those cases where nanomaterials are a final product by themselves, such as nanocatalysts). Nanotechnology has greatly progressed during the last two decades and, nowadays, we can find applications for almost any type of nanomaterial in any type of matrix.

A nanocomposite is a multiphase solid material that has at least one of the phases in the nanoscale. The main difference between nanocomposites and traditional composites is the high surface of contact between the phases in the first case. The addition of nanomaterials to solid matrices produces materials with enhanced, or even completely new, attributes, such as conductive polymeric matrices, electroluminescent metals, semiconductor ceramics or photo-luminescent textiles. The properties and quality of the resulting nanocomposite depend on the constituents of the composite, but also on the degree of dispersion and homogeneity of the different phases, which depend on the compatibilization between the phases and the mixing/addition methods.

In addition, nanomaterials can be also added to the surface of a material to obtain new or improved surface properties. This surface addition can be done by in situ nanocomposite formation, such as surface treatment of ceramic tiles with a solution of nanosized titania and a resin that is later dried [121]; by physical or chemical attachment of the nanomaterial to the surface, such as textiles with silver nanoparticles bonded to the fibres [122]; or by direct deposition of a thin layer over the surface, such as solar cells of nanosized TiO_2 prepared by direct CVD over the cell surface [123].

Other types of nano-additivated formulations include nanomaterial dispersion in emulsions, such as paints or cosmetics, or mixtures of non-consolidated solids, such as catalytic mixtures for gas emission treatment.

The literature on this field is very broad (8,280; 88,000 and 75,000 results in Google Books, Google Scholar and Web of Science, respectively, when looking for *all in title: nanocomposites*).[1] Due to the scope of this book, this chapter will only provide an overview on the production processes for the main types of nanocomposites: polymeric, ceramic, metallic and textiles. If needed, the reader can expand this information in some of the existing reviews [124–126].

3.3.1 Polymeric-Matrix Nanocomposites

Research on polymeric nanocomposites has exponentially grown in the last decades and this has been reflected in an increase in the number of products launched to the market based on such materials, from conductive polymeric materials to artificial tissues.

Polymeric matrix nanocomposites can be synthesized by different techniques that can be divided in three major groups: solution casting, melt blending and *in situ* polymerization [127–130]. Solution casting consists in the dissolution of the polymer and dispersion of the nanomaterial in a solvent (usually using ultrasonication). Then, the nanocomposite is obtained by removing the solvent. In the melt blending method, the polymer and the nanomaterial are intensively mixed in an extruder or a mixer at a temperature that allows polymer mobility. *In situ* polymerization consists on the mixture of the nanomaterial and monomers (in solution or not) under conditions that favour the polymerization. Polymerization can be catalyzed by the nanomaterial itself (i.e. silicate layers promote intercalated monomer polymerization) or by the addition of polymerization catalysts. Moreover, the nanomaterial may be coated with vinyl moieties where polymerization can start.

The choice of the synthesis method and its conditions completely depend on the polymer and on the type of nanomaterial. A good compatibility of the polymer and the nanofiller is critical for a homogeneous physical-chemical behaviour of the composite and to reduce nanomaterial release in following life cycle phases [131]. The most common strategies used to improve such compatibility are the use of additives that act as a surfactant between the nanomaterial and the polymer [132–135], and the surface modification of the nanomaterial to make it more compatible with the polymer [135–138].

[1] Search done on the 21st January 2015.

3.3.2 Ceramic-Matrix Nanocomposites

Nano-additivation of ceramic materials has resulted in the development of new materials with enhanced properties. The most important disadvantage of ceramic materials is their fragility, and the addition of nanomaterials is mainly used for the reinforcement of the ceramics that allows their use in new applications (i.e. armours, surgery materials or artificial bones). In addition, nanomaterials can also confer other properties that make these materials useful in fields such as optical, electronic or sensing [128, 139]. Ceramic matrices were traditionally reinforced with metallic particles [140–142], but nowadays one can find in the literature ceramics reinforced with carbon based nanomaterials [143, 144], nanometal oxides [145, 146] or quantum dots [147].

Three methodologies are basically used in the processing of ceramic matrix nanocomposites: powder process, polymer precursor process and sol-gel process [128]. Powder process consists of the mixing of the different materials that are thoroughly milled together in wet conditions; later the mixture is dried and consolidated, usually by pressure or moulding. This process is simple but results on a heterogeneous material. Polymer precursor process is similar, but the nanomaterial precursor is added to a polymer that is later pyrolized. Sol-gel process consists in the hydrolysis and condensation of molecular precursors dissolved in organic media to form a sol-gel, which is later dried and consolidated.

3.3.3 Metal-Matrix Nanocomposites

Particulate reinforced metal-matrix composites have been used for decades [148], but the reinforcement with nanomaterials has been developed recently and metal-matrix nanocomposites are still in their infancy [149]. The main advantages of metal matrices are their inherent thermal stability, resistance to abrasion, and thermal and electrical conductivities. But their development was strained by their cost and the difficulties of preparation [128, 149, 150]. The nano-additivation of metal matrices confers a combination of ceramic and metal properties to the material. This makes the material ideal for multiple applications, such as structural materials in the aeronautic industry or in light energy conversion.

Several methods for metal-matrix nanocomposites processing are described in the literature, including vapour phase processing, spray pyrolysis, powder metallurgy, solidification, chemical and deformation processes. The most used and cheapest method is solidification, which consists on the melting of the metal and the nano-reinforcement and rapid solidification of the melt by different processes. Liquid infiltration is similar, but in this case only the metal is melt and surrounds the nanomaterial. The homogeneity of the mixture can be improved by ultrasounds. The other methods are similar to the ones used in nanomaterial synthesis (sol-gel synthesis, CVD, spray pyrolysis, etcetera) (see Sect. 3.2).

3.3.4 Nano-additivated Textiles

Nanomaterials can be integrated in the textiles in different phases of their fabrication, which leads to different types of nano-additivated textiles: (i) Nanotextiles, when the nanomaterial is added once the fabric is produced, most of the products falls in this category; (ii) Nanocomposite textiles, when the material used to make the fibres is a nanocomposite; and (iii) Nanofibrous materials, when they are made from nanofibres (woven or non-woven) [151]. Nanomaterials are added to textiles to provide new or improved properties. The most common ones are antimicrobial activity and UV-filtering, but they are also used as flame retardants, water repellent, static protection, electrical conductivity, enhanced resistance or strength, photo- or electro-luminescent, self-cleaning, etcetera.

Several methods have been used for surface modification of fibre-based materials, such as textiles and membranes [152–155]. Usually they involve small modifications during the fabric processing; the nanomaterial is added as any other additive by methods such as impregnation, roll-to-roll and pad-dry-cure. The main problem of these methods is that the nanomaterials are usually not well fixed to the fabric and majorly released during the washing process [156–161]. In order to minimize the release of nanomaterials, binders and functionalized particles are used to improve particle affinity for the textiles. Ultrasounds, UV irradiation, plasma-treatment and ion-beam-assisted deposition are very effective for the surface modification of textiles, but impractical for large scale manufacturing (several preparatory steps, time-consuming and costly).

Nanocomposite fibres have emerged in the last decade as a very interesting material for nano-enabled textiles processing. In this case, nanocomposite material is produced as any other polymer-matrix nanocomposite (see Sect. 3.3.2). The main challenge is to get nano-reinforced polymers that can be processed as fibres and that, later, do not reduce the mechanical properties of the fabric.

Textiles and other non-woven products (such as filtering membranes) can be made of nanofibres. Nanofibres are produced by electrospinning (see Sect. 3.2.5) and can contain pure polymer(s) or nanocomposites. At the moment, nanofibres are not woven at industrial scale and are usually used as additive over other fibres [151]. Non-woven nanofibres are used as layer and barrier materials [115, 117, 118, 162, 163].

3.3.5 Occupational and Environmental Exposure During Nanocomposite Production

Most of the available studies on occupational exposure to nanomaterials focused on the nano-additivated material preparation. Nanomaterial synthesis is usually done in close reactors and workers are basically exposed during the nanomaterial recovery and during cleaning and maintenance. In contrast, weighting, pouring, and mixing of nanomaterial and bulk materials (common steps during nano-additivation of materials) are usually done in open conditions, and can involve big amounts of nanomaterials, so that the exposure during this step is potentially high [10, 12, 164–166]. This has been corroborated by exposure monitoring campaigns in workplaces as described in different reviews [167, 168] and other later studies [169–175].

On the other hand, nanomaterial release to the environment during this step is usually considered unlikely. Once the process is finished, the nanomaterials are embedded in a matrix or a mixture, so their recovery is easier and also the waste treatment [16].

3.4 Product Manufacturing

Product manufacturing involves a series of processes to convert the nano-additivated material into the final product. Machining is necessary to obtain final products with specified dimensions, surface finishing and tolerances. Most of the machining processes are physically aggressive and can lead to nanomaterial release. Although these processes are carried out by machines, they usually need an operator, sometimes in close and long contact to the material (i.e. sewing). Some examples of machining processes include soldering, welding, cutting, sewing, grinding, shredding, sanding, punching and drilling. Moreover, one has to consider than several of these processes may be necessary to get the final product, which may mean the product manufacturing divided in several phases that can even occur in different companies or locations.

Nanomaterial release from nanocomposites during the machining processes has received special attention in the literature in comparison to other processes during manufacturing and use stages. Indeed, almost half of the papers identified in a recent review on nanomaterial release from nanocomposites focused on machining processes [176].

Most of these studies are focused on CNT- and NMOx-based nanocomposites, with almost no attention to nanometal-based nanocomposites. Regarding the matrix, most of the studies focus on polymeric nanocomposites, probably because they are the ones with highest production volumes (Fig. 3.2) [176, 177].

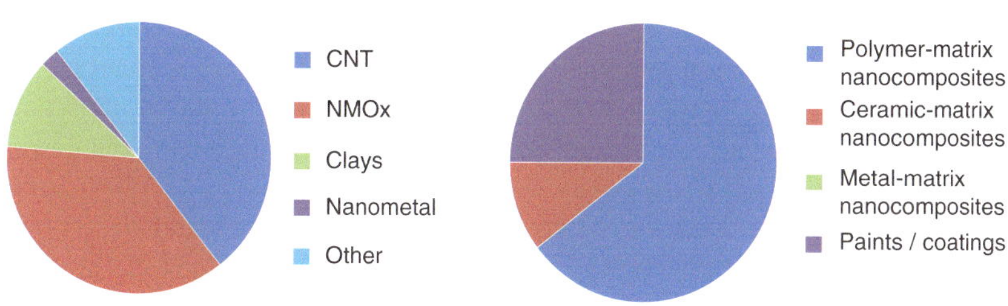

Fig. 3.2 Summary of reviewed papers by nano-reinforcement, *left*, and by base material, *right*, submitted to machining processes (Based on Froggett et al. 2014)

Nanomaterial release from nanocomposites during the machining processes can be studied simulating real operations or using standardized protocols. Non-standard studies of cutting/sawing [120, 178, 179], grinding [120, 180], shredding [181], sanding [65, 120, 182–186], and drilling [187–190] under different conditions (wet/dry, hot/cold, etcetera) are found in the literature. Studies based on standard protocols usually focus on abrasion, using a Taber abraser [131, 185, 191–198]. Regardless on the type of simulation, most studies analyze the released material and usually conclude that part of the matrix released contains nanomaterials embedded. Only four publications report significant release of isolated nanomaterial [120, 131, 186, 192]. It is important to notice that most of these sanding/abrasion studies do not clearly distinguish between abrasions due to aging or industrial processes.

From the publications mentioned in this section, it can be concluded that the matrix play a more important role than the nano-reinforcement on the overall degradation caused by the machining processes. Moreover, good dispersion of the nanofillers in the matrix could reduce the release of isolated nanomaterials [131].

3.5 Use Phase

At the moment, the major usage of nanomaterials is considered to be at the industrial level. For example, they are used as catalysts, membranes, and as additives or technical components of materials in various application fields. In addition, some nano-enabled products are addressed to professionals and consumers. There is no doubt that the diversity of applications of ENMs in commercial products has grown extensively over the past decade, and continues to grow rapidly [199]. However, the actual distribution of nanomaterials over different product categories is largely unknown. According to the (US) Nanotechnology Consumer Products Inventory [200], which has been updated very recently, the number of consumer products that are claimed to contain nanomaterials has increased from 54 products in 2005 to 1628 products in 2013. Although these numbers are likely to reflect real trends, their accuracy is questionable because tracking products that contain nanomaterials is rather challenging. With a few exceptions, current labelling regulations do not require that the nanomaterial be listed specifically as an ingredient. On the other hand, some products on the market with the claim of "nano" may neither contain nanomaterials nor be produced with nanotechnology. Depending on the area of application, interest in reporting the use of nanomaterials can differ, which could result on biased estimations on the main area of application if based on reported use.

Such lack of information regarding the real use of nanomaterials in consumer products may change in the coming years. First, some regulations, such as those affecting cosmetics and food ingredients in the EU are currently already requesting producers to label nanomaterials in their products [201, 202]. And second, some countries established compulsory registries of nano-enabled products.

Table 3.1 Distribution of sectors of use among the total declared in 2014

Code	Descriptor	Occurrence	Percentage
SU1	Agriculture, forestry, fishery	6417	58.28
SU10	Formulation [mixing] of preparations and/or re-packaging (excluding alloys)	2131	19.36
SU0	Other	877	7.97
SU17	General manufacturing, e.g. machinery, equipment, vehicles, other transport equipment	330	3.00
SU4	Manufacture of food products	233	2.12
SU24	Scientific research and development	227	2.06
SU11	Manufacture of rubber products	161	1.46
SU12	Manufacture of plastics products, including compounding and conversion	161	1.46
SU9	Manufacture of fine chemicals	119	1.08

Adapted from Ministère de l'Écologie du Développement durable et de l'Énergie (2014)

France was the first European country to require the identification of 'substances with nanoparticle status' that are produced, imported, distributed, or formulated from the 1st of January 2013 (Article 185 of the French Environmental Code [3]). Since June 2014, the Danish EPA also requests the reporting into the nano-product register of mixtures and articles that are intended for sale to the general public and which contain nanomaterials. They did, however, limit the type of products that should be reported on the basis of their potential to represent a risk to the user or the environment. Therefore, reporting is only requested for products where the nanomaterial itself is released under normal or reasonably foreseeable use or where the nanomaterial itself is not released but substances in soluble form that are classified as carcinogen/mutagen/reprotoxic (CMRs) or environmentally dangerous substances are released from the nanomaterial. In addition, some type of products (mostly those covered by specific product risk assessments, such as medicines or cosmetics) are also exempt [2].

A recent report outlines the results of the two first declaration periods in France (up to 1st June 2014) [9, 203]. Table 3.1 includes the sectors of use with more than 100 declarations in 2014. The sectors with the highest number of declarations were agriculture, forestry and fishery, and formulation [mixing] of preparations and/or re-packaging (excluding alloys) with 58 and 19 of the declarations, respectively. Regarding chemical product categories, the most commonly reported are: (1) coatings and paints, thinners, paint removers, (2) cosmetics and personal care products, and (3) plant protection products, altogether accounting for almost 70 % of the chemical product categories registered (Table 3.2). Finally, among the registered articles, the most frequently reported categories were rubber articles (AC10), machinery, mechanical appliances, and electrical/electronic articles (AC2), plastic articles (AC13), vehicles (AC1), and other articles with intended release of substances (AC30).

Future updates of this registry and other registries will provide more realistic estimates of the global production of nanomaterials and their main applications.

Table 3.2 Distribution of chemical product categories among the total declared in 2014

Code	Descriptor	Occurrence	Percentage
PC9a	Coatings and paints, thinners, paint removers	631	24.0
PC39	Cosmetics, personal care products	605	23.0
PC27	Plant protection products	575	21.9
PC13	Fuels	216	8.2
PC32	Polymer preparations and compounds	160	6.1

Adapted from Ministère de l'Écologie du Développement durable et de l'Énergie (2014)

3.5.1 Common Nanomaterial Applications and Potential for Release of Nanomaterials

Some applications involve the intended release of NM, either to result on an intended human exposure (e.g., application of nano-enabled sunscreen onto the skin), or to application on other surfaces (e.g., generation of a nanocoating by spraying into a glass surface). In these cases, the estimation of the direct release of NMs is rather straightforward. However, understanding which fraction of it reaches its target application point, and which is the fate of such fraction after application is still largely unknown.

In other many applications, NM are part of the product matrix and are not intended to become released during use. Nevertheless, some of the normal use processes for some products may result on such unintended release. These can be mechanical processes, such as washing, wearing, tearing, breaking, and drilling, or physical-chemical degradation processes, such as weathering and chemical abrasion. The amount of the NM released from the matrix during the use stage will depend on several factors: the amount of NM in the product, the product lifetime, the way the NM are incorporated in the material (surface applications or in matrix), the surface contact area of the product that is affected by the process inducing release, the transfer factor of the NM within the matrix, the thickness of the product, and the frequency and duration of use.

During the last years, an increasing interest has resulted on research on the release of NM during the use phase of nano-enabled products. Indeed, it is assumed that unintended emissions from diffuse sources are one of the most important sources of NM releases to the environment [14]. Nevertheless, the number of studies evaluating release of NM from solid nanocomposites is still very low (Table 3.3) [176].

In general, weathering studies with polymeric nanocomposites have shown the degradation of the polymeric matrices due to photo- and chemical degradation. As a consequence, the nanoparticles tend to accumulate in the degraded zone, at the surface of the nanocomposite. However, free released NM are barely detected and rarely freed from the matrix in which they were included, even when weathering experiments have been combined with secondary mechanical forces [204, 205].

The release of nanomaterials (embedded in organic binder, as aggregates, or as single particles) from conventional paints during run-off events has been reported [195, 206–209]. However, the amounts released greatly differ among studies.

Table 3.3 Summary of the current literature on release of nanomaterials from solid nanocomposites

	Weathering	Washing	Contact
Textiles/Fabrics	0	7	2
Thermoset	8	0	0
Termoplastic	6	0	2
Paints/Coatings	9	0	1
Cement	2	0	0
Dental glass	0	0	1
Ceramic	0	2	0
Total	25	9	6

Adapted from Froggett et al. [176]

A considerable number of studies have also focused on the release of nanomaterials (mostly silver) from textiles during washing processes [157, 160, 161, 210–212]. High releases have often been reported during the first washing event [211]. All these studies suggest that the silver particles in the textile dissolve to silver ions in the water and form secondary particles. A similar process seems to occur when textiles containing silver nanoparticles are immersed in artificial sweat [213–215].

The available research still provides a rather partial view of the potential release of nanomaterials or dissolved ions from consumer products. And further research is needed to understand and model which factors and how determine release under different processes.

3.6 End of Life

Products containing nanomaterials will eventually reach the end of their useful lives and, unless recycled, be discarded. In addition, waste materials containing nanomaterials are being generated during the manufacture of nanotechnology products. These waste streams generated during the life cycle of products containing nanomaterials are potential sources of nanomaterials into the environment. The handling, treatment and disposal of such wastes will determine the resulting environmental releases of nanomaterials. Therefore, the development of appropriate end-of-life management strategies for waste streams containing nanomaterials is critical.

This section provides an overview of the most common recycling and end-of-life processes for products and waste streams containing nanomaterials.

3.6.1 Recycling

Two categories of waste can be considered in terms of recycling processes. First, waste streams that are treated as broad waste categories, such as plastics or paper. These are typically highly heterogeneous mixtures of different products that could include multitude of different nanomaterials. And second, narrower categories, such as PET bottles, tyres, and Li-ion batteries, that are comparatively much more homogeneous. Regardless of the category, current recycling processes will handle products with and without nanomaterials in unknown proportions. Research is needed to estimate the type and quantity of nanomaterials in different material flows entering recycling systems, and on how the presence of nanomaterials alters the quality of the recycled material. Indeed some research has been published on the performance of recycled composites containing nanomaterials, and results show that the presence of nanomaterials may negatively affect the quality of the recycled composites [216, 217]. Such information could result on changes in the optimal applications for the recycled materials or on changes in the recycling processes per se. In addition, information is also needed on the potential release of nanomaterials during these processes and on technical measures that could be used for minimizing them [218]. The generation of such information is necessary to evaluate potential negative impacts on workers or the environment [219].

3.6.2 Incineration

Incineration is a thermal treatment, through which waste is combusted in an oxidizing ambient at temperatures in the range of 850–1200 °C [220]. There are different types of plants, which mostly differ in the off-gas treatment section. Materials (including nanomaterials) that enter an incineration plant can be totally or partially combusted or remain unaltered, depending on the local conditions in the combustion chamber, the melting point and reactivity of the materials, and additional matrix materials in which they are present. Unaltered or partly combusted materials can end up in the slag/bottom ash, retained in the particle control filters and becoming part of the fly ash, or go through such filters and be released to the environment.

Nanomaterials in the waste streams entering an incinerator may exist as free particles (i.e. a powder) or dispersed in a liquid or solid material. Based on theoretical thermodynamic considerations and on some experimental data, it is generally assumed that most nanomaterials in waste would end up in bottom ash. This would be the case for particle aggregates or particles that do not totally combust. A smaller fraction, mainly free particles and some partly combusted materials, would reach the air filtration systems, where a proportion of those would be retained [17, 220–225]. Some experimental data suggest that state-of-the-art flue gas cleaning systems (such as electrostatic precipitators and wet scrubbers) would effectively retain nanomaterials, but the efficiency of current filter techniques is still controversial [221, 222]. Further experiments are needed to fully substantiate these assumptions on the fate of nanomaterials in an incineration plant, and quantify efficiency of filter techniques for different type of nanomaterials. In addition, it remains unclear in what form the NM are present in the bottom ash. Treatment of the bottom ash depends on regional legislation, but it is usually disposed in landfills, unless originated from special waste streams that justify its further confinement. It is assumed that most of the waste streams containing nanomaterials will be considered domestic wastes, resulting in less strict regulations on the fate of resulting bottom ashes. Therefore, understanding in which form the nanomaterials are usually present in the bottom ash (i.e. whether or not enclosed in vitrified fragments) is important to understand their possible later mobility [225]. This information could be used to evaluate if current treatments are appropriate for the resulting ashes.

3.6.3 Landfilling

Landfill is a system of waste disposal that is based on burial of municipal solid waste (MSW) in specifically designed sites. Although landfill is one of the most exploited treatments for MSW end of life, it is not yet clear how NM behave during disposal. If NM are able to be transported through waste, then the potential for release from landfills to the surrounding environment increases. Existing studies show some degree of mobility for different NM, which depends on the NM and the composition of the leachate (organic composition, ionic strength, pH) [226, 227]. Another concern about the presence of NM in landfills is related to their capacity to influence biological activity. Very few data is available on this issue, and so far this indicates no effects on the overall biological activity [228, 229], although bacterial community structure has been shown to be sensitive to some nanomaterials [229].

3.6.4 Waste Water Treatment

Domestic (and some industrial) waste water containing nanomaterials will end up in sewage treatment plants and industrial waste water treatment plants.

Concerns are related to the impact of nanomaterials on the biological systems within such treatments, and on their fate. Several studies have investigated such processes (see recent review by Neale et al. [230]), but available information is still rather partial. Part of the sewage sludge, when metal concentrations are below established maximum limits, is applied on land as supplemental fertilizer of landfill cover. Current regulations establish metal content limits without consideration of particle size. Yang et al. estimated the proportion of nano-TiO_2 present in a landfill and concluded that it represented around 0.1–0.2 % of the total Ti [231]. However, these values could vary regionally and with changing trends in the production of nano-TiO_2. Further knowledge on the mechanisms of metal transport in soils and effects of environmental conditions and particle size are needed to evaluate the potential impact of applying sewage sludge containing nanomaterials on soils.

3.6.5 Current Practise and Regulations

Altogether, there is very limited information on the possible risks associated to the presence of nanomaterials in wastes. In the lack of specific evidence for concern, no specific processes are required for wastes containing nanomaterials in neither

Europe nor the United States [232, 233]. In Europe, wastes are classified as hazardous or non-hazardous based on Regulation No. 1272/2008 on Classification, Labelling and Packaging of Substances and Mixtures [234]. This regulation does not include specific requirements for nanomaterials. Therefore, it is likely that nanomaterials will be classified in the same categories as their bulk form, and nano-specific hazards may be overlooked. The classification of waste as hazardous or non-hazardous is a key step as it leads to different requirements under the Waste Framework Directive. For example, mixing restrictions, labelling, and record keeping do not apply to wastes containing nanomaterials, unless they have been classified as hazardous [235].

Even when nanomaterials would be classified as hazardous, they may still be appropriate for use in some consumer products. In those cases, it is unlikely that their classification would result on specific end-of-life treatments for consumer products containing them. However, this is an issue that also applies to other type of hazardous substances.

More details on how current regulations affect wastes containing nanomaterials (and associated gaps) can be found in previous review reports [232, 233, 236].

References

1. Luther W, Zweck A (2013) Safety aspects of engineered nanomaterials. doi: 10.4032/9789814364867
2. BEK (2014) BEK nr 644 af 13/06/2014. Bekendtgørelse om register over blandinger og varer, der indeholder nanomaterialer samt producenter og importørers indberetningspligt til registeret. BEK, Denmark
3. JORF (2010) Article 185. Prévention des risques pour la santé et l'environnement résultant de l'exposition aux substances à l'état nanoparticulaire. JORF n°0160 du 13 juillet 2010 page 12905, France
4. Robichaud CO, Uyar AE, Darby MR et al (2009) Estimates of upper bounds and trends in nano-TiO 2 production as a basis for exposure assessment. Environ Sci Technol 43:4227–4233. doi:10.1021/es8032549
5. Schmid K, Riediker M (2008) Use of nanoparticles in Swiss industry: a targeted survey. Environ Sci Technol 42:2253–2260. doi:10.1021/es071818o
6. Hendren CO, Mesnard X, Dröge J, Wiesner MR (2011) Estimating production data for five engineered nanomaterials as a basis for exposure assessment. Environ Sci Technol 45:2562–2569. doi:10.1021/es103300g
7. Piccinno F, Gottschalk F, Seeger S, Nowack B (2012) Industrial production quantities and uses of ten engineered nanomaterials in Europe and the world. J Nanopart Res 14:1109. doi:10.1007/s11051-012-1109-9
8. Gao Y, Luo Z, He N, Wang M (2013) Metallic nanoparticle production and consumption in China between 2000 and 2010 and associative aquatic environmental risk assessment. J Nanopart Res 15:1681. doi:10.1007/s11051-013-1681-7
9. ANSES (2015) R-nano.fr - Déclaration des substances à l'état nanoparticulaire
10. Sánchez Jiménez A, Brouwer D, van Tongeren M (2014) Workplace inhalation exposure to engineered nanomaterials. Detection, measurement, and assessment. In: Monteiro-Riviere NA, Tran CL (eds) Nanotoxicology prog. towar. nanomedicine, 2 edn. CRC Press, London, pp. 77–96

11. Brouwer D, Duuren-Stuurman B, Berges M et al (2009) From workplace air measurement results toward estimates of exposure? Development of a strategy to assess exposure to manufactured nano-objects. J Nanopart Res 11:1867–1881. doi:10.1007/s11051-009-9772-1
12. Vogel U, Savolainen K, Wu Q, et al. (2014) Handbook of nanosafety. Handb nanosafety. doi: 10.1016/B978-0-12-416604-2.00002-0
13. Brouwer D (2010) Exposure to manufactured nanoparticles in different workplaces. Toxicology 269:120–127. doi:10.1016/j.tox.2009.11.017
14. Gottschalk F, Nowack B (2011) The release of engineered nanomaterials to the environment. J Environ Monit 13:1145–1155. doi:10.1039/c0em00547a
15. Nowack B, Ranville JF, Diamond S et al (2012) Potential scenarios for nanomaterial release and subsequent alteration in the environment. Environ Toxicol Chem 31:50–59. doi:10.1002/etc.726
16. Nowack B, David RM, Fissan H et al (2013) Potential release scenarios for carbon nanotubes used in composites. Environ Int 59:1–11. doi:10.1016/j.envint.2013.04.003
17. Keller AA, Mcferran S, Lazareva A, Suh S (2013) Global life cycle releases of engineered nanomaterials. J Nanopart Res 15:1692–1703. doi:10.1007/s11051-013-1692-4
18. Biswas A, Bayer IS, Biris AS et al (2012) Advances in top-down and bottom-up surface nanofabrication: techniques, applications & future prospects. Adv Colloid Interface Sci 170:2–27. doi:10.1016/j.cis.2011.11.001
19. Varshney K (2014) Carbon nanotubes: a review on synthesis, properties and applications. Int J Eng Res Gen Sci 2:660–677
20. Prasek J, Drbohlavova J, Chomoucka J et al (2011) Methods for carbon nanotubes synthesis—review. J Mater Chem 21:15872. doi:10.1039/c1jm12254a
21. Purohit R, Purohit K, Rana S et al (2014) Carbon nanotubes and their growth methods. Procedia Mater Sci 6:716–728. doi:10.1016/j.mspro.2014.07.088
22. Rao C, Govindaraj A (2011) Nanotubes and nanowires, 2nd edn. doi:10.1039/9781847552525
23. Zhao X, Ohkohchi M, Wang M et al (1997) Preparation of high-grade carbon nanotubes by hydrogen arc discharge. Carbon N Y 35:775–781. doi:10.1016/S0008-6223(97)00033-X
24. Shimotani K, Anazawa K, Watanabe H, Shimizu M (2014) New synthesis of multi-walled carbon nanotubes using an arc discharge technique under organic molecular atmospheres. Appl Phys 73 A:451–454. doi:10.1007/s003390100821
25. Parkansky N, Boxman RL, Alterkop B et al (2004) Single-pulse arc production of carbon nanotubes in ambient air. J Phys D Appl Phys 37:2715–2719. doi:10.1088/0022-3727/37/19/015
26. Tsai YY, Su JS, Su CY, He WH (2009) Production of carbon nanotubes by single-pulse discharge in air. J Mater Process Technol 209:4413–4416. doi:10.1016/j.jmatprotec.2008.10.049
27. Saito Y, Nishikubo K, Kawabata K, Matsumoto T (1996) Carbon nanocapsules and single-layered nanotubes produced with platinum-group metals (Ru, Rh, Pd, Os, Ir, Pt) by arc discharge. J Appl Phys 80:3062. doi:10.1063/1.363166
28. Journet C, Maser WK, Bernier P et al (1997) Large-scale production of single-walled carbon nanotubes by the electric-arc technique. Nature 388:756–758. doi:10.1038/41972
29. Zhang Y, Gu H, Iijima S (1998) Single-wall carbon nanotubes synthesized by laser ablation in a nitrogen atmosphere. Appl Phys Lett 73:3827. doi:10.1063/1.122907
30. Muñoz E, Maser W, Benito A et al (2000) Gas and pressure effects on the production of single-walled carbon nanotubes by laser ablation. Carbon N Y 38:1445–1451. doi:10.1016/S0008-6223(99)00277-8
31. Arepalli S, Scott C (1999) Spectral measurements in production of single-wall carbon nanotubes by laser ablation. Chem Phys Lett 302:139–145. doi:10.1016/S0009-2614(99)00098-6
32. Kokai F, Takahashi K (2000) Laser ablation of graphite-Co/Ni and growth of single-wall carbon nanotubes in vortexes formed in an Ar atmosphere. J Phys Chem B 104:6777–6784. doi:10.1021/jp000359+
33. Yudasaka M, Kokai F, Takahashi K et al (1999) Formation of single-wall carbon nanotubes: comparison of CO2 laser ablation and Nd: YAG laser ablation. J Phys Chem B 103:3576–3581. doi:10.1021/jp990072g

34. Zhang Y, Iijima S (1999) Formation of single-wall carbon nanotubes by laser ablation of fullerenes at low temperature. Appl Phys Lett 15:3087. doi:10.1063/1.125239
35. Niu K, Sun J, Yang J, Du X (2012) The synthesis of carbon nanotubes by pulsed-laser ablation of a nickel/carbon composite target in ethanol or ambient air. Sci Adv Mater 4:463–466. doi:10.1166/sam.2012.1302
36. Hu B, Wang K, Wu L, Yu S (2010) Engineering carbon materials from the hydrothermal carbonization process of biomass. Adv Sci 22:813–828. doi:10.1002/adma.200902812
37. Skrabalak S (2009) Ultrasound-assisted synthesis of carbon materials. Phys Chem Chem Phys 11:4930–4942. doi:10.1039/B823408F
38. Manafi S, Rahimipour M, Mobasherpour I, Soltanmoradi A (2012) The synthesis of peculiar structure of springlike multiwall carbon nanofibers/nanotubes via mechanothermal method. J Nanomater 803546:8. doi:10.1155/2012/803546
39. Sen LH, Nainar M, Begum S (2014) Model, synthesis and applications of graphene oxide–a review. Nanomater. Energy 3:61–65. doi:10.1680/nme.13.00031
40. Liu W, Chai S, Mohamed A, Hashim U (2014) Synthesis and characterization of graphene and carbon nanotubes: a review on the past and recent developments. J Ind Eng Chem 20:1171–1185. doi:10.1016/j.jiec.2013.08.028
41. Geim AK (2009) Graphene: status and prospects. Science 324:1530–1534. doi:10.1126/science.1158877
42. Choi W, Lahiri I (2010) Synthesis of graphene and its applications: a review. Crit Rev Solid State Mater Sci 35:52–71. doi:10.1080/10408430903505036
43. Zhang Y, Petibone D, Xu Y et al (2014) Toxicity and efficacy of carbon nanotubes and graphene: the utility of carbon-based nanoparticles in nanomedicine. Drug Metab Rev. doi:10.3109/03602532.2014.883406
44. Muñoz R, Gómez-Aleixandre C (2013) Review of CVD synthesis of graphene. Chem Vap Depos 19:297–322. doi:10.1002/cvde.201300051
45. Coleman J (2012) Liquid exfoliation of defect-free graphene. Acc Chem Res 46:14–22. doi:10.1021/ar300009f
46. Khan U, O'Neill A, Lotya M et al (2010) High-concentration solvent exfoliation of graphene. Small 6:864–871. doi:10.1002/smll.200902066
47. Hernandez Y, Nicolosi V, Lotya M (2008) High-yield production of graphene by liquid-phase exfoliation of graphite. Nat Nanotechnol 3:563–568. doi:10.1038/nnano.2008.215
48. Kaniyoor A, Baby T, Ramaprabhu S (2010) Graphene synthesis via hydrogen induced low temperature exfoliation of graphite oxide. J Mater Chem 20:8467–8469. doi:10.1039/C0JM01876G
49. Zhu Y, Stoller M, Cai W et al (2010) Exfoliation of graphite oxide in propylene carbonate and thermal reduction of the resulting graphene oxide platelets. ACS Nano 4:1227–1233. doi:10.1021/nn901689k
50. Stankovich S, Piner R, Nguyen S, Ruoff R (2006) Synthesis and exfoliation of isocyanate-treated graphene oxide nanoplatelets. Carbon N Y 44:3342–3347. doi:10.1016/j.carbon.2006.06.004
51. Cano-Márquez A (2009) Ex-MWNTs: graphene sheets and ribbons produced by lithium intercalation and exfoliation of carbon nanotubes. Nano Lett 9:1527–1533. doi:10.1021/nl803585s
52. Erlanger B, Chen B, Zhu M, Brus L (2001) Binding of an anti-fullerene IgG monoclonal antibody to single wall carbon nanotubes. Nano Lett 1:465–467. doi:10.1021/nl015570r
53. McDevitt M, Chattopadhyay D, Kappel B et al (2007) Tumor targeting with antibody-functionalized, radiolabeled carbon nanotubes. J Nucl Med 48:1180–1189. doi:10.2967/jnumed.106.039131
54. Juárez B, Klinke C, Kornowski A, Weller H (2007) Quantum dot attachment and morphology control by carbon nanotubes. Nano Lett 7:3564–3568. doi:10.1021/nl071225b
55. Zanella R, Basiuk E, Santiago P et al (2005) Deposition of gold nanoparticles onto thiol-functionalized multiwalled carbon nanotubes. J Phys Chem B 109:16290–16295. doi:10.1021/jp0521454

56. Dahm M, Evans D (2012) Occupational exposure assessment in carbon nanotube and nanofiber primary and secondary manufacturers. Ann Occup Hyg 56:542–556. doi:10.1093/annhyg/mer110
57. Lee JH, Lee S-B, Bae GN et al (2010) Exposure assessment of carbon nanotube manufacturing workplaces. Inhal Toxicol 22:369–381. doi:10.3109/08958370903367359
58. Lam C, James J, McCluskey R et al (2006) A review of carbon nanotube toxicity and assessment of potential occupational and environmental health risks. Crit Rev Toxicol 36:189–217. doi:10.1080/10408440600570233
59. Liu Y, Zhao Y, Sun B, Chen C (2013) Understanding the toxicity of carbon nanotubes. Acc Chem Res 46:702–713. doi:10.1021/ar300028m
60. Kolosnjaj-Tabi J, Szware H, Moussa F (2012) In vivo toxicity studies of pristine carbon nanotubes: a review. Deliv Nanoparticles. doi:10.5772/34201
61. Aschberger K, Johnston H (2010) Review of carbon nanotubes toxicity and exposure-appraisal of human health risk assessment based on open literature. Crit Rev Toxicol 40:759–790. doi:10.3109/10408444.2010.506638
62. Wang J, Xu Y, Yang Z et al (2013) Toxicity of carbon nanotubes. Curr Drug Metab 14:891–899
63. Donaldson K, Aitken R, Tran L (2006) Carbon nanotubes: a review of their properties in relation to pulmonary toxicology and workplace safety. Toxicol Sci 92:5–22. doi:10.1093/toxsci/kfj130
64. Boczkowski J, Lanone S (2012) Respiratory toxicities of nanomaterials – a focus on carbon nanotubes. Adv Drug Deliv Rev 64:1694–1699. doi:10.1016/j.addr.2012.05.011
65. Cena LG, Peters TM (2011) Characterization and control of airborne particles emitted during production of epoxy/carbon nanotube nanocomposites. J Occup Environ Hyg 8:86–92. doi:10.1080/15459624.2011.545943
66. Han JH, Lee EJ, Lee JH et al (2008) Monitoring multiwalled carbon nanotube exposure in carbon nanotube research facility. Inhal Toxicol 20:741–749. doi:10.1080/08958370801942238
67. Fujitani Y, Kobayashi T, Arashidani K et al (2008) Measurement of the physical properties of aerosols in a fullerene factory for inhalation exposure assessment. J Occup Environ Hyg 5:380–389. doi:10.1080/15459620802050053
68. Masala O, Seshadri R (2004) Synthesis Routes for large volumes of nanoparticles. Annu Rev Mat Res 34:41–81. doi:10.1146/annurev.matsci.34.052803.090949
69. Panacek A, Kvítek L, Prucek R et al (2006) Silver colloid nanoparticles: synthesis, characterization, and their antibacterial activity. J Phys Chem B 110:16248–16253. doi:10.1021/jp063826h
70. Alex S, Tiwari A (2015) Functionalized gold nanoparticles: synthesis, properties and applications—a review. J Nanosci Nanotechnol 15:1869–1895. doi:10.1166/jnn.2015.9718
71. Chen A, Holt-Hindle P (2010) Platinum-based nanostructured materials: synthesis, properties, and applications. Chem Rev 110:3767–3804. doi:10.1021/cr9003902
72. Nemamcha A (2006) Synthesis of palladium nanoparticles by sonochemical reduction of palladium (II) nitrate in aqueous solution. J Phys Chem B 110:383–387. doi:10.1021/jp0535801
73. Xiong Y, Xia Y (2007) Shape-controlled synthesis of metal nanostructures: the case of palladium. Adv Mater 19:3385–3391. doi:10.1002/adma.200701301
74. Viau G, Brayner R, Poul L et al (2003) Ruthenium nanoparticles: size, shape, and self-assemblies. Chem Mater 15:486–494. doi:10.1021/cm0212109
75. Mévellec V, Nowicki A, Roucoux A et al (2006) A simple and reproducible method for the synthesis of silica-supported rhodium nanoparticles and their investigation in the hydrogenation of aromatic compounds. New J Chem 30:1214–1219. doi:10.1039/B605893K
76. Stowell C, Korgel B (2005) Iridium nanocrystal synthesis and surface coating-dependent catalytic activity. Nano Lett 5:1203–1207. doi:10.1021/nl050648f
77. Lista M, Liu D, Mulvaney P (2014) Phase transfer of noble metal nanoparticles to organic solvents. Langmuir 30:1932–1938. doi:10.1021/la404569h
78. Tao AR, Habas S, Yang P (2008) Shape control of colloidal metal nanocrystals. Small 4:310–325. doi:10.1002/smll.200701295

79. Philippot K, Chaudret B (2003) Organometallic approach to the synthesis and surface reactivity of noble metal nanoparticles. C R Chimie 6:1019–1034. doi:10.1016/j.crci.2003.07.010
80. Chaudhuri RG, Paria S (2011) Core/shell nanoparticles: classes, properties, synthesis mechanisms, characterization, and applications. Chem Rev 112:2373–2433. doi:10.1021/cr100449n
81. Park J, Kwak BK, Bae E et al (2009) Characterization of exposure to silver nanoparticles in a manufacturing facility. J Nanopart Res 11:1705–1712. doi:10.1007/s11051-009-9725-8
82. Zimmermann E, Derrough S, Locatelli D et al (2012) Results of potential exposure assessments during the maintenance and cleanout of deposition equipment. J Nanopart Res 14:1209. doi:10.1007/s11051-012-1209-6
83. Debia M, Beaudry C, Weichenthal S et al (2013) Report R-777. Characterization and control of occupational exposure to nanoparticles and ultrafine particles. IRSST, Montréal
84. Ling M-P, Lin W-C, Liu C-C et al (2012) Risk management strategy to increase the safety of workers in the nanomaterials industry. J Hazard Mater 229–230:83–93. doi:10.1016/j.jhazmat.2012.05.073
85. Lee JH, Ahn K, Kim SM et al (2012) Continuous 3-day exposure assessment of workplace manufacturing silver nanoparticles. J Nanopart Res 14:1134. doi:10.1007/s11051-012-1134-8
86. Lee J, Kwon M, Ji J, Kang C (2011) Exposure assessment of workplaces manufacturing nanosized TiO2 and silver. Inhal Toxicol 23:226–236. doi:10.3109/08958378.2011.562567
87. Gottschalk F, Sun T, Nowack B (2013) Environmental concentrations of engineered nanomaterials : review of modeling and analytical studies. Environ Pollut. doi:10.1016/j.envpol.2013.06.003
88. Fan Z, Lu J (2005) Zinc oxide nanostructures: synthesis and properties. J Nanosci Nanotechnol 5:1561–1573. doi:10.1166/jnn.2005.182
89. Chen X, Selloni A (2014) Introduction: titanium dioxide (TiO2) nanomaterials. Chem Rev 114:9281–9282. doi:10.1021/cr500422r
90. Lu A-H, Salabas EL, Schüth F (2007) Magnetic nanoparticles: synthesis, protection, functionalization, and application. Angew Chem Int Ed Engl 46:1222–1244. doi:10.1002/anie.200602866
91. Chen Y, Lv S, Chen C, Qiu C (2014) Controllable synthesis of ceria nanoparticles with uniform reactive {100} exposure planes. J Phys Chem C 118:4437–4443. doi:10.1021/jp410625n
92. Chen H, Mazzolini J, Ayers J, et al. (2013) Synthesis and characterization of nano ceria for biological applications. In: Kang SW, Park SH, Lee LP, et al (eds) Proc. SPIE 8879, Nano-bio sensing, imaging, spectrosc. Jeju, Republic of Korea p 887910. doi: 10.1117/12.2018564
93. Chen G, Guo C, Qiao H et al (2013) Well-dispersed sulfated zirconia nanoparticles as high-efficiency catalysts for the synthesis of bis(indolyl) methanes and biodiesel. Catal Commun 41:70–74. doi:10.1016/j.catcom.2013.07.006
94. Yin M, Wu C, Lou Y et al (2005) Copper oxide nanocrystals. J Am Chem Soc 127:9506–9511. doi:10.1021/ja050006u
95. Cai W, Yu J, Anand C et al (2011) Facile synthesis of ordered mesoporous alumina and alumina-supported metal oxides with tailored adsorption and framework properties. Chem Mater 23:1147–1157. doi:10.1021/cm102512v
96. Rahman I, Padavettan V (2012) Synthesis of silica nanoparticles by sol-gel: size-dependent properties, surface modification, and applications in silica-polymer nanocomposites—a review. J Nanomater 132424:15. doi:10.1155/2012/132424
97. Wu S, Mou C, Lin H (2013) Synthesis of mesoporous silica nanoparticles. Chem Soc Rev 42:3862–3875. doi:10.1039/C3CS35405A
98. Tang F, Li L, Chen D (2012) Mesoporous silica nanoparticles: synthesis, biocompatibility and drug delivery. Adv Mater 24:1504–1534. doi:10.1002/adma.201104763
99. Yang J, Liu H, Martens W, Frost RL (2009) Synthesis and characterization of cobalt hydroxide, cobalt oxyhydroxide, and cobalt oxide nanodiscs. J Phys Chem C 114:111–119. doi:10.1021/jp908548f

100. Kango S, Kalia S, Celli A et al (2013) Surface modification of inorganic nanoparticles for development of organic–inorganic nanocomposites—a review. Prog Polym Sci 38:1232–1261. doi:10.1016/j.progpolymsci.2013.02.003
101. Curwin B, Bertke S (2011) Exposure characterization of metal oxide nanoparticles in the workplace. J Occup Environ Hyg 8:580–587. doi:10.1080/15459624.2011.613348
102. Leppänen M, Lyyränen J, Järvelä M et al (2012) Exposure to CeO(2) nanoparticles during flame spray process. Nanotoxicology 6:643–651. doi:10.3109/17435390.2011.600838
103. Yang Y, Mao P, Wang Z, Zhang J (2012) Distribution of nanoparticle number concentrations at a nano-TiO2 plant. Aerosol Air Qual Res 12:934–940. doi:10.4209/aaqr.2012.02.0047
104. Methner M, Hodson L, Dames A, Geraci C (2010) Nanoparticle Emission Assessment Technique (NEAT) for the identification and measurement of potential inhalation exposure to engineered nanomaterials–Part B: Results from 12 field studies. J Occup Environ Hyg 7:163–176. doi:10.1080/15459620903508066
105. Bera D, Qian L, Tseng T-K, Holloway PH (2010) Quantum dots and their multimodal applications: a review. Materials (Basel) 3:2260–2345. doi:10.3390/ma3042260
106. Valizadeh A, Mikaeili H, Samiei M et al (2012) Quantum dots: synthesis, bioapplications, and toxicity. Nanoscale Res Lett 7:480. doi:10.1186/1556-276X-7-480
107. Evans C, Cass L, Knowles K (2012) Review of the synthesis and properties of colloidal quantum dots: the evolving role of coordinating surface ligands. J Coord Chem 65:2391–2414. doi:10.1080/00958972.2012.695019
108. Deng Z, Samanta A (2012) Robust DNA-functionalized core/shell quantum dots with fluorescent emission spanning from UV–vis to near-IR and compatible with DNA-directed self-assembly. J Am Chem Soc 134:17424–17427. doi:10.1021/ja3081023
109. Goswami N, Giri A, Kar S, Bootharaju M (2012) Protein-directed synthesis of NIR-emitting, tunable HgS quantum dots and their applications in metal-ion sensing. Small 8:3175–3184. doi:10.1002/smll.201200760
110. Wang D, Qian J, Cai F et al (2012) "Green"-synthesized near-infrared PbS quantum dots with silica–PEG dual-layer coating: ultrastable and biocompatible optical probes for in vivo animal imaging. Nanotechnology 23:245701. doi:10.1088/0957-4484/23/24/245701
111. Lee J, Kwon B, Park H et al (2013) Solar cells: exciton dissociation and charge-transport enhancement in organic solar cells with quantum-dot/N-doped CNT hybrid nanomaterials. Adv Mater 25:2104. doi:10.1002/adma.201370088
112. Hardman R (2006) A toxicologic review of quantum dots: toxicity depends on physicochemical and environmental factors. Environ Health Perspect 114:165–172. doi:10.1289/ehp.8284
113. Chen N, He Y, Su Y et al (2012) The cytotoxicity of cadmium-based quantum dots. Biomaterials 33:1238–1244. doi:10.1016/j.biomaterials.2011.10.070
114. Tsoi K, Dai Q (2012) Are quantum dots toxic? Exploring the discrepancy between cell culture and animal studies. Acc Chem Res 46:662–671. doi:10.1021/ar300040z
115. Huang Z-M, Zhang Y-Z, Kotaki M, Ramakrishna S (2003) A review on polymer nanofibers by electrospinning and their applications in nanocomposites. Compos Sci Technol 63:2223–2253. doi:10.1016/S0266-3538(03)00178-7
116. Wu H, Pan W, Lin D, Li H (2012) Electrospinning of ceramic nanofibers: fabrication, assembly and applications. J Adv Ceram 1:2–23. doi:10.1007/s40145-012-0002-4
117. Panda PK, Sahoo B (2013) Synthesis and applications of electrospun nanofibers-a review. In: Navani KN, Sinha S, Govil JN (eds) Nanotechnology, vol. 1: Fundamental applications. Studiun Press LLC, Houston, TX, USA, pp 399–416
118. Lee YS, Im JS (2010) Preparation of functionalized nanofibers and their applications. In: Kumar A (ed) Nanofibers. InTech, Rijeka, Croatia, p 121–140
119. Dai Y, Liu W, Formo E (2011) Ceramic nanofibers fabricated by electrospinning and their applications in catalysis, environmental science, and energy technology. Polym Adv Technol 22:326–338. doi:10.1002/pat.1839

120. Methner M, Crawford C, Geraci C (2012) Evaluation of the potential airborne release of carbon nanofibers during the preparation, grinding, and cutting of epoxy-based nanocomposite material. J Occup Environ Hyg 9:308–318. doi:10.1080/15459624.2012.670790
121. Daoud WA (2013) Self-cleaning materials and surfaces: a nanotechnology approach. Wiley, New York
122. Rai M, Yadav A, Gade A (2009) Silver nanoparticles as a new generation of antimicrobials. Biotechnol Adv 27:76–83. doi:10.1016/j.biotechadv.2008.09.002
123. Thelakkat M, Schmitz C, Schmidt H (2002) Fully vapor-deposited thin-layer titanium dioxide solar cells. Adv Mater 14:577–581. doi:10.1002/1521-4095(20020418)14:8<577::AID-ADMA577>3.0.CO;2-S
124. Davim JP, Charitidis CA (2013) Nanocomposites: materials, manufacturing and engineering. De Gruyter, Berlin
125. Kumar CSSR (2010) Nanocomposites. John Wiley & Sons, Inc., Hoboken, NJ, USA. doi: 10.1002/9781119096122
126. Twardowski TE (2007) Introduction to nanocomposite materials: properties, processing, characterization, 1st edn. DEStech Publications Inc, Lancaster
127. Anandhan S, Bandyopadhyay S (2011) Polymer nanocomposites: from synthesis to applications. In: Cuppoletti J (ed) Nanocomposites polym. with anal. methods. InTech, Rijeka, Croatia, p 3–28
128. Camargo PHC, Satyanarayana KG, Wypych F (2009) Nanocomposites : synthesis, structure, properties and new application opportunities. Mater Res 12:1–39. doi:10.1590/S1516-14392009000100002
129. Hussain F (2006) Review article: polymer-matrix nanocomposites, processing, manufacturing, and application: an overview. J Compos Mater 40:1511–1575. doi:10.1177/0021998306067321
130. Mittal V (2010) Polymer nanocomposites: synthesis, microstructure, and properties. In: Mittal V (ed) Optim. polym. nanocomposite prop. Wiley-VCH Verlag GmbH & Co. KGaA, Weinheim, Germany, p 1–20. ISBN:978-3-527-32521-4
131. Golanski L, Guiot A, Pras M et al (2012) Release-ability of nano fillers from different nanomaterials (toward the acceptability of nanoproduct). J Nanopart Res 14:962. doi:10.1007/s11051-012-0962-x
132. Phua Y, Lau N, Sudesh K et al (2014) A study on the effects of organoclay content and compatibilizer addition on the properties of biodegradable poly (butylene succinate) nanocomposites under natural weathering. J Compos Mater. doi:10.1177/0021998314527328
133. Mistretta M, Fontana P, Ceraulo M et al (2015) Effect of compatibilization on the photooxidation behaviour of polyethylene/polyamide 6 blends and their nanocomposites. Polym Degrad Stab 112:192–197. doi:10.1016/j.polymdegradstab.2015.01.002
134. García-López D, Picazo O, Merino JC, Pastor JM (2003) Polypropylene–clay nanocomposites: effect of compatibilizing agents on clay dispersion. Eur Polym J 39:945–950. doi:10.1016/S0014-3057(02)00333-6
135. Taguet A, Cassagnau P, Lopez-Cuesta J (2014) Structuration, selective dispersion and compatibilizing effect of (nano)fillers in polymer blends. Prog Polym Sci 39:1526–1563. doi:10.1016/j.progpolymsci.2014.04.002
136. Sengupta R, Bhattacharya M (2011) A review on the mechanical and electrical properties of graphite and modified graphite reinforced polymer composites. Prog Polym Sci 36:638–670. doi:10.1016/j.progpolymsci.2010.11.003
137. Li D, You Y, Li R, Deng X (2013) Effects of nanometer-TiO2 surface modification and concentration on the mechanical performances of polypropylene/polyamide maleic anhydride-grafted. J Reinf Plast Compos 32:1807–1820. doi:10.1177/0731684413493341
138. Beyou E, Akbar S, Chaumont P, Cassagnau P (2013) Polymer nanocomposites containing functionalised multiwalled carbon nanoTubes: a particular attention to polyolefin based materials. Synth Appl Carbon Nanotub Their Compos. doi:10.5772/50710
139. Low I (2014) Advances in ceramic matrix composites. Woodhead Publishing Limited, Cambridge, UK
140. Chawla K (1998) Ceramic matrix composites, 2nd edn. Springer
141. Yeomans J (2008) Ductile particle ceramic matrix composites—Scientific curiosities or engineering materials? J Eur Ceram Soc 28:1543–1550. doi:10.1016/j.jeurceramsoc.2007.12.009

142. Rosso M (2006) Ceramic and metal matrix composites: Routes and properties. J Mater Process Technol 175:364–375. doi:10.1016/j.jmatprotec.2005.04.038
143. Samal S, Bal S (2008) Carbon nanotube reinforced ceramic matrix composites-A review. J Miner Mater Charact Eng 7:355–370
144. Peigney A, Laurent C, Flahaut E, Rousset A (2000) Carbon nanotubes in novel ceramic matrix nanocomposites. Ceram Int 26:677–683. doi:10.1016/S0272-8842(00)00004-3
145. Zhao Z, Sun R, Xin G (2013) A review: application of nanomaterials in concrete. Appl Mech Mater 405–408:2881–2884. doi:10.4028/www.scientific.net/AMM.405-408.2881
146. Mukhopadhyay A, Basu B (2007) Consolidation–microstructure–property relationships in bulk nanoceramics and ceramic nanocomposites: a review. Int Mater Rev 52:257–288. doi:10.1179/174328007X160281
147. Hedayati M, Faupel F, Elbahri M (2014) Review of plasmonic nanocomposite metamaterial absorber. Materials (Basel) 7:1221–1248. doi:10.3390/ma7021221
148. Ibrahim I, Mohamed F, Lavernia E (1991) Particulate reinforced metal matrix composites—a review. J Mater Sci 26:1137–1156. doi:10.1007/BF00544448
149. Rohatgi P, Schultz B (2007) Lightweight metal matrix nanocomposites–stretching the boundaries of metals. Mater Matters 2:16–19
150. He F, Han Q, Jackson M (2008) Nanoparticulate reinforced metal matrix nanocomposites–a review. Int J Nanoparticles 1:301–309. doi:10.1504/IJNP.2008.026473
151. Haydon B (2013) Nanomaterials and their applications in textiles, standards: domestic standardization for Canadian Manufacturers and Importers and International Standardization Developments. Waterloo, Ontario, Canada. ISBN:978-1-100-21089-6
152. Dastjerdi R, Montazer M (2010) A review on the application of inorganic nano-structured materials in the modification of textiles: focus on anti-microbial properties. Colloids Surf B Biointerfaces 79:5–18. doi:10.1016/j.colsurfb.2010.03.029
153. Zille A, Almeida L, Amorim T et al (2014) Application of nanotechnology in antimicrobial finishing of biomedical textiles. Mater Res Express 1:032003. doi:10.1088/2053-1591/1/3/032003
154. Coyle S, Diamond D (2010) Smart nanotextiles: materials and their application. In: Encycl. Mater. Elsevier Ltd., Atlanta, GA, USA, p 8. doi: 10.1016/B978-008043152-9.02220-X
155. Coyle S, Wu Y, Lau K, Rossi DD (2007) Smart nanotextiles: a review of materials and applications. MRS Bull 32:434–442. doi:10.1557/mrs2007.67
156. Som C, Wick P, Krug H, Nowack B (2011) Environmental and health effects of nanomaterials in nanotextiles and façade coatings. Environ Int 37:1131–1142. doi:10.1016/j.envint.2011.02.013
157. Geranio L, Heuberger M, Nowack B (2009) The behavior of silver nanotextiles during washing. Environ Sci Technol 43:8113–8118. doi:10.1021/es9018332
158. Farkas J, Peter H, Christian P et al (2011) Characterization of the effluent from a nanosilver producing washing machine. Environ Int 37:1057–1062. doi:10.1016/j.envint.2011.03.006
159. Windler L, Lorenz C, von Goetz N et al (2012) Release of titanium dioxide from textiles during washing. Environ Sci Technol 46:8181–8188. doi:10.1021/es301633b
160. Lorenz C, Windler L, von Goetz N et al (2012) Characterization of silver release from commercially available functional (nano)textiles. Chemosphere 89:817–824. doi:10.1016/j.chemosphere.2012.04.063
161. El-Rafie MH, Ahmed HB, Zahran MK (2014) Characterization of nanosilver coated cotton fabrics and evaluation of its antibacterial efficacy. Carbohydr Polym 107:174–181. doi:10.1016/j.carbpol.2014.02.024
162. Chronakis I (2005) Novel nanocomposites and nanoceramics based on polymer nanofibers using electrospinning process—a review. J Mater Process Technol 167:283–293. doi:10.1016/j.jmatprotec.2005.06.053
163. Leong S, Razmjou A, Wang K (2014) TiO2 based photocatalytic membranes: a review. J Membr Sci 472:167–184. doi:10.1016/j.memsci.2014.08.016
164. Nowack B, Brouwer C, Geertsma RE et al (2013) Analysis of the occupational , consumer and environmental exposure to engineered nanomaterials used in 10 technology sectors. Nanotoxicology 7:1152–1156. doi:10.3109/17435390.2012.711863
165. Wohlleben W, Kuhlbusch TAJ, Schnekenburger J, Lehr C-M (2014) Safety of nanomaterials along their lifecycle: release, exposure, and human hazards. CRC Press, London

166. Vance ME, Marr LC (2014) Exposure to airborne engineered nanoparticles in the indoor environment. Atmos Environ. doi:10.1016/j.atmosenv.2014.12.056
167. Kuhlbusch TA, Asbach C, Fissan H et al (2011) Nanoparticle exposure at nanotechnology workplaces: a review. Part Fibre Toxicol 8:22. doi:10.1186/1743-8977-8-22
168. Pietroiusti A, Magrini A (2014) Engineered nanoparticles at the workplace: current knowledge about workers' risk. Occup Med (Lond) 64:319–330. doi:10.1093/occmed/kqu051
169. Gomez V, Irusta S, Balas F et al (2014) Unintended emission of nanoparticle aerosols during common laboratory handling operations. J Hazard Mater 279:75–84. doi:10.1016/j.jhazmat.2014.06.064
170. Heitbrink WA, Lo L-M, Dunn KH (2015) Exposure controls for nanomaterials at three manufacturing sites. J Occup Environ Hyg 12:16–28. doi:10.1080/15459624.2014.930559
171. Kim B, Lee J, Choi B, Park S (2013) Ultrafine particle characteristics in a rubber manufacturing factory. Ann Occup Hyg 57:728–739. doi:10.1093/annhyg/mes102
172. Kim B, Kim H, Yu IJ (2014) Assessment of nanoparticle exposure in nanosilica handling process: including characteristics of nanoparticles leaking from a vacuum cleaner. Ind Health 52:152–162. doi:10.2486/indhealth.2013-0087
173. Göhler D, Stintz M (2014) Granulometric characterization of airborne particulate release during spray application of nanoparticle-doped coatings. J Nanopart Res 16:2520. doi:10.1007/s11051-014-2520-1
174. Brouwer DH, Duuren-Stuurman B, Berges M et al (2013) Workplace air measurements and likelihood of exposure to manufactured nano-objects, agglomerates, and aggregates. J Nanopart Res 15:2090. doi:10.1007/s11051-013-2090-7
175. Voliotis A, Bezantakos S, Giamarelou M et al (2014) Nanoparticle emissions from traditional pottery manufacturing. Environ Sci Process Impacts 16:1489–1494. doi:10.1039/c3em00709j
176. Froggett SJ, Clancy SF, Boverhof DR, Canady RA (2014) A review and perspective of existing research on the release of nanomaterials from solid nanocomposites. Part Fibre Toxicol 11:17. doi:10.1186/1743-8977-11-17
177. Duncan TV (2015) Release of engineered nanomaterials from polymer nanocomposites: the effect of matrix degradation. ACS Appl Mater Interfaces 7:20–39. doi:10.1021/am5062757
178. Hsu L-Y, Chein H-M (2006) Evaluation of nanoparticle emission for TiO2 nanopowder coating materials. J Nanopart Res 9:157–163. doi:10.1007/s11051-006-9185-3
179. Bello D, Wardle BL, Yamamoto N et al (2008) Exposure to nanoscale particles and fibers during machining of hybrid advanced composites containing carbon nanotubes. J Nanopart Res 11:231–249. doi:10.1007/s11051-008-9499-4
180. Ogura I, Kotake M, Shigeta M et al (2013) Potential release of carbon nanotubes from their composites during grinding. J Phys Conf Ser 429:12049. doi:10.1088/1742-6596/429/1/012049
181. Raynor PC, Cebula JI, Spangenberger JS et al (2012) Assessing potential nanoparticle release during nanocomposite shredding using direct-reading instruments. J Occup Environ Hyg 9:1–13. doi:10.1080/15459624.2012.633061
182. Koponen IK, Jensen KA, Schneider T (2009) Sanding dust from nanoparticle-containing paints: physical characterisation. J Phys Conf Ser 151:012048. doi:10.1088/1742-6596/151/1/012048
183. Koponen IK, Jensen KA, Schneider T (2010) Comparison of dust released from sanding conventional and nanoparticle-doped wall and wood coatings. J Expo Sci Environ Epidemiol 21:408–418. doi:10.1038/jes.2010.32
184. Göhler D, Stintz M, Hillemann L, Vorbau M (2010) Characterization of nanoparticle release from surface coatings by the simulation of a sanding process. Ann Occup Hyg 54:615–624. doi:10.1093/annhyg/meq053
185. Wohlleben W, Brill S, Meier MW et al (2011) On the lifecycle of nanocomposites: comparing released fragments and their in-vivo hazards from three release mechanisms and four nanocomposites. Small 7:2384–2395. doi:10.1002/smll.201002054
186. Huang G, Park JH, Cena LG et al (2012) Evaluation of airborne particle emissions from commercial products containing carbon nanotubes. J Nanopart Res. doi:10.1007/s11051-012-1231-8
187. Irfan A, Sachse S, Njuguna J et al (2013) Assessment of nanoparticle release from polyamide 6- and polypropylene-silicon composites and cytotoxicity in human lung A549 cells. J Inorg Organomet Polym Mater 23:861–870. doi:10.1007/s10904-013-9856-3

188. Sachse S, Silva F, Zhu H et al (2012) The effect of nanoclay on dust generation during drilling of PA6 nanocomposites. J Nanomater 2012:1–8. doi:10.1155/2012/189386
189. Sachse S, Silva F, Irfan A et al (2012) Physical characteristics of nanoparticles emitted during drilling of silica based polyamide 6 nanocomposites. IOP Conf Ser Mater Sci Eng 40:12012. doi:10.1088/1757-899X/40/1/012012
190. Bello D, Wardle BL, Zhang J et al (2010) Characterization of exposures to nanoscale particles and fibers during solid core drilling of hybrid carbon nanotube advanced composites. Int J Occup Environ Health 16:434–450. doi:10.1179/107735210799159996
191. Wohlleben W, Meier MW, Vogel S et al (2013) Elastic CNT-polyurethane nanocomposite: synthesis, performance and assessment of fragments released during use. Nanoscale 5: 369–380. doi:10.1039/c2nr32711b
192. Schlagenhauf L, Chu BTT, Buha J et al (2012) Release of carbon nanotubes from an epoxy-based nanocomposite during an abrasion process. Environ Sci Technol 46:7366–7372. doi:10.1021/es300320y
193. Guiot A, Golanski L, Tardif F (2009) Measurement of nanoparticle removal by abrasion. J Phys Conf Ser 170:12014. doi:10.1088/1742-6596/170/1/012014
194. Vorbau M, Hillemann L, Stintz M (2009) Method for the characterization of the abrasion induced nanoparticle release into air from surface coatings. J Aerosol Sci 40:209–217. doi:10.1016/j.jaerosci.2008.10.006
195. Zuin S, Gaiani M, Ferrari A, Golanski L (2013) Leaching of nanoparticles from experimental water-borne paints under laboratory test conditions. J Nanopart Res 16:2185. doi:10.1007/s11051-013-2185-1
196. Zhou L, Zhang Z, Xia S et al (2014) Effects of suspended titanium dioxide nanoparticles on cake layer formation in submerged membrane bioreactor. Bioresour Technol 152:101–106. doi:10.1016/j.biortech.2013.11.006
197. Golanski L, Guiot A, Braganza D, Tardif F (2010) New method for the characterization of abrasion-induced nanoparticle release into air from nanomaterials. In: NSTI-Nanotech 2010. Anaheim, CA, USA, p 720–723
198. Biswal M, Mohanty S, Nayak SK, Kumar PS (2013) Effect of functionalized nanosilica on the mechanical, dynamic-mechanical, and morphological performance of polycarbonate/nanosilica nanocomposites. Polym Eng Sci 53:1287–1296. doi:10.1002/pen.23388
199. Grieger KD, Laurent A, Miseljic M et al (2012) Analysis of current research addressing complementary use of life-cycle assessment and risk assessment for engineered nanomaterials: have lessons been learned from previous experience with chemicals? J Nanopart Res 14:958. doi:10.1007/s11051-012-0958-6
200. Project on Emerging Nanotechnologies (2014) Consumer Products Inventory
201. EP (2009) Regulation (EC) No 1223/2009 of the European Parliament and of the Council of 30 November 2009 on cosmetic products. Europe
202. EP (2011) Regulation (EU) No 1169/2011 of the European Parliament and the Council of 25 October 2011 on the provision of food information to consumers. Europe
203. Ministère de l'Écologie du Développement durable et de l'Énergie (2014) Éléments issus des déclarations des substances à l'état nanoparticulaire – exercice 2014. Agence nationale de sécurité sanitaire de l'alimentation, de l'environnement et du travail, Maisons-Alfort, France
204. Göhler D, Nogowski A, Fiala P, Stintz M (2013) Nanoparticle release from nanocomposites due to mechanical treatment at two stages of the life-cycle. J Phys Conf Ser 429:12045. doi:10.1088/1742-6596/429/1/012045
205. Hirth S, Cena L, Cox G et al (2013) Scenarios and methods that induce protruding or released CNTs after degradation of nanocomposite materials. J Nanopart Res 15:1504. doi:10.1007/s11051-013-1504-x
206. Kaegi R, Sinnet B, Zuleeg S et al (2010) Release of silver nanoparticles from outdoor facades. Environ Pollut 158:2900–2905. doi:10.1016/j.envpol.2010.06.009
207. Al-Kattan A, Wichser A, Vonbank R et al (2013) Release of TiO2 from paints containing pigment-TiO2 or nano-TiO2 by weathering. Environ Sci Process Impacts 15:2186–2193. doi:10.1039/c3em00331k

208. Kaegi R, Ulrich A, Sinnet B et al (2008) Synthetic TiO2 nanoparticle emission from exterior facades into the aquatic environment. Environ Pollut 156:233–239. doi:10.1016/j.envpol.2008.08.004
209. Al-Kattan A, Wichser A, Zuin S et al (2014) Behavior of TiO(2) released from Nano-TiO(2)-containing paint and comparison to pristine Nano-TiO(2). Environ Sci Technol 48:6710–6718. doi:10.1021/es5006219
210. Benn TM, Westerhoff P (2008) Nanoparticle silver released into water from commercially available sock fabrics. Environ Sci Technol 42:4133–4139
211. Benn T, Cavanagh B, Hristovski K et al (2010) The release of nanosilver from consumer products used in the home. J Environ Qual 39:1875. doi:10.2134/jeq2009.0363
212. Mitrano DM, Rimmele E, Wichser A et al (2014) Presence of nanoparticles in wash water from conventional silver and nano-silver textiles. ACS Nano 8:7208–7219. doi:10.1021/nn502228w
213. Kulthong K, Srisung S, Boonpavanitchakul K et al (2010) Determination of silver nanoparticle release from antibacterial fabrics into artificial sweat. Part Fibre Toxicol 7:8. doi:10.1186/1743-8977-7-8
214. von Goetz N, Lorenz C, Windler L et al (2013) Migration of Ag- and TiO2-(Nano)particles from textiles into artificial sweat under physical stress: experiments and exposure modeling. Environ Sci Technol 47:9979–9987. doi:10.1021/es304329w
215. Yan Y, Yang H, Li J et al (2012) Release behavior of nano-silver textiles in simulated perspiration fluids. Text Res J 82:1422–1429. doi:10.1177/0040517512439922
216. Sánchez C, Hortal M, Aliaga C et al (2014) Recyclability assessment of nano-reinforced plastic packaging. Waste Manag 34:2647–2655. doi:10.1016/j.wasman.2014.08.006
217. Touati N, Kaci M, Bruzaud S, Grohens Y (2011) The effects of reprocessing cycles on the structure and properties of isotactic polypropylene/cloisite 15 A nanocomposites. Polym Degrad Stab 96:1064–1073. doi:10.1016/j.polymdegradstab.2011.03.015
218. Struwe J, Schindler E, Pfirrmann O (2012) Relevance of nanomaterials in waste recycling. Hans-Böckler Foundation, Düsseldorf, Germany, AP270
219. Caballero-Guzman A, Sun T, Nowack B (2015) Flows of engineered nanomaterials through the recycling process in Switzerland. Waste Manag 36:33–43. doi:10.1016/j.wasman.2014.11.006
220. Holder AL, Vejerano EP, Zhou X, Marr LC (2013) Nanomaterial disposal by incineration. Environ Sci Process Impacts 15:1652–1664. doi:10.1039/c3em00224a
221. Mueller NC, Buha J, Wang J et al (2013) Modeling the flows of engineered nanomaterials during waste handling. Environ Sci Process Impacts 15:251. doi:10.1039/c2em30761h
222. Roes L, Patel MK, Worrell E, Ludwig C (2012) Preliminary evaluation of risks related to waste incineration of polymer nanocomposites. Sci Total Environ 417–418:76–86. doi:10.1016/j.scitotenv.2011.12.030
223. Walser T, Limbach LK, Brogioli R et al (2012) Persistence of engineered nanoparticles in a municipal solid-waste incineration plant. Nat Nanotechnol 7:520–524. doi:10.1038/nnano.2012.64
224. Vejerano EP, Leon EC, Holder AL, Marr LC (2014) Characterization of particle emissions and fate of nanomaterials during incineration. Environ Sci Nano. doi:10.1039/c3en00080j
225. Massari A, Beggio M, Hreglich S et al (2014) Behavior of TiO2 nanoparticles during incineration of solid paint waste: a lab-scale test. Waste Manag 34:1897–1907. doi:10.1016/j.wasman.2014.05.015
226. Lozano P, Berge ND (2012) Single-walled carbon nanotube behavior in representative mature leachate. Waste Manag 32:1699–1711. doi:10.1016/j.wasman.2012.03.019
227. Khan IA, Berge ND, Sabo-attwood T et al (2013) Single-walled carbon nanotube transport in representative municipal solid waste landfill conditions. Env Sci Technol 47:8425–8433. doi:10.1021/es401748f
228. Bolyard SC, Reinhart DR, Santra S, States U (2013) Behavior of engineered nanoparticles in landfill leachate. Environ Sci Technol 47:8114–8122. doi:10.1021/es305175e
229. Gitipour A, El Badawy A, Arambewela M et al (2013) The impact of silver nanoparticles on the composting of municipal solid waste. Environ Sci Technol 47:14385–14393. doi:10.1021/es402510a

230. Neale PA, Jämting ÅK, Escher BI, Herrmann J (2013) A review of the detection, fate and effects of engineered nanomaterials in wastewater treatment plants. Water Sci Technol 68:1440–1453. doi:10.2166/wst.2013.388
231. Yang Y, Wang Y, Westerhoff P et al (2014) Metal and nanoparticle occurrence in biosolid-amended soils. Sci Total Environ 485-486:441–449. doi:10.1016/j.scitotenv.2014.03.122
232. Ganzleben C, Pelsy F, Hansen SF, et al (2011) Review of environmental legislation for the regulatory control of nanomaterials. Directorate-General for Environment-European Commision, Brussels, Belgium, p. 210
233. Beaudrie CEH, Kandlikar M, Satterfield T (2013) From cradle-to-grave at the nanoscale: gaps in U.S. regulatory oversight along the nanomaterial life cycle. Environ Sci Technol 47:5524–5534. doi:10.1021/es303591x
234. EP (2008) Regulation (EC) No 1272/2008 of the European Parliament and of the Council of 16 December 2008 on classification, labelling and packaging of substances and mixtures, amending and repealing. Europe
235. EP (2008) Directive 2008/98/EC of the European Parliament and of the Council of 19 November 2008 on waste and repealing certain Directives. 3–30
236. Breggin LK, Pendergrass J (2007) Does the nano go? End-of-life regulation of nanotechnologies. Project on Emerging Nanotechnologies at the Woodrow Wilson International Center for Scholars, Washington, DC

Chapter 4
From Dose to Response: In Vivo Nanoparticle Processing and Potential Toxicity

Uschi M. Graham, Gary Jacobs, Robert A. Yokel, Burtron H. Davis, Alan K. Dozier, M. Eileen Birch, Michael T. Tseng, Günter Oberdörster, Alison Elder, and Lisa DeLouise

Abstract Adverse human health impacts due to occupational and environmental exposures to manufactured nanoparticles are of concern and pose a potential threat to the continued industrial use and integration of nanomaterials into commercial products. This chapter addresses the inter-relationship between dose and response and will elucidate on how the dynamic chemical and physical transformation and breakdown of the nanoparticles at the cellular and subcellular levels can lead to the *in vivo* formation of new reaction products. The dose-response relationship is complicated by the continuous physicochemical transformations in the nanoparticles induced by the dynamics of the biological system, where dose, bio-processing, and response are related in a non-linear manner. Nanoscale alterations are monitored using high-resolution imaging combined with in situ elemental analysis and emphasis is placed on the importance of the precision of characterization. The result is an in-depth understanding of the starting particles, the particle transformation in a biological environment, and the physiological response.

Keywords Nanotechnology • Electron Microscopy • Transformation • Subcellular • Nanoparticle Instability

---

U.M. Graham (✉)
University of Kentucky, Lexington, KY, USA

CDC/NIOSH DART, Cincinnati, OH, USA
e-mail: graham@topasol.com

G. Jacobs • R.A. Yokel • B.H. Davis
University of Kentucky, Lexington, KY, USA

A.K. Dozier • M.E. Birch
CDC/NIOSH DART, Cincinnati, OH, USA

M.T. Tseng
University of Louisville, Louisville, KY, USA

G. Oberdörster • A. Elder • L. DeLouise
University of Rochester, Rochester, NY, USA

L. Tran et al. (eds.), *Modelling the Toxicity of Nanoparticles*, Advances in Experimental Medicine and Biology, Vol. 947, DOI 10.1007/978-3-319-47754-1_4

4.1 Introduction

Nanotechnology is a key modernization driver that balances innovations in material synthesis with the need for novel solutions that impact all energy sectors, emerging medical fields, and rapidly evolving electronics applications [3]. It also offers environmental technology breakthroughs by integrating nanotechnology products and synthetic biology and offers opportunities that focus on human health and animal welfare. The field of nanomaterials is a multidisciplinary area in which material science is explored at the nano-scale, but the concepts behind nanoscience are not new. In his celebrated lecture at Caltech, in 1959, physicist Richard Feynman described the process of manipulating and controlling individual atoms, molecules and nanoparticles, and he anticipated an "enormous number of technical applications" through the creation of novel materials and compounds [22, 76, 78]. More than half a century later, scientists and engineers are finding various ways to produce a wide range of nanoparticles [1]. Importantly, the fast exploration and deployment of nanomaterials must also incorporate exposure, toxicity and risk assessment studies in order to balance the successful integration of nanomaterials into everyday life with any potential safety and environmental issues [7, 17, 18, 48, 58, 69, 72, 78]. This is critically important in determining which parts of life may be enriched with the assistance of nanomaterials and which parts may suffer.

Manufactured nanoparticles (MNPs) typically range in size from 1–100 nm [23]. They exhibit unique properties compared with those of their larger-sized "macro" counterparts. The differences are due to vastly increased surface-to-bulk ratios and because of the distinct structures of MNPs [13, 27, 80, 88]. Nanotechnology and the application of nanoparticles in consumer products has become an integral part of today's life and require safety assessments [4, 12, 20, 27, 32, 34, 50, 59, 66, 73]. The growing rate of nanoparticle-based product developments has raised worldwide apprehension regarding the release of MNPs into the environment and their subsequent uptake. There are several uptake pathways for MNPs, which complicates the issue of modelling exposure risks tremendously [5, 15, 30, 40, 54, 60, 83]. Nanosafety studies have seen an exponential rise over the past two decades, but the effects and dangers of nanoparticles, either for animals, humans, or cell structures, are still not clearly defined [10, 43]. Safety concerns have led industrial and academic researchers to adopt strategies to make MNPs more biocompatible, by employing techniques such as capping with various functional groups and also by exploring new synthesis routes [64], but the ultimate fate of the MNPs after uptake remains unresolved [28, 56, 85]. This is, in part, heightened by additional effects from nanoparticles that come from sources other than controlled manufacturing labs, such as pollution-derived nanoparticles where the composition, size ranges and effects are often unknown. Another important issue is the environmental significance of natural colloids and nanoparticles that govern elemental mobility and bioavailability [33], where much of the environmental pool of nanoparticles consists of breakdown products from both organic and inorganic sources such as cellulose fragments and clays that may be in the same size range as MNPs. There is also the influx

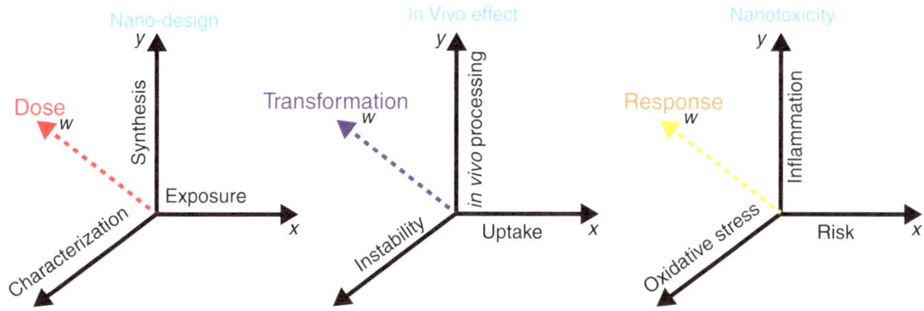

Fig. 4.1 Schematic illustration of nanoparticle dose-dependent reactivity and physiological response

of pollution-derived nanoparticles produced in urban settings from industrial effluents and auto exhausts, which are an important part of risk assessment models and have been linked to major health problems [11, 19, 88].

Adverse human health effects due to occupational and also environmental exposures to nanoparticles are of worldwide concern. Concepts of nanoparticle dose metric and response metric are of paramount importance [42, 43] and can provide key insights into relationships between the nanoparticles' synthetic identity and chemical reactivity, their biological activity which involves aggregation, protein interactions, protective surface coatings as well as migration and, lastly, their stability, all of which contribute individually and collectively to dose-dependent toxicity outcomes [79]. An in-depth understanding of biokinetics is vital to obtaining meaningful risk assessment protocols for MNPs [71, 78, 87]. This has to include information on the biodistribution and clearance of MNPs as a function of the exposure route [43]. Furthermore, it has to include information on uptake, transport and transformation of MNPs as a function of dose and epithelial route of entry (including but not limited to gastro-intestinal, dermal and respiratory ports-of-entry). It also requires thorough data collection on the biotransformation of MNPs within target tissues and cells [27]. The cellular and subcellular interactions of nanoparticles are a function of the physiological environment which can only respond to a certain number of invader nanoparticles or reactive surface area (smaller nanoparticles contribute higher surface areas and in this regard, also contribute different surface properties such as charge, composition, structures, porosity, redox-state and reactivity). This is sometimes referred to as the "surface area dose-response relationship" [61] and affects the short and long term fate of nanoparticles after uptake. Dose and nanoparticle properties (nano-design) will undoubtedly influence the transport and bioprocessing (*in vivo* effect) of the MNPs and their derivatives (break-down products) which leads to a dose-dependent reactivity and physiological response (nanotoxicity) (Fig. 4.1).

A dose-dependent instability of synthesized MNPs after exposure and cellular uptake leads to *in vivo* processing and transformation, which may be followed by a certain response (oxidative stress and inflammation) and ultimately results in nanotoxicity (Fig. 4.1). Clearly, a nanoscale substance might potentially be toxic for a biological system when the "dose" or concentration exceeds an adverse threshold. The response "effect" could be initiated by a single "acute" dose, or, by repeated low "chronic" dose that occurs over an extended time frame. Careful dose evaluations are necessary for meaningful risk assessments of nanoparticles and play a major role in regulatory processes to help determine health-relevant limits [43]. For example, instillation studies are typically carried out with high MNPs doses and it is impossible to know whether effects are caused by overload conditions or due to the MNPs' inherent effects. Inhalation studies can offer insights at lower dose, yet they too are met with inflammatory responses which have been determined more often than not to be independent of the nature of MNPs that are inhaled [43, 88]. Because of these difficulties in dose-response studies, the mechanisms that induce toxicity from respiratory exposures are poorly understood and thus hinder the building of predictive models. Similarly, low dose response studies for MNP exposures to skin and gastrointestinal epithelial tissues are lacking as is our understanding of how differences in the local biological milieu effects microenvironment around a nanoparticle and vice-versa.

4.1.1 In Vivo Processing and Transformation of Nanoparticles

The issue that will be addressed in this chapter is the relationship between dose and potential *in vivo* processing of nanoparticles (response) shown in Fig. 4.1. The issue includes nanoparticle uptake, transport and transformation as a function of dose and uptake routes. *In vivo* processing is defined here as the dynamic chemical and/or physical breakdown of nanoparticles at the cellular and subcellular level [27]. The process can be followed *by in vivo* formation of new reaction products including ions, nuclei and growth of second generation nanoparticles all of which may be set in motion by the breakdown of the original nanomaterials. Such *in vivo* biotransformation processes are known to occur with implanted orthopedic materials that can lead to both pathologic and beneficial patient outcomes. For example, nanoparticle wear debris formed from articulating prosthetic surfaces can lead to osteolysis [29], whereas the successful adherence and osteoinduction of amorphous bioactive glass results from a dissolution and re-precipitation reaction and induces a material phase change to crystalline hydroxyapatite [35, 39].

Uptake and transport of nanoparticles to different regions in the body have been extensively studied and are generally linked to certain pathology and toxicity [40, 43, 51, 52, 56, 61]. However, the *in vivo* breakdown and processing of MNPs that leads to formation of new reaction products with different properties is not very clear and obfuscates the issue of exposure risk and related outcomes. It also makes the design of meaningful predictive models significantly more challenging. The breakdown mechanism of MNPs in cells depends on the material composition, surface coatings, ports of entry and the organs they invade (Fig. 4.2). The instability of nanoparticles in cells then initiates another cascade of responses that yet have to be defined. In this Chapter, we describe applications of advanced electron microscopy methods to the

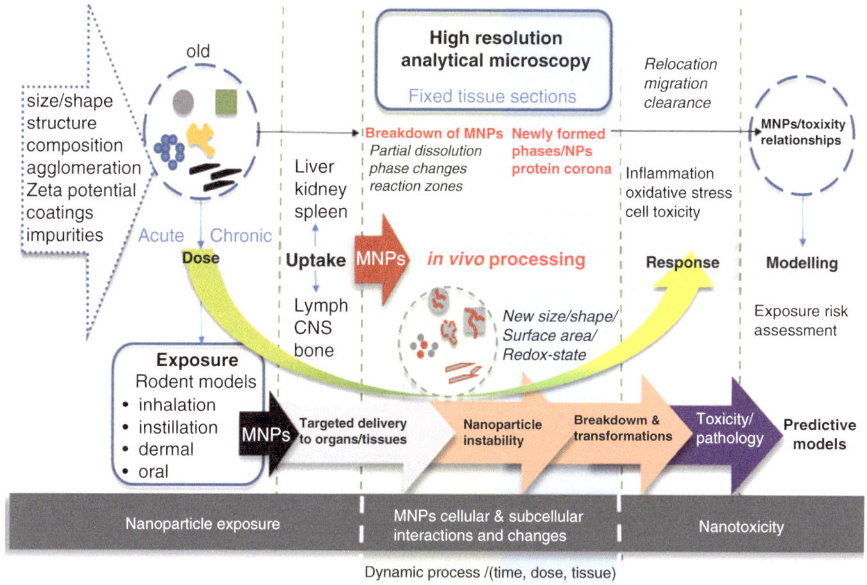

Fig. 4.2 From dose to response: MNPs characterization, followed by dose-dependent exposure in rodent model, uptake into organs/tissues, *in vivo* processing and nanotoxicity assessment and modelling

analysis of fixed tissue sections, which provides critical information on material phase changes and the oxidation states of MNPs [27]. Specifically, we discuss use of high resolution (nanometer) transmission electron microscopy (HRTEM) coupled with simultaneous elemental analysis for the investigation of the *in vivo* processing of nanoparticles as a function of dose and uptake route (Fig. 4.2). The *in vivo* processing evidence can then be used for more comprehensive modelling of the potential exposure risks for nanoparticles. The goal of these studies is to investigate cellular and subcellular interactions of MNPs using advanced imaging and analysis of the retained particles and correlate these interactions with biological and toxicological effects. The data are important to build meaningful predictive models that are based on the dynamic interaction of nanoparticles at the cellular and subcellular levels after uptake. A thorough understanding of nanoparticle processing in biological systems as a function of dose is vital in making determinations of the long-term toxicological effects. This requires studies to determine *in vivo* solubility (nanoparticle dissolution), size and shape changes in response to the original dose and nanoparticle retention time (Fig. 4.2). A possible increase or decrease in protein corona around the MNPs and their cell associations also must be examined [14, 25]. *In vivo* processing of MNPs is a function of dose and residence time in particular tissues or cells. The *in vivo* processed nanoparticles and any "newly" formed phases and reaction products must be compared with the original MNPs (Fig. 4.2). The nanoparticle's characteristic "fingerprints" before and after *in vivo* processing are based on composition, geometrical parameters and physio-chemical, structural and spectroscopic properties. Without this data, it would be impossible to build meaningful models that correlate nanoparticle dose and exposure risks. However, a deep understanding of biokinetics is also central to obtain an all-encompassing exposure risk assessment and involves identification of target organs following different routes of exposure. One has to evaluate the *in vivo* processing of MNPs not only in the regions where uptake first occurs (portal-of-entry-organs), but also must observe any particle breakdown or processing in secondary and further ancillary target tissues while considering the original dose and residence time of the particles. Often the experimental characterization of nanoparticles that is obtained before exposure is directly linked to cellular-based assays. This means that risk assessment models typically assume that the "invader" nanoparticles that cause toxicity are exactly the same as those that were used in the exposure experiments. Unfortunately that is an oversimplification. To date we know that nanoparticles are processed *in vivo* [28] and the extent to which they are processed needs to be systematically studied so that this information can be incorporated into advanced risk assessment models. Future studies will need to evaluate the *in vivo* processing of MNPs in portal-of entry organs and also in secondary target tissues and evaluate any modifications/transformations of MPNs with regards to their physicochemical changes as a function of the route and duration of exposure. Only then can predictive models be designed to better forecast nanoparticle-dose-toxicity relationships. State of the art microscopy methods can be applied to obtain needed *in vivo* processing data, and several examples of this approach are presented in the remainder of this Chapter.

4.2 The Role of Cellular Breakdown and In Vivo Processing of Nanoparticles

The study of *in vivo* induced changes to nanoparticles is an emerging area of investigation. In the case of highly soluble materials such as nano-copper and nano-silver any dissolution and particle breakdown after uptake into biological media can be expected and has been demonstrated [6]. However, the *in vivo* breakdown and transformation mechanisms of essentially poorly soluble particles (PSP) like ceria (CeO_2) on a cellular and subcellular level are not well understood. The breakdown mechanism of CeO_2 nanoparticles in the liver of rats was recently demonstrated for the first time [27]. These findings confirm that nanoparticle uptake and sequestration in peripheral organs can lead to the formation of secondary particles with different physiochemical properties including altered reactivity and effects that result in varying degrees of toxic effects over long periods of exposure. Furthermore, ceria *in vivo* transformation can progress to a toxic, more benign, or potentially beneficial state [32]. In this regard, CeO_2 nanoparticles after prolonged residence time of 90 days inside liver undergo *in vivo* processing that causes a shift towards smaller particle size and an increased reactive surface area with enhanced free radical scavenging potential of the new *in vivo* formed ultrafine particles [27]. This work also

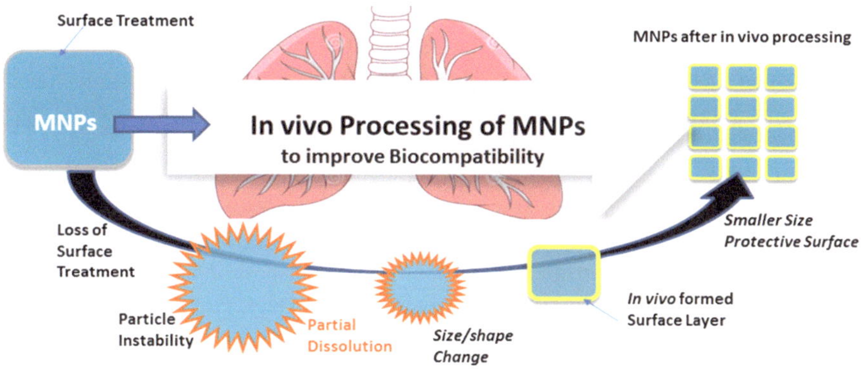

Fig. 4.3 Nanoparticle breakdown can lead to improved biocompatibility

showed with the help of high resolution imaging and analysis that essentially insoluble CeO_2 nanoparticles experience partial dissolution and reformation inside the liver. Breakdown and redistribution after inhalation of ceria nanoparticles could be a possible coping mechanism of biological systems and a step towards improving nanoparticle biocompatibility as illustrated in Fig. 4.3.

Because CeO_2 is basically insoluble under laboratory controlled conditions, one has to question what drives dissolution of CeO_2 and similar nanoparticles in the liver, lung and possibly other regions *in vivo* and whether enzyme activity and other factors need to be incorporated into risk assessment models. This is particularly important for more soluble nanoparticles such as amorphous silica (SiO_2), alumina (Al_2O_3), titania (TiO_2) and iron oxides (Fe_2O_3 and Fe_3O_4) which constitute the vast volumes of MNPs used today in consumer products and medical imaging. Faster dissolution rates could lead to rapid particle breakdown and transformation. Clearly, how to obtain insights into biotransformation routes of nanoparticles and their *in vivo* processing response depends on well-designed experimental studies that provide dose-controlled nanoparticle uptake, i.e., via instillation, inhalation (lung, olfactory system), oral intake (stomach, GI) or dermal uptake (skin: intact versus injured) (Fig. 4.2). This has to be followed by a systematic comparison of the *in vivo* transformed particles with the pristine precursor materials by examining morphological changes, size variations, dissolution patterns and the presence or absence of secondary reaction zones (new precipitates) in the vicinity of the transforming nanoparticles. Further detailing the physio-chemical changes during bioprocessing of nanoparticles may be an effective tool in understanding their subcellular and temporal fate that controls toxicity. These analyses depend on advanced imaging methods. High-resolution electron microscopy applications allow the use of fixed tissues to examine nanoparticle location, size and composition immediately after deposition and also after prolonged residence time. Nanoparticle-cell interaction and dose-dependent inflammatory response raises the question about underlying cellular mechanisms that produce nanoparticle instability (Fig. 4.2). Therefore, dose-dependent toxicity that is caused by *in vivo* processing of nanoparticles needs to be considered in risk assessment models. Also, it is important to model nanotoxicity as a function of the nanoparticle instability, transformation, mobility and potential *in vivo* reformation (precipitation) at the cellular and subcellular level. Nanoparticle instability *in vivo* is a function of the particle's inherent composition, size, molecular structure and surface chemistry among other properties, but also a function of the complex cellular condition such as protein corona, inflammatory responses (chronic vs. acute), upregulation of inflammatory defense mechanisms and availability of enzymatic catalysts just to name some. Mobility of individual nanoparticles may be controlled by both physical transport of the intact particles, and also by a sequence of dissolution and reformation steps. High resolution analysis of the reaction zones around dissolving nanoparticles in phagolysosomes show breakdown patterns, void spaces and pore-formation, suggesting that there are continuous processes that release and relocate molecules during the nanoparticle transformation. This information is important in creating government regulations for nanoparticle exposure to workers and consumers. One very important aspect for obtaining nanoscale structural and chemical information to be able to study the breakdown and processing of MNPs, of course, is the preparation and conditioning of tissue materials which precedes all of the advanced imaging and analysis techniques. We refer here to previous works that give excellent overviews of the tissue preparation techniques [49, 55, 74].

The following sections will discuss the importance and application of advanced imaging methods to help identify the various processes involved during *in vivo* nanoparticle transformation and give three specific examples for: (1) high and low dose inhaled amorphous silica (SiO_2) nanoparticles that are deposited, transformed and relocated inside rat lung; (2) ceria (CeO_2) nanoparticle dispersion and *in vivo* processing in spleen after a single high dose instillation; and (3) discuss the spatial and temporal relationship of *in vivo* synthesized ferritin nanoparticles (iron oxyhydroxide Fe-OOH) as a direct response to the uptake and processing of invader SiO_2 and CeO_2 nanoparticles, and suggest mechanisms at the cellular and subcellular levels.

4.3 Advanced Imaging and Analysis of Nanoparticles in Tissue Sections

The study of nanoparticle *in vivo* processing is compounded by the number of variables in play when it comes to biotransformation, such as composition, morphology, size, and exposure mechanism or route of entry. The situation is made even more complex for researchers in that specialized methods of investigation are required to observe nanoparticle transformations in biological systems. Typical methods employed in biological research can only partly reveal nanoparticle transformations or information about the mechanisms involved due to the small size range. These methods include fluorescence, confocal, and polarized microscopy, electron micrographs, radiological tracing, and measurements of biological toxic response indicators. Traditional material characterization methods used by material engineers/scientists need to be employed that allow resolution and analysis at the nanoscale. Methods that have been employed so far are the standard electron microscopy techniques used in materials characterization such as, electron diffraction [2, 27, 46], and scanning transmission electron microscopy (STEM) [81] with the associated analytical techniques energy dispersion spectroscopy (EDS) [36, 84], and electron energy loss spectroscopy (EELS) [16, 26, 80]. Also, x-ray photoelectron spectroscopy (XPS) has been used [27]. Aberration corrected STEM allows imaging at the atomic scale and will be instrumental in determining the structures and composition of *in vivo* formed nanoparticles that are only a few nanometers or possibly sub-nano size [57]. The use of these methods is complicated by the nanoparticles being hosted inside a biological matrix. This requires modifications to the standard biological sample preparation techniques [49, 55, 74].

4.3.1 High Resolution Analytical Microscopy

Electron microscopy has been fundamental in gaining knowledge about biological systems since the 1950's and was instrumental in developing insights into cellular ultrastructure [21]. As electron microscopes evolved, the imaging needs of the biologist and that of material scientist diverged. The biologist needed high contrast, wide field, and low accelerating voltages, whereas the material scientist needed high resolution imaging, high accelerating voltages, and high brightness through the use of a field emission electron source. This resulted in differing classes of electron microscopes being manufactured such as the Philips 201 and CM-10 for biological applications versus the Philips 300, 400, and CM-12 for material characterization. This has resulted in major medical research centers having biologically oriented electron microscopes with an inability to apply what are now common material characterization techniques. Multidisciplinary collaborations between medical researchers and material scientists can overcome this. High-resolution transmission electron microscopy (HRTEM), coupled with advanced detectors allows one to probe materials in unprecedented detail, providing both local chemical information and also structural properties.

An analytical electron microscope today can image and obtain compositional and electronic information down to the angstrom level. This ability provides highly local information from surface-environment interactions such as *in vivo* nanoparticles. Other material characterization techniques work for bulk samples and have resolutions larger than several nanometers. Thus, for the study of *in vivo* transformation of nanoparticles there is only one choice, an HRTEM designed for materials analysis [28, 84]. A typical HRTEM used in materials characterization will have both TEM and STEM capability with EDS and EELS being incorporated with the use of computer technology to allow the acquisition of elemental line profiles and maps acquired in STEM mode. This allows not only elemental analysis but also acquisition of material phase changes and oxidative states via the EELS data [16, 53]. Combining these observations with material phase databases such as the Materials Project (www.materialsproject.org) and computation from first principles using spectroscopy oriented software such as FEFF9 [67] in principle, allows the identification of phases and electronic states. Because a standard non-aberration corrected field emission electron microscope designed for materials analysis will typically be able to achieve a STEM spot size of 0.2 nm, changes in nanoparticle surfaces versus their main bodies can be analyzed [27]. This data combined with material phase structure data and spectroscopy computation can, in principle, provide information on structural and electronic changes in nanoparticles in tissue. This is the type of information needed to understand the interaction of a nanoparticle with its local environment in order to gain an understanding of the mechanisms behind *in vivo* transformation and how this relates to toxicity.

4.3.2 Example I: Amorphous Silica (SiO_2) Inside Lung Tissue

Analysis of the clearance kinetics using modelling of retained lung burden of SiO_2-MNPs showed a significant *in vivo* solubility which raises questions about underlying cellular mechanisms that result in the instability of the SiO_2-MNPs and related toxicity [24]. This was the stimulus to use HRTEM applications and to look for evidence of particle breakdown and mobility in the lung tissue at both cellular and subcellular levels. The principal objective for HRTEM is to examine any nano-scale alteration, dissolution and processing of SiO_2-MNPs after inhalation by comparing the translocated particles with the precursor SiO_2-MNPs. A dose and time controlled inhalation study involved groups of rats that were exposed to aerosols containing amorphous SiO_2-MNPs for 4 h/day, 5 days/week for 4 weeks with a 27 day post-exposure observation period at three different concentrations and dose-dependent pulmonary inflammation in the rats, and data was collected in relation to the exposure time and corresponding dose that was used [62]. In this particular study, subchronic inhalation exposures of the SiO_2-MNPs were investigated using an approach of dosimetric modelling to determine the mechanisms for clearance of these nanoparticles from the lung. Both mechanical clearance and partial dissolution have to be considered as potential pathways. Determining the *in vivo* bioprocessing mechanisms of the nanoparticles will be important towards risk characterization and to better assess possible health effects caused by the transformation, translocation and clearance of the particles after exposure. The HRTEM observation of the precursor (as synthesized) SiO_2-MNPs particles showed a typical size of ~20–50 nm with a corresponding spherical morphology of the individual aerosolized SiO_2-MNPs components (Fig. 4.4). Furthermore, after 27 days post-exposure the phagocytosed SiO_2-MNPs that were sequestered in alveolar macrophages in the fixed tissue sections were imaged using high resolution Dark Field STEM. The STEM images show clear indication of significant *in vivo* breakdown and transformation (Fig. 4.5). There is also structural evidence in the Dark Field STEM (Fig. 4.5) that a portion of SiO_2-MNPs had been completely dissolved out. The degree of *in vivo*

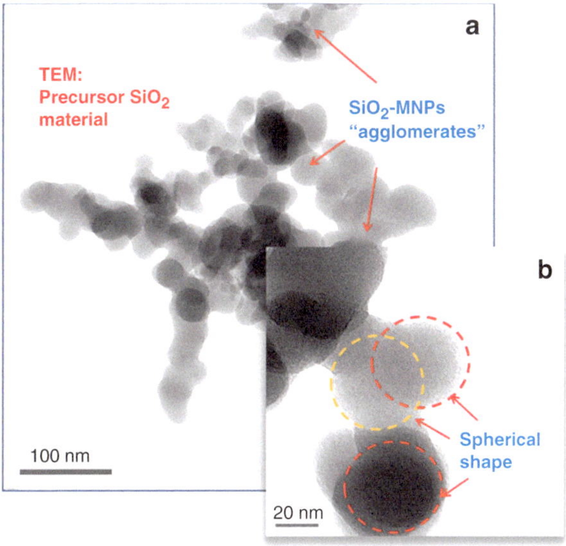

Fig. 4.4 illustrates precursor SiO_2-MNPs prior to dose controlled inhalation into lungs. (**a**) Large agglomerates. (**b**) Precursor SiO_2-MNPs with amorphous nanostructures and predominantly *spherical shape*. Some *overlapping spheres* are marked with *circles*

processing of the particles and partial dissolution most likely depends on the residence time, dose, and synthetic identity of the original inhaled SiO_2-MNPs. Most of the SiO_2-MNPs particles lost their original spherical morphology after prolonged lung retention and are now displaying various dissolution patterns (pitting), void formation and secondary outward growth that results in the formation of multiple reaction zones. To gain greater insights into what controls particle transformation and determine if there are any relationships with subcellular components, one has to perform detailed elemental mapping of the regions of interest. As an example, elemental EDS maps of O, Si, S and P (Fig. 4.6) are obtained from a region that is illustrated in the Dark Field STEM image in Fig. 4.5. The EDS elemental maps of O, Si, S, and P require the use of a 1 nm STEM probe to have enough counting statistics within a reasonable dwell time as illustrated in (Fig. 4.6).

Typically a 1–2 s dwell time is used depending on the signal strength. EELS mapping can be done with a 0.2 nm probe and dwell times as small as 0.1 second, depending on the elemental edge being mapped. Higher edges require longer dwell times. When doing simultaneous EDS and EELS mapping a compromise must be worked out to have a long enough time for a good EDS count and short enough so as not to overload the EELS CCD detector [27]. Other EELS acquisition parameters such as dispersion and y-binning can be adjusted to obtain a satisfactory EELS signal [16, 26]. After performing EDS mapping of a region that seems to have undergone *in vivo* processing of SiO_2-MNPs in alveolar macrophages the O and Si signals clearly follow the outline of the SiO_2-MNPs (Fig. 4.6). However, the Si signal furthermore is indicating that some Si is present in the close neighborhood of the SiO_2-MNPs, while the O signal is predominantly confined to the outline of the alveolar macrophage-entrapped nanoparticles and not seen in the immediate subcellular surroundings.

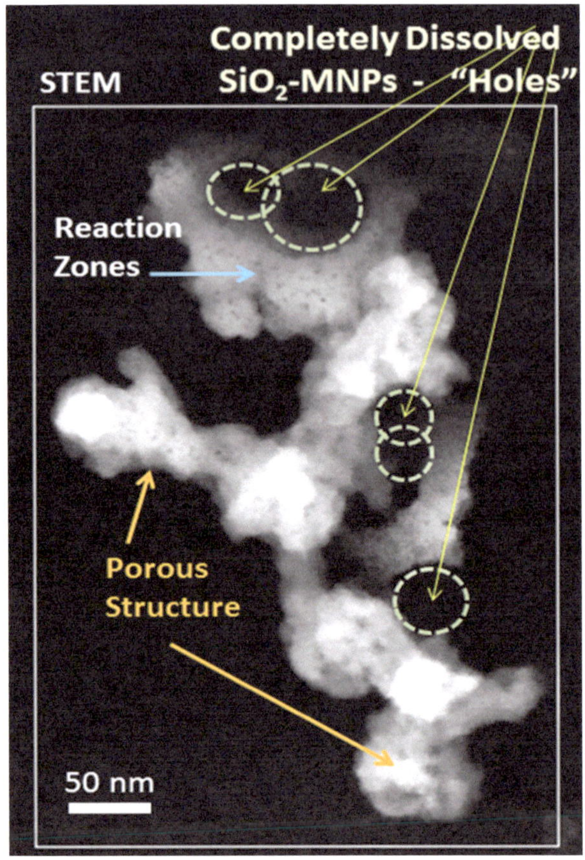

Fig. 4.5 Dark field STEM imaging of lung section after repeated dose inhalation and 27 days post treatment. SiO_2-MNPs show pores and significant *in vivo* processing. Almost all of the original spherical morphology has disappeared after 27 days post treatment exposure. SiO_2-MNPs show dissolution patterns, void/pore formation (*yellow arrows*) and significant outward growth of reaction zones (secondary growth shown by *blue arrow*)

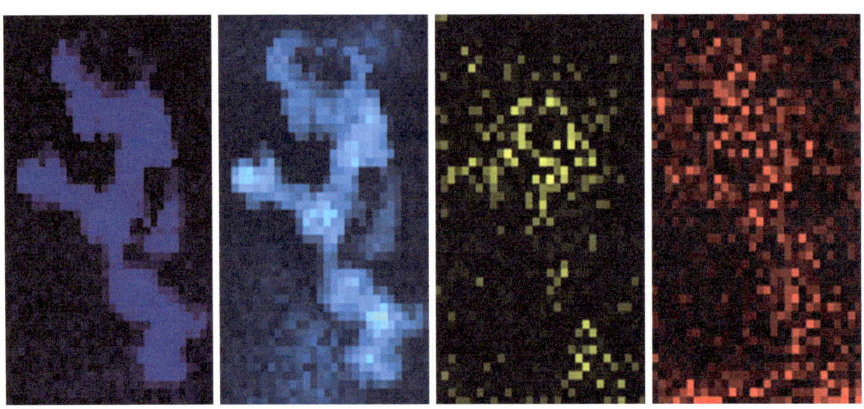

Fig. 4.6 Elemental EDS maps of O, Si, S, and P taken from a region in Fig. 4.4

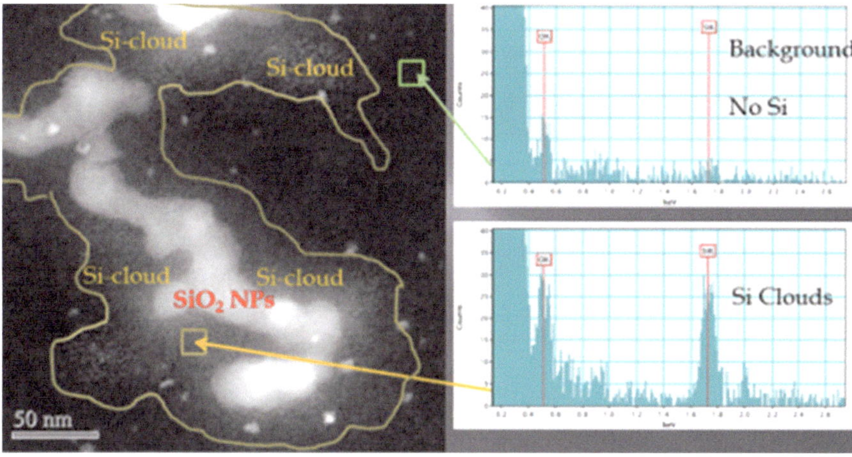

Fig. 4.7 Dark Field STEM imaging and EDS spot analyses show the bioprocessing of SiO_2-MNPs in alveolar macrophages. Migration of Si occurred outwards and led to a secondary reaction zone "Si-cloud" between SiO_2-MNPs and yellow line. The small very bright spots in the Si-cloud region are ferritin nanoparticles

Some O may be present in –(Si-O-Si)- forming anionic silanol components within the tissue and this could be a critical mechanism for Si transport and new precipitates and studies are needed at the molecular level to determine the processes involved in Si mobility after processing of the MNPs. The elemental scans for S and P were included here to show that, surprisingly, the signals are shadowing the location of the SiO_2-MNPs. This opens chief questions for future work including queries into the underlying *in vivo* processing mechanisms that guide nanoparticle delivery to certain cellular and subcellular locations and chemical environments after uptake and also how this may be affected by dose. The dose variations (high vs. low) all resulted in significant *in vivo* processing of the SiO_2-MNPs after inhalation which probably is based on the relatively high solubility of amorphous silica [9, 24, 31]. Importantly, *in vivo* processing gives rise to second generation nanoparticles and reaction zones containing Si- phase within the vicinity of the bio-transformed SiO_2-MNPs, which suggests that migration and relocation processes take place at the cellular and subcellular levels as determined in elemental mapping (Fig. 4.6). Moreover, high resolution STEM coupled with EDS confirms that release of Si ions and relocation and precipitation of secondary Si- phases in the alveolar macrophages results in generation of Si-rich halos "Si-clouds" at the outskirts of partially dissolving SiO_2-MNPs (Fig. 4.7). An analogous cloud-formation process was shown for the first time to take place when poorly soluble ceria (CeO_2) nanoparticles bioprocess in liver tissue [27].

Detailing all of the physio-chemical changes that take place during bioprocessing of SiO_2-MNPs in alveolar macrophages and other tissue locations as a function of dose is still under development. This may be an effective tool in understanding their subcellular and temporal fate and how this factors into controlling a toxic

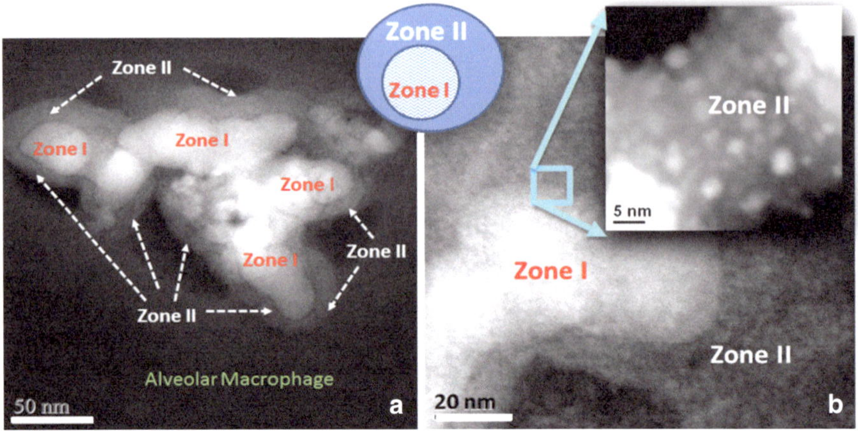

Fig. 4.8 (**a**) Dark Field STEM image shows *in vivo* breakdown of SiO_2-MNPs in alveolar macrophage (Zone I) and formation of Zone II. (**b**) Magnified region shows small nanoparticles in Zone II

response after environmental uptake of nanoparticles. More work is needed to study the dose effects on the extent of SiO_2-MNPs breakdown and the relocation of Si as a function of saturation levels. An example of further detailing the chemical and structural content of the Si-Cloud contents is shown in the Dark Field STEM images that demonstrate a greater extent of *in vivo* processed SiO_2-MNPs in the alveolar macrophage and the development of nanozone formation (Fig. 4.8). The chemical breakdown of the SiO_2-MNPs (Zone I) leads to pitting in the original particles with subsequent material migration and relocation into satellite zones (Zone II in Fig. 4.8) which hosts much smaller particles that are highly dispersed. This is the reason why Zone II appears less dense and concentrated in the Dark Field STEM image. It will be of paramount importance to apply aberration corrected STEM and 3D–imaging to probe the chemical composition of the matrix of Zone II that engulfs the very small SiO_2-MNPs. If Zone II matrix is chemically distinct from other nanoparticle-free regions in the alveolar macrophage that hosts the SiO_2-MNPs it can help determine if protein formation or encapsulation helps stabilize the SiO_2-MNPs and make them more biocompatible after *in vivo* processing.

The examples above show that *in vivo* processing of nanoparticles can occur and that a materials-oriented electron microscope can reveal some aspects of the changes that are occurring. This coupled with toxicological response monitoring could provide information as to whether nanoparticle dose-dependent changes reduce or increase toxic effects. Much work remains to be done in determining the *in vivo* properties of the many different types of nanoparticles and how variables such as particle morphology, size, surface treatments, and composition effect *in vivo* processing. The application of aberration corrected electron microscopes to the study of *in vivo* processing would most likely be very fruitful [65, 86]. These microscopes have the resolution to determine if a cloud surrounding a nanoparticle is composed of single molecules or very small clusters as in the silica examples above. It could also determine if the composition of such clusters is that of the precursor particle or if a reaction has occurred thereby modifying the clusters and resulted in new compounds. In addition aberration-corrected STEM could provide high resolution maps of the surface layers of nanoparticles and corresponding surrounding tissue to better understand the mechanisms behind *in vivo* processing. Importantly, aberration-corrected electron optical sectioning can give insights into any potential protein corona formation at the exterior of the nanoparticles. Recently, RAMAN mapping has also become commercially available. This opens the possibility of identifying molecular changes in the tissue surrounding nanoparticles *in vivo* as a function of dose.

4.3.3 Example II: Ceria (CeO_2) Nanoparticles Inside Spleen Tissue

EELS analysis was performed on a 200 kV JEOL 2100F TEM/STEM and spectrum images were collected to investigate morphologies, size distribution and oxidation states of ceria nanoparticles (CeO_2-MNPs) in rat spleen tissue. A therapeutic dose (4 g/kg) of ~15–20 nm CeO_2-MNPs was used and instilled four times over a 2 week time period. The hydrothermal synthesis procedure for the CeO_2-MNPs resulted in a narrow size range ~20 nm [47]. The particle surfaces were capped using a citrate coating (10 %) in 5 % aqueous dispersion. The (CeO_2-MNPs) synthetic identity included size, surface charge (Zeta potential: −40 mV at pH 7.3) and structural characterization using HRTEM/STEM analyses and EELS in the spleen tissue (Fig. 4.9).

Previously, the *in vivo* processing, transformation and subcellular effects of CeO_2-MNPs in a rat model using a single high dose (85 mg/kg) was presented with corresponding effects on oxidative stress increases and decreases and internalized CeO_2-MNPs were shown to cause distinct cellular responses and oxidative stress, but also presented significant *in vivo* processing which releases smaller CeO_2-MNPs clouds with much improved ROS potential [27, 32]. Since CeO_2-MNPs can do both generate and scavenge free radical oxygen species (ROS), it is important to distinguish CeO_2-MNPs that contribute to either ROS production or ROS scavenging in subcellular levels [41]. Example II shows how ceria MNPs translocate to spleen tissue and in the spleen the original CeO_2-MNPs produced Ce-clouds (Fig. 4.9). In this particular case a therapeutic dose (4 g/kg) of CeO_2-MNPs not only bio-accumulated in spleen which can be demonstrated with the help of HRTEM and Dark Field STEM imaging, but the original ceria nanoparticles were also structurally altered and second generation plumes of ultra-fine (<3 nm) ceria nanoparticles formed close by, which can be seen as clouds next to the *in vivo* processed ceria precursors (Fig. 4.9). Corresponding EELS analyses along the EELS-trace line in Fig. 4.9 compare the redox state of the precursor and newly precipitated ceria clouds. The high angle STEM analysis along the EELS line profile used a small probe size (0.2 nm) to minimize any fixed tissue sample damage that could occur under the prolonged electron beam exposure. The oxidation states of Ce were determined by the fine structures of M4,5 edges in EELS as described

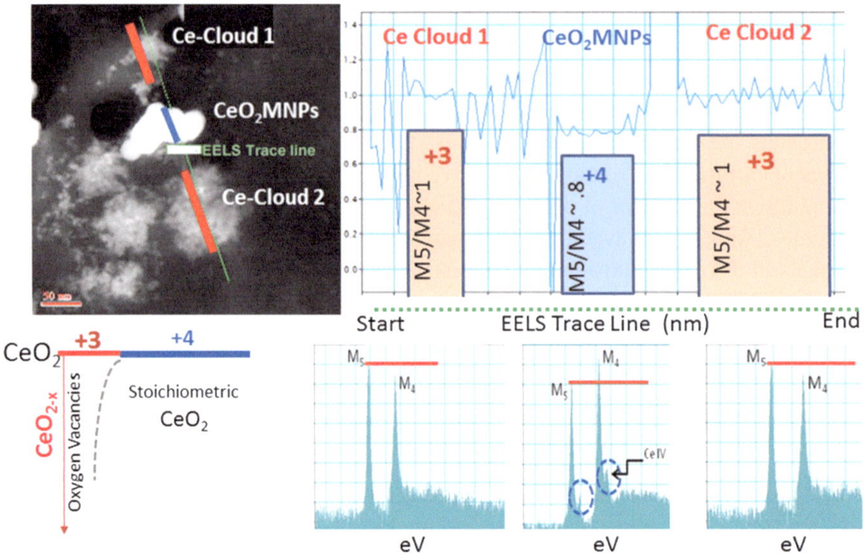

Fig. 4.9 STEM and EELS analyses of CeO_2-MNPs in spleen after a therapeutic dose and 14 days residence time. Analysis of the ceria M5/M4 ratio along the line profile from an EELS spectrum image

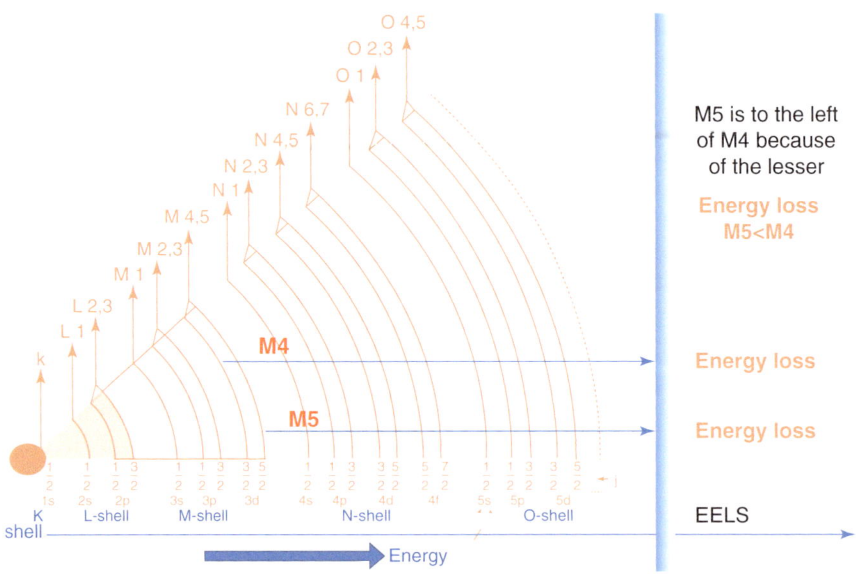

Fig. 4.10 Schematic illustration of energy loss for computing EELS edges for Ce

elsewhere in details [77]. The schematic in Fig. 4.10 gives some insights on how to compute the EELS edges for Ce and, in particular, the energy loss for Ce M_4 versus Ce M_5. The greater the contribution of Ce M5, the higher is the ceria reduction potential [77]. Interestingly, the same kind of Ce-cloud formation was also shown

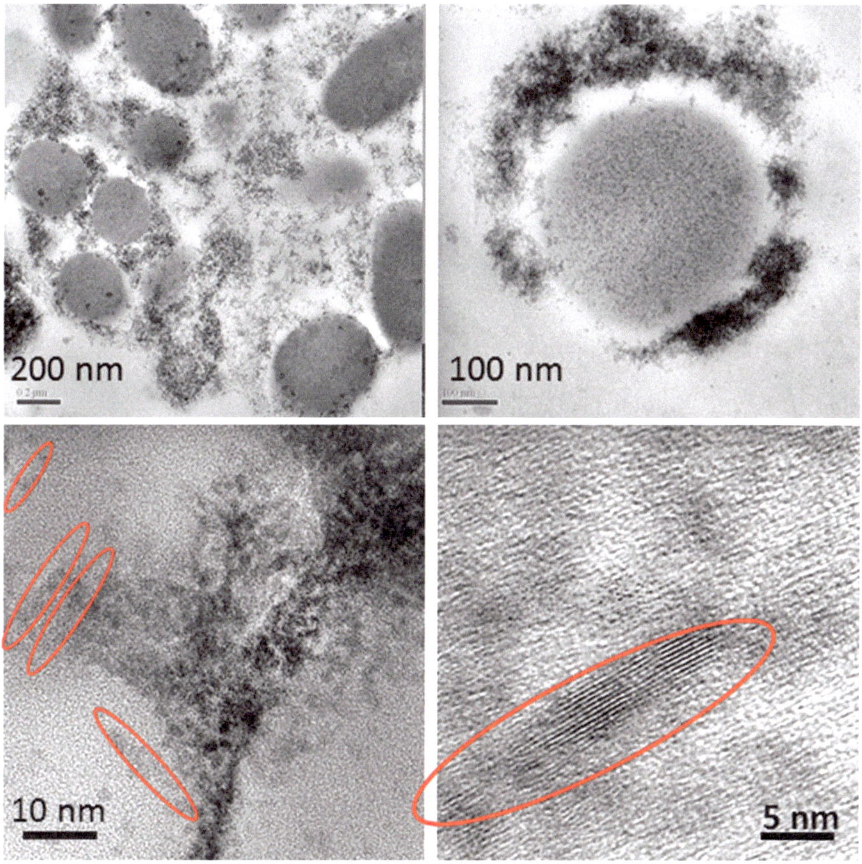

Fig. 4.11 TEM and HRTEM images with increasing magnification show the presence of CeO_2-MNPs and Ce-phosphate after *in vivo* processing and leads to a local arrangement. Many MNPs are self-aligning to form needle- shaped structures indicted with *red arrows*

previously for a high-dose of CeO_2-MNPs after intravenous uptake and sequestration of CeO_2-MNPs in liver and was associated with a much improved ROS potential [27]. Different valance states of the ultra-small CeO_2-MNPs needles are characterized by core loss EELS to have very high Ce^{+3} signatures (corresponding to oxygen vacancies) as evidenced by the greater Ce M_5 contributions obtained via the EELS analyses and are similar to those in the Ce-clouds (Fig. 4.9). Both HRTEM and Dark Field STEM demonstrate that Ce-phosphates formed in the spleen and this typically occurs in lysosomal regions where ultra-small ceria particles transform/reform (the mechanism of transformation/reformation is not known at this time) (Figs. 4.11 and 4.12).

Elemental maps can be produced from EDS and EELS spectrum images (in high resolution STEM mode) and in the example below span across the regions where the Ce-nanoparticles accumulate. This information can then be used to build a thorough understanding of the temporal, structural and cellular relationships involving

Fig. 4.12 CeO2-MNPs are imaged using low angle dark field STEM condition showing cellular structures in the spleen. The CeO_2-MNPs localize around cellular inclusions

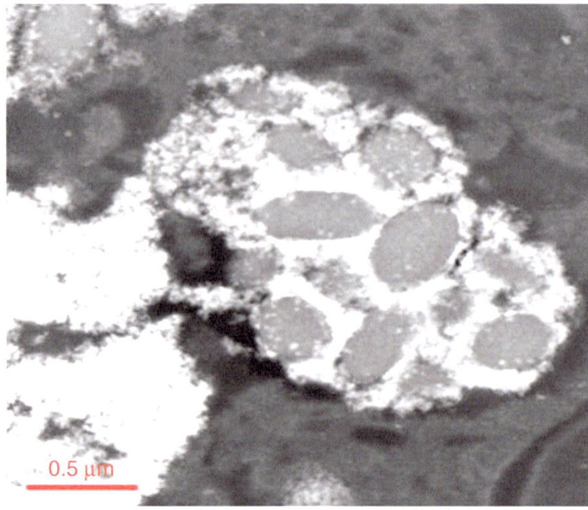

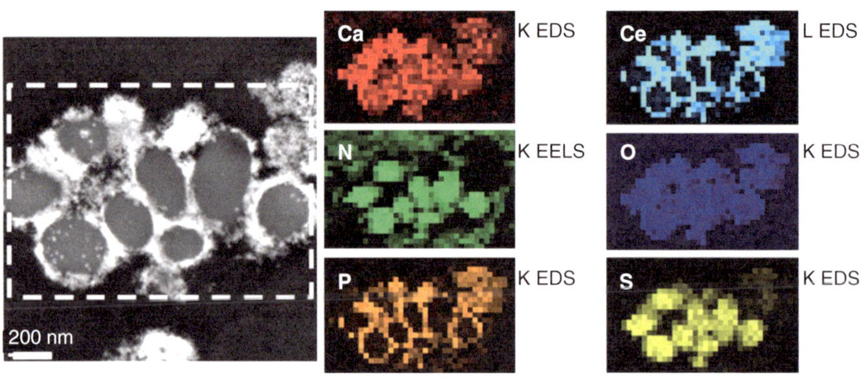

Fig. 4.13 STEM spectrum imaging and elemental maps of Ce-MNPs localized around spleen inclusions

tissue composition and location of nanoparticles. An example is shown in Fig. 4.13 where elemental maps were generated over a select region that is illustrated using dark field STEM in Fig.4.12. These are fairly low EELS pixel count maps in order not to destroy the tissue structures during the prolonged electron beam scanning. Therefore, a low angle dark field STEM condition has to be selected to bring out the cellular structures in the spleen (Fig. 4.13) while simultaneously analyzing the relative elemental composition and spectral signatures of for example Ce, P, C, Ca, N and O in the same region using STEM spectrum imaging (Fig. 4.13).

Elemental maps obtained from the EDS and EELS spectrum images allow for a comparison of the elemental distribution that is associated with the cellular structures and that of the accumulated nanoparticles. The low pixel count maps take about 30 min to acquire. Higher pixel count maps can be obtained but the required time increases rapidly and this can affect the beam/sample interaction and lead to artifacts. Elemental imaging may also be accomplished using an EELS imaging filter. Typically an in-column filter is more successful on biological specimens due to the smaller beam dose required. When using post column filters the sample stability may be impacted due to the high brightness required (electron intensity). The elemental imaging for Ce-MNPs that seem to preferentially locate around globular lipid-based components (Figs. 4.12 and 4.13) clearly shows that the Ca and N are highly enriched as part of the internal composition of the lipid structures and they have an outer shell or corona that is phosphor rich (P signal is high at the outside of the globular structures in Fig. 4.13). The Ce signal completely overlaps with the P signal suggesting, at the least, a spatial relationship. In case of the preferential Ce deposition at the outside of the lipid structures it would suggest that there is either a mechanism that controls the delivery of the Ce-MNPs to that particular P-rich location or, that Ce ions migrated to that region and formed new Ce-oxides, Ce-hydroxides or Ce-phosphates. The O-signal clearly shadows the areas of both, Ce and P signals (Fig. 4.13). The elemental map of the S-signal shows that it is confined to the lipid structures only. At this time it is not known how certain regions in cells govern nanoparticle delivery and accumulation, but there seems to be an underlying chemical control that needs to be considered. Much work will be required to understand the relationship between tissue components and nanoparticles, but the use of HRTEM/STEM and EELS is certainly a tool that will be very useful towards that goal. How the MNPs' port-of-entry, dose, exposure duration and post-exposure time factor into the transport phenomena, particle transformation and *in vivo* processing mechanisms is not known at this time.

4.3.4 Example III: Ferritin Nanoparticles Inside Lung and Liver Tissue

The uptake and sequestration of MNPs, both silica (amorphous SiO_2) and ceria (CeO_2) results in the partial breakdown and *in vivo* processing of the original MNP-particles as discussed in Examples I and II in earlier sections in this Chapter. Remarkably, there is additional evidence one can gain from HRTEM: the presence of different MNPs (SiO_2, CeO_2) in different organs (lung, liver, spleen) after being delivered via different uptake routes (inhalation, intravenous), have at least one response in common, specifically, the simultaneous formation of ferritin nanoparticles in the vicinity of the invader MNPs. Ferritins represent bio-mineralized iron nanoparticles that are typically 5–8 nm in size and trapped inside the cage of the iron storage protein [8]. They occur immediately juxtaposed to the cell-invading and inflammation-inducing MNPs [27]. Moreover, the ferritin nanoparticles are highly concentrated next to the MNPs when compared to tissue regions that are not affected by inflammation, as shown in the Dark Field STEM images in Figs. 4.14 and 4.15. The individual solitary bright white spots surrounding the invader MNPs (inside the lysosomal regions) each represent one ferritin nanoparticle of 5–8 nm size (Figs.

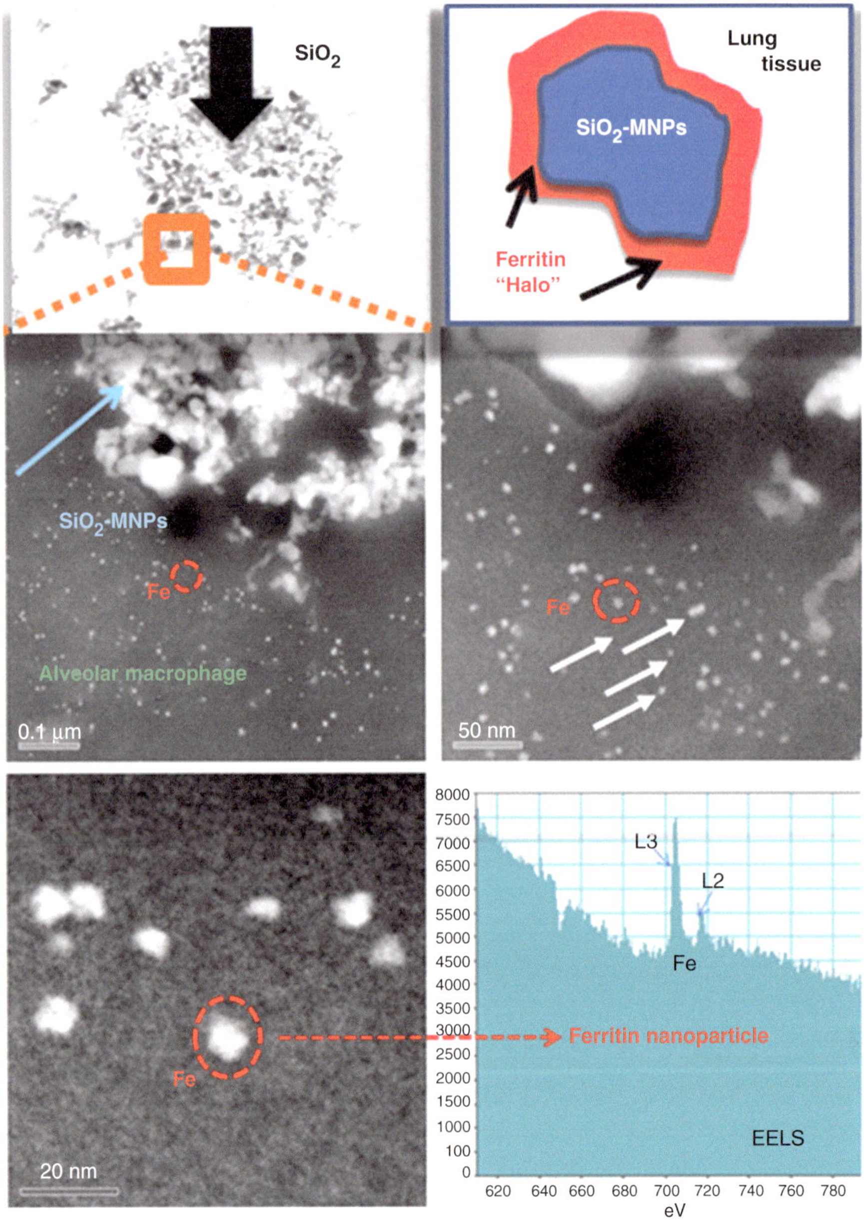

Fig. 4.14 SiO_2-MNPs in TEM and dark field STEM showing cellular structure of the alveolar macrophage, location. The SiO_2-MNPs are surrounded by ferritin nanoparticle halos identified using EELS spot analysis

4.14 and 4.15). It is well established that ferritin nanoparticles form during the biomineralization of ferrous (reduced) iron. A conserved iron-binding site, the ferroxidase center of the ferritin protein regulates iron storage in iron metabolism [38]. It is

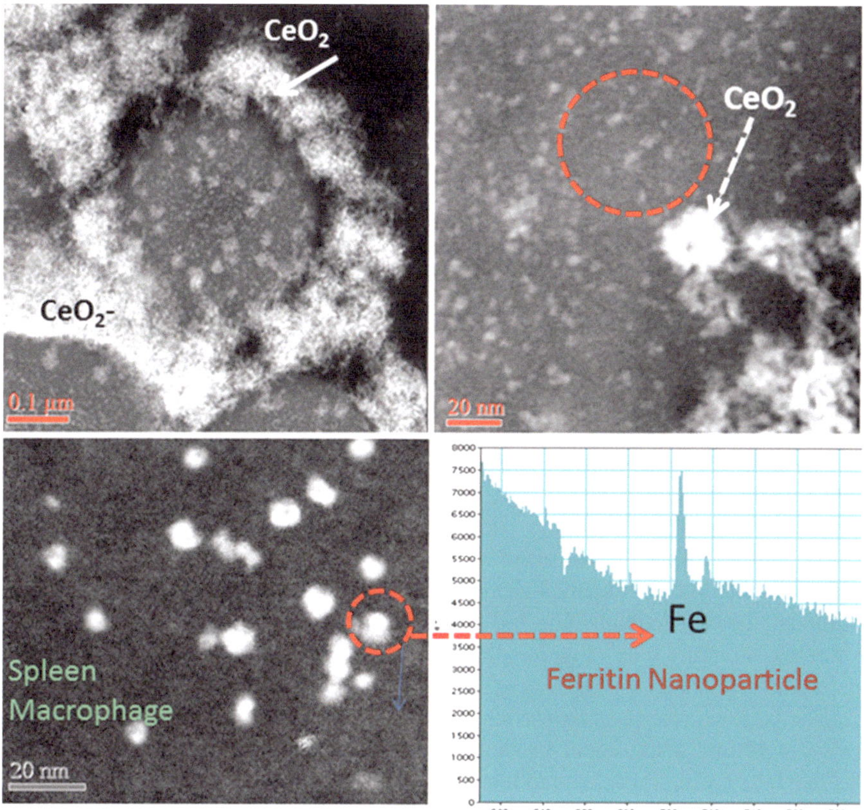

Fig. 4.15 CeO_2-MNPs in dark field STEM showing cellular structure in the spleen macrophage. The CeO_2-MNPs localize around cellular inclusions and are surrounded by ferritin nanoparticles identified using EELS spot analysis

generally assumed that ferrous iron Fe(II) binds the ferroxidase center and the oxidized iron Fe(III) spontaneously enters the ferritin cage. High resolution imaging as well as spectroscopic and kinetic studies of ferritins (family of 24 iron storage proteins), suggest many common characteristics, including highly symmetric subunits of a cavity-engulfing protein coating in which the iron bio-mineralization takes place. Furthermore, there are four channel passages through the protein shell that help facilitate ingress and egress of ions which results in an iron core with eight subunits rather than a single dense sphere [75]. There are catalytic sites at the inner shell "ferroxidase center" which control the oxidation of Fe(II). The mechanisms of bio-mineralization of iron that result in ferritin nanoparticles like the ones shown in Figs. 4.14 and 4.15 are described elsewhere [8, 38], but the association (close locality) with invader MNPs is novel and requires a thorough investigation of the subcellular mechanisms and participation of iron as a redox mediator to counter the effects of invader MNPs. The significance of iron in biological systems is due to its ability to engage in redox reactions, including the scavenging of free radicals [27]. In general, iron forms a labile iron pool that includes iron atoms, but free Fe(II) must be managed either by use in hemoglobin or inside of the iron storage protein, ferritin. Otherwise, reduced Fe(II) iron can participate in the Fenton reaction and cause free radical formation [44]. There are numerous transferrin receptors, all of which are proteins that participate in iron transport at the cellular and subcellular levels. Once Fe(II) is sequestered in the ferric form within the ferritin protein shell, this particular iron will not participate in free radical formation. In fact, the oxidation of one Fe(II) to Fe(III) releases an electron that can neutralize a free radical species and, thereby, act as an anti-oxidant. Each ferritin cavity can hold up to 4500 oxidized iron atoms [63] and each one had to release an electron while being oxidize. It is this catalytic process that provides ferritin with the anti-oxidant property. The physical characterization of the ferritin particles includes details of the protein shell as well as characterization of the mineralized iron oxide core. In the high resolution dark field images of the lung and spleen thin sections only the iron oxide core is visible due to the comparatively high atomic number and close packing of the iron atoms, while the surrounding ferritin-protein shell has about the same density and general chemical make up as the cellular matrix and, therefore, is difficult to distinguish (Figs. 4.14 and 4.15). The dense iron core allows z-contrast imaging using HAADF-STEM (Figs. 4.14 and 4.15). When using aberration corrected STEM this technique provides insights into the detailed morphologies and structures of the iron core [70]. Although the exact composition and stoichiometry of the core is not well understood yet, most literature today suggests that the core is composed of a ferrihydrite (iron-oxyhydrite) and also approximates this structure in human liver [63].

The copious ferritin nanoparticles that form halos around sequestered MNPs (Figs. 4.14 and 4.15) occur in such a high concentration that it, unmistakably, seems to be a direct response to the presence (invasion) of the MNPs in either the lung or spleen (Figs. 4.14 and 4.15). Ferritin nanoparticles are typically present throughout cells, but not in the particularly high concentration that is shown inside the ferritin-halos around MNPs (Figs. 4.14 and 4.15). Unexpectedly, the elevated ferritin nanoparticle accumulation seems independent of the nature of the MNPs (amorphous SiO_2, CeO_2 and others not shown in this Chapter). The mechanisms that control the abundant *in vivo* formation of ferritin nanoparticles next to the invader MNPs need to be further investigated, but it seems to indicate that the cellular and subcellular response mechanism(s) trigger an upregulation of iron immediately juxtaposed to the MNPs. This is very important since MNPs are linked to inflammatory processes and possible cell toxicity, which results in formation of free radicals [44]. Either the MNPs or the free radicals, or both, initiate mechanisms that trigger the upregulation of iron in the same regions. Consequently, ferritin nanoparticles that form as a result of the oxidation of Fe(II) to Fe(III) can participate in free radical scavenging processes as mentioned earlier and provide the needed anti-oxidant response to counteract invader MNPs. This can explain the ferritin-rich halos that are observed in the HAADF-STEM images around the MNPs (Figs. 4.14 and 4.15). At this time there is no available data on a nanoparticle induced dose-dependent ferritin response, but it seems logical that the higher the MNPs dose, the greater the ferritin nanoparticle concentrations would be in the affected tissue regions.

4.4 Synchrotron Analysis: Dose-Dependent Nanoparticle Signatures in Tissue

X-ray absorption spectroscopy methods making use of synchrotron radiation, such as XANES and EXAFS, may prove useful in providing information on the electronic and local atomic structure of elements in nanotoxicology. A screening of selected tissue samples for elements of interest may provide information regarding the incorporation of such elements from exposure to nanoparticles as a function of a particular dose that led to a certain pathological response. An important analogy is when an X-ray absorption spectroscopy survey was conducted on coal samples to determine the chemical nature and structure of elemental impurities. In that case, the researchers were faced with a similar staggering problem. Coal contains nearly the entire periodic table as impurity elements, and many of the impurity elements were of concentrations 1000 ppm or less, which could not be confidently characterized by conventional microscopic, spectroscopic, or diffraction techniques [82]. Using predominantly fluorescence mode, significant and important information was obtained on numerous trace elemental impurities in coal using X-ray absorption spectroscopy [37]. Just below the edge energy and prior to the single scattering region, pre-edge features provide useful information on site symmetry (e.g., a sharp feature is typically produced with tetragonal symmetry, while octahedral generally produces a faint signal), the white line region provides information on the oxidation state of the material, and the multi-scattering region provides information on the immediate environment of neighboring atoms. The higher energy region provides a wealth of information on the identity of neighboring atoms, their interatomic distances, and their degree of coordination. Although XANES in principal can also be obtained in high resolution STEM mode using electron microscope applications, elemental dispersion over larger tissue areas is not possible using large magnification settings and needs to be done at a synchrotron source. Two examples related to catalyst particles (iron oxide and ceria nanoparticles) that are often examined with regards to their nano-toxic response are provided below. In the first case [68] the role of the element K in promoting the carburization rate of iron oxide in Fischer-Tropsch synthesis catalysts was explored by XANES and EXAFS spectroscopy. The XANES spectra were recorded with the catalyst heated in flowing carbon monoxide (Fig. 4.16a). Changes in the white line are evident (Fig. 4.16a), and in comparing the spectra to those of reference compounds, reduction was found to proceed by way of Fe_2O_3 to Fe_3O_4 to FeO to Fe carbides. Simultaneous EXAFS spectra were recorded (Fig. 4.16b). The low distance- peak of Fe-O coordination and the high distance peak of Fe-Fe coordination in Fe_2O_3 change to match the distances of Fe-O and Fe-Fe coordination in Fe_3O_4. At the end of the trajectory, Fe-C bonds in Fe carbides are clearly observed in the intermediary range of distance. Thus, the two techniques (XANES and EXAFS) provided similar information on the chemical changes occurring, but simultaneously and from two different perspectives. Biomineralized iron oxides are very often present at the cellular and subcellular levels and it is important to distinguish oxidation states and also to observe whether iron

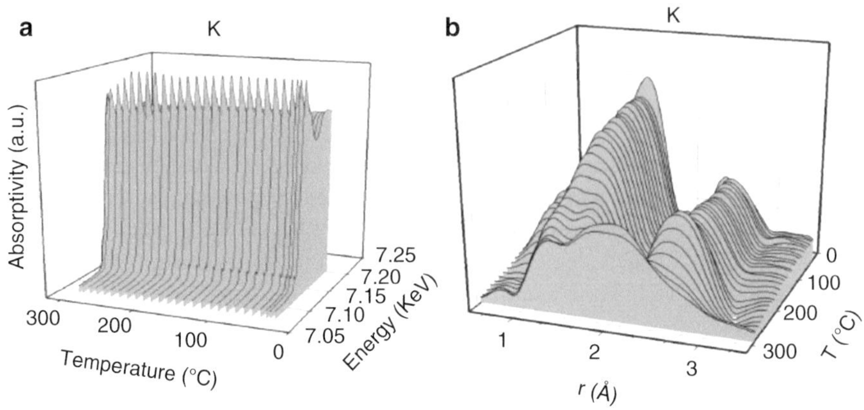

Fig. 4.16 (**a**) XANES spectra for iron oxide; (**b**) EXAFS recorded during heating

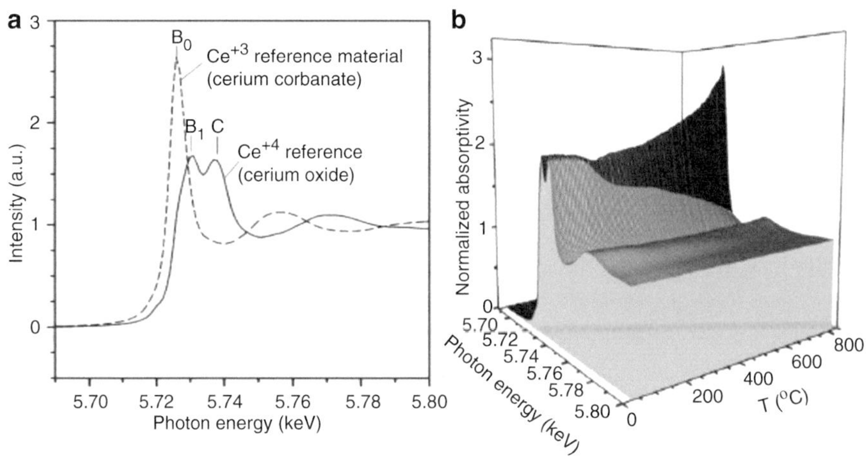

Fig. 4.17 (**a**) XANES for Ce+3 and Ce+4 and (**b**) XANES as a function of heating

may be coordinated to carbon, sulfur or phosphor and if variations occur as a function of dose (nanoparticle exposure).

A second example [45], on the doping of nanosized ceria (CeO_2) domains with platinum doping of 0.5 % (by weight) and 50 % (by mole) calcia, is described below. The XANES patterns of Ce^{4+} and Ce^{3+} are very different (Fig. 4.17a).

where Ce^{4+} contains two very broad peaks, as well as additional features, while Ce^{3+} exhibits a sharp distinct peak, B_0. This is based on changes in the electronic structure of ceria, and its effect on the allowed electronic transitions. When the same CeO_2 nanoparticle catalyst was heated in hydrogen (Fig. 4.17b) to activate the surface by reduction, the addition of the dopants (Pt and Ca) facilitated surface shell reduction to ~200°C (from 450 to 500°C for undoped ceria), and bulk reduction commenced at ~400°C (rather than >700°C for undoped ceria). This clearly demonstrates the sensitivity of the analysis tool towards elemental, structural and thermal changes as a function of reduction potential. Thus, using high intensity X-rays generated at the synchrotron can provide a wealth of information regarding the identity, chemical state, and local atomic structure of nanoparticles, and may provide key information in any survey of tissue samples for nanotoxicology.

4.5 Synopsis

In this chapter the inter-relationship between dose, nanoparticle uptake, cellular and subcellular interactions and nanotoxicity has been discussed with examples of means of observation of *in vivo* bio-processing and response. Emphasis is placed on the importance of the precision of characterization of the starting particles, the particles in a biological environment, and the physiological response. The relationship between dose, bio-processing, and response is an area of active research as all three may be related in a non-linear manner. It is pointed out that relatively insoluble materials like CeO_2 have been observed *in vivo* to undergo significant changes in shape, size, material phase and electronic structure. Because of this, the modelling and prediction of dose versus toxicity over time becomes a non-linear problem because the initial particles can transform over time and initiate different responses that evolve as the dynamic system undergoes further transformations. The examples in this chapter illustrate two advanced materials characterization methods that are useful in the characterization of nanoparticles, before and after introduction in the biological environment, and in observing specific types of physiological response. These methods are advanced analytical electron microscopy (STEM/EELS) and x-ray absorption near edge spectroscopy (XANES). In conclusion the dose response relationship is complicated by the physicochemical transformations in the nanoparticles induced by the biological system producing an altered response. Thus, the modelling and prediction of dose-response-toxicity relationships has to take into account non-linear dependencies when attempting to predict a dose versus toxicity response relationship. This has to be especially considered when predictive modelling of nanomaterials utilizes *in vitro* models. Therefore, the long-term goal is to develop cellular in vitro models that can support dynamic processing of nanoparticles for exposure risk assessment.

References

1. Altavilla C, Ciliberto E (eds) (2010) Inorganic Nanoparticles Synthesis, Applications, and Perspectives. CRC Press, Boca Raton, Florida, p 576. Print ISBN: 978-1-4398-1761-2, eBook ISBN: 978-1-4398-1762-9
2. Anderson SC, Birkeland CR, Anstis GR, Cockayne DJH (1997) An approach to quantitative compositional profiling at near atomic resolution using high-angle anular dark-field imaging. Ultramicroscopy 69:83–103
3. Bates ME, Keisler JM, Zussblatt NP, Plourde KJ, Wender BA, Linkov I (2015) Balancing research and funding using value of information and portfolio tools for nanomaterial risk classification. Nat Nanotechnol 9:249

4. Baisch BL, Corson NM, Wade-Mercer P, Gelein R, Kennell AJ, Oberdörster G, Elder A (2014) Equivalent titanium dioxide nanoparticle deposition by intratracheal instillation and whole body inhalation: the effect of dose rate on acute respiratory tract inflammation. Part Fibre Toxicol 11(5):16 . PMCID:PMC3905288
5. Bermudez E, Mangum JB, Wong BA, Asgharian B, Hext PM, Warheit DB, Everitt JI (2004) Pulmonary responses of rats, mice, and hamsters to subchronic inhalation of ultrafine titanium dioxide particles. Toxicol Sci 77:347–357
6. Bilberg K, Hovgaard MB, Besenbacher F, Baatrup E (2012) In vivo toxicity of silver nanoparticles and silver ions in zebrafish (*Danio rerio*). J Toxicol 2012:9 Article ID 283784
7. Bonner JC, Silva RM, Taylor AJ, Brown JM, Hilderbrand SC, Castranova V, Porter D, Elder A, Oberdörster G, Harkema J, Bramble L, Kavanagh TJ, Botta D, Nel A, Pinkerton KE (2013) Interlaboratory evaluation of rodent pulmonary responses to engineered nanomaterials. Environ Health Perspect 121:676–682 PMC3672912
8. Bradley JM, Moore GR, LeBrun NE (2014) Mechanisms of iron mineralizaiton in ferritins: one size does not fit all. J Biol Inorg Chem 19(6):775–785. doi:10.1007/s00775-014-1136-3
9. Brunner TJ, Wick P, Manser P, Spohn P, Grass RN, Limbach LK, Bruinink A, Stark WJ (2006) In vitro cytotoxicity of oxide nanoparticles: comparison to asbestos, silica, and the effect of particle solubility. Environ Sci Technol 15(4D(14)):4374–4381
10. Burello E, Worth AP (2011) A theoretical framework for predicting the oxidative stress potential of oxide nanoparticles. Nanotoxicol 5(2):228–235
11. Buzea C, Blandino IIP, Robbie K (2007) Nanomaterials and nanoparticles: sources and toxicity. Biointerphases 2(4):MR17–MR172
12. Crosera M, Bovenzi M, Maina G, Adami G, Zanette C, Florio C, Larese FF (2009) Nanoparticle dermal absorption and toxicity: a review of the literature. Int Arch Occup Environ Health 82(9):1043–1055
13. DeSouza PM, Neto RCR, Borges LEP, Jacobs G, Davis BH, Graham UM, Resasco D, Noronha FB (2015) Effect of zirconia morphology on hydrogenation of phenol over Pd/ZrO2. ACS Catal 5(12):7358
14. Docter D, Strieth S, Westmeier D, Hayden O, Gao M, Knauer SK, Stauber RH (2015) No king without a crown – impact of the nanomaterial-protein corona on nanobiomedicine. Nanomedicine (London) 10(3):503–519
15. Driscoll KE (1996) Role of inflammation in the development of rat lung tumors in response to chronic particle exposure. Inhal Toxicol 8(Suppl):139–153
16. Egerton RF (2011) Electron energy loss in the electron microscope, 3rd ed. Springer, New York. ISBN 978-1-4419-9582-7
17. Elder ACP, Gelein R, Finkelstein JN, Cox C, Oberdörster G (2000) The pulmonary inflammatory response to inhaled ultrafine particles is modified by age, ozone exposure, and bacterial toxin. Inhal Toxicol 12(Suppl. 4):227–246
18. Elder A, Gelein R, Finkelstein JN, Driscoll KE, Harkema J, Oberdörster G (2005) Effects of subchronically inhaled carbon black in three species. I. Retention kinetics, lung inflammation, and histopathology. Toxicol Sci 88(2):614–629
19. Elder A, Couderc JP, Gelein R, Eberly S, Cox C, Xia X, Zareba W, Hopke P, Watts W, Kittelson D, Frampton M, Utell M, Oberdörster G (2007) Effects of on-road highway aerosol exposures on autonomic responses in aged, spontaneously hypertensive rats. Inhal Toxicol 19:1–12
20. Elder A, Vidyasagar S, DeLouise L (2009) Physicochemical factors that affect metal and metal oxide nanoparticle passage across epithelial barriers. Wiley Interdiscip Rev Nanomed Nanobiotechnol 1(4):434–450 .4004356
21. Fawcett DW (1966) An atlas of fine structure: the cell, its organelles, and inclusions. W.B. Saunders, Philidelphia
22. Feynman RP (1960) There's Plenty of Room at the Bottom. Eng Sci 23(5):22–36
23. Foroozandeh P, Aziz AA (2015) Merging worlds of nanomaterials and biological environment: factors governing protein corona formation on nanoparticles and its biological consequences. Nanoscale Res Lett 10:221 4437989

24. Fruijtier-Polloth C (2012) The toxicological mode of action and the safety of synthetic amorphous silica – A nanostructured material. Toxicology 284(2):61–70. doi:10.1016/j.tox.2012.02.001
25. Gebauer JS, Malissek M, Simon S, Knauer SK, Maskos M, Stauber RH, Peukert W, Treuel L (2012) Impact of the nanoparticle-protein corona on colloidal stability and protein structure. Langmuir 28(25):9673–9679
26. Graham UM, Khatri RA, Dozier A, Jacobs G, Davis BH (2009) 3D ridge-valley structure of a Pt ceria catalyst: HRTEM and EELS spectrum imaging. J Cat Lett 132(2):335–341
27. Graham UM, Tseng MT, Jasinski JB, Yokel RA, Unrine JM, Davis BH, Dozier AK, Hardas SS, Sultana R, Grulke EA, Butterfield DA (2014) In vivo processing of ceria nanoparticles inside liver: impact on free-radical Scavenging activity and oxidative stress. Chempluschem 79(8):1083–1088 .4551665
28. Graham UM, Dozier AK, Wang C, Tseng MT, Fernbeck JE, Birch ME, Davis BH (2015) Observations of in vivo processing of metal oxide nanoparticles by analytical TEM/STEM. Microsc Microanal:2287, 21, (Suppl 3)
29. Grosse S, Haugland HK, Lilleng P, Ellison P, Hallan G, Hol PJ (2015) Wear particles and ions from cemented and uncemented titanium-based hip prostheses-a histological and chemical analysis of retrieval material. J Biomed Mater Res B Appl Biomater 103(3):709–717 . 4413358
30. Gulson B, McCall MJ, Bowman DM, Pinheiro T (2015) A review of critical factors for assessing the dermal absorption of metal oxide nanoparticles from sunscreens applied to humans, and a research strategy to address current deficiencies. Arch Toxicol 89(11):1909–1930
31. Hajarul AAW, Nor DZ, Aziz AA, Razak KA (2012) Properties of amorphous silica entrapped isoniazid drug delivery system. Adv Mat Res 364:134–138
32. Hardas SS, Sultana R, Warrier G, Dan M, Wu P, Grulke EA, Tseng MT, Unrine JM, Graham UM, Yokel RA, Butterfield DA (2013) Rat hippocampal responses up to 90 days after a single nanoceria dose extends a hierarchical oxidative stress model for nanoparticle toxicity. Nanotoxicol 8(Suppl. 1):155–166
33. Hartland A, Lead JR, Slaveykova VI, O'Carroll D, Valsami-Jones E (2013) The environmental significance of natural nanoparticles. Nat Educ Knowl 4(8):7
34. Heinrich U, Fuhst R, Rittinghausen S, Creutzenberg O, Bellmann B, Koch W, Levsen K (1995) Chronic inhalation exposure of Wistar rats and two different strains of mice to diesel engine exhaust, carbon black, and titanium dioxide. Inhal Toxicol 7:533–556
35. Hellmann R, Cotte S, Cadel E, Malladi S, Karlsson LS, Lozano-Perez S, Cabie M, Seyeux A (2015) Nanometre-scale evidence for interfacial dissolution-reprecipitation control of silicate glass corrosion. Nat Mater 14(3):307–311
36. Hirsch PB, Howie A, Whelan MJ (1962) On the production of X-rays in thin metal foils. Philos Mag 7:2095–2100
37. Hodgson KO, Hedman B, Penner-Hahn JE (eds) (1984) EXAFS and near edge structure III, Proc. In Physics 2, Springer, Berlin/Heidelberg GmbH
38. Honarmand Ebrahimi K, Bill E, Hagedoorn PL, Hagen WR (2012) The catalytic center of ferritin regulates iron storage via Fe(II)-Fe(III) displacement. Nat Chem Biol 8(11):941–948. doi:10.1038/nchembio.1071
39. Jones JR (2013) Review of bioactive glass: from Hench to hybrids. Acta Biomater 9(1):4457–4486
40. Keller J, Wohlleben W, Ma-Hock L, Strauss V, Gröters S, Küttler K, Wiench K, Herden C, Oberdörster G, Van Ravenzwaay B, Landsiedel R (2014) Time course of lung retention and toxicity of inhaled particles: short-term exposure to nano-ceria. Arch Toxicol 88(11): 2033–2059
41. Kim CK, Kim T, Choi IY, Soh M, Kim D, Kim YJ, Jang H, Yang HS, Kim JY, Park HK, Park S, Yu T, Yoon BW, Lee SH, Hyeon T (2012) Ceria nanoparticles that can protect against ischemic stroke. Angew Chem Int Ed Eng 51:11039
42. Kreyling WG, Semmler-Behnke M, Seitz J, Scymczak W, Wenk A, Mayer P, Takenaka S, Oberdörster G (2009) Size dependence of the translocation of inhaled iridium and carbon

nanoparticle aggregates from the lung of rats to the blood and secondary target organs. Inhal Toxicol 21(Suppl. 1):55–60
43. Krug HF (2014) Nanosafety research – are we on the right track. J Angew Chemie 53(46):12304–12319
44. Lemire JA, Harrison JJ, Turner RJ (2013) Box 3: the Fenton reaction, free radical chemistry and metal poisoning. Nat Rev Microbiol 11:371–384. doi:10.1038/nrmicro3028
45. Linganiso L, Pendyala V, Jacobs G, Davis B, Cronauer D, Kropf A, Marshall C (2011) Low-temperature water-gas-shift: doping ceria improves reducibility and mobility of O-bound species and catalyst activity. Cat Letters 141(12):1723
46. Loane RF, Kirkland EJ, Silcox J (1988) Visibility of single heavy atoms on thin cyrstalline silicon in simulated annular dark-field STEM images. Acta Crystallogr A 44:912–927
47. Mai HX, Sun LD, Zhang YW, Si R, Feng W, Zhang HP, Liu HC, Yan CH (2005) Shape-selective synthesis of oxygen storage behavior of ceria nanopolyhedra, nanorods and naocubes. J Phys Chem B 109(51):24380–24385. doi:10.1021/jp055584b
48. Maynard AD, Baron PA, Foley M, Shvedova AA, Kisin ER, Castranova V (2004) Exposure to carbon nanotube material: aerosol release during the handling of unrefined single-walled carbon nanotube material. J Toxicol Environ Health Part A 67(1):87–107
49. Massie I, Dziasko M, Kureshi A, Levis HJ, Morgan L, Neale M, Sheth R, Tovell VE, Vernon AJ, Funderburgh JL, Daniels JT (2015) Advanced imaging and tissue engineering of the human limbal epithelial stem cell niche. Methods Mol Biol 1235:179–202. doi:10.1007/978-1-4939-1785-3_15
50. Mercer RR, Scabilloni JF, Hubbs AF, Wang L, Battelli LA, McKinney W, Castranova V, Porter DW (2013) Extrapulmonary transport of MWCNT following inhalation exposure. Part Fibre Toxicol 10(38):13
51. Mortensen LJ, Ravichandran S, Delouise LA (2013) The impact of UVB exposure and differentiation state of primary keratinocytes on their interaction with quantum dots. Nanotoxicology 7:1244–1254 3779483
52. Mortensen LJ, Faulknor R, Ravichandran S, Zheng H, DeLouise LA (2015) UVB dependence of quantum dot reactive oxygen species generation in common skin cell models. J Biomed Nanotechnol 11(9):1644–1652 PMCID: PMC4625909
53. Muller DA, Sorsch T, Moccio S, Baumann FH, Evans-Lutterodt K, Timp G (1999) The electronic structure at the atomic scale of ultrathin gate oxides. Nature 399:758–761
54. Naess EM, Hofgaard A, Skaug V, Gulbrandsen M, Danielsen TE, Grahnstedt S, Skogstad A, Holm JO (2015) Titanium dioxide nanoparticles in sunscreen penetrate the skin into viable layers of the epidermis: a clinical approach. Photodermatol Photoimmunol Photomed 32(1):48–51
55. Nagashima K, Zheng J, Pamiter D, Patri AK (2011) Biological tissue and cell culture specimen preparation for TEM nanoparticle characterization. Methods Mol Biol 697:83–91. doi:10.1007/978-1-60327-198-1_8
56. Nel A, Xia T, Mädler L, Li N (2006) Toxic potential of materials at the nanolevel. Science 311:622–627
57. Nellist PD, Pennycook SJ (2000) The principles and interpretation of annular dark-field Z contrast imaging. Advances Imaging Electron Phys 113:147–203
58. Oberdörster G, Finkelstein JN, Johnston C, Gelein R, Cox C, Baggs R, Elder ACP (2000) Acute pulmonary effects of ultrafine particles in rats and mice. Health Effects Institute, Cambridge, pp. 1–74
59. Oberdörster G (2002) Toxicokinetics and effects of fibrous and nonfibrous particles. Inhal Toxicol 14:29–56
60. Oberdörster G, Sharp Z, Atudorei V, Elder A, Gelein R, Kreyling W, Cox C (2004) Translocation of inhaled ultrafine particles to the brain. Inhal Toxicol 16(6–7):437–445
61. Oberdörster G, Oberdörster E, Oberdörster J (2007) Concepts of nanoparticle dose metric and response metric. Environ Health Perspect 115(6):A290 . PMC1892118

62. Oberdörster G (2015) Predictive modeling of nanomaterial risk: combining or replacing *in vivo* with in vitro studies? CompNanoTox, 4 Nov, Malaga, Spain
63. Pan Y-H, Sader K, Powell JJ, Bleloch A, Gass M, Trinick J, Warley A, Brydson AR, Brown A (2009) 3D morphology of the human hepatic ferritin mineral core: new evidence for a subunit structure revealed by single particle analysis of HAADF-STEM images. J Struct Biol 166(1):22–31. doi:10.1016/j.jsb.2008.12.001
64. Pan H, Myerson JW, HU L, Marsh JN, Hou K, Scott MJ, Allen JS, Hu G, Roman SS, Lanza GM, Schreiber RD, Schlesinger PH, Wickline SA (2013) Programmable nanoparticle functionalization for in vivo targeting. FASEB 27(1):255–264
65. Pennycook SJ, Varela M (2011) New views of materials through aberration-corrected scanniung transmission electron microscopy. Microscopy 60(1):5213–5223
66. Ravichandran S, Mortensen LJ, DeLouise LA (2011) Quantification of human skin barrier function and susceptibility to quantum dot skin penetration. Nanotoxicol 5(4):675–686 PMID: 21142716
67. Rehr JJ, Kas JJ, Vila FD, Prange MP, Jorissen K (2010) Parameter-free calculations of X-ray spectra with FEFF9. Phys Chem Chem Phys 12(21):5503–5513. doi:10.1039/b926434e
68. Ribeiro MC, Jacobs G, Davis BH, Cronauer DC, Kropf AJ, Marshall CL (2010) Fischer-Tropsch synthesis: deactivation as a function of potassium promoter loading for precipitated iron catalyst. J Phys Chem C 114:7895–7903
69. Rushton EK, Jiang J, Leonard SS, Eberly S, Castranova V, Biswas P, Elder A, Han X, Gelein R, Finkelstein J, Oberdorster G (2010) Concept of assessing nanoparticle hazards considering nanoparticle dosemetric and chemical/biological response metrics. J Toxicol Environ Health A 73(5):445–461
70. Sader K, Pan Y, Bleloch AL, Brydson R, Brown A (2008) Structural characterization of protein-caged iron minerals in biological systems. J. Physics: Conference Series 126(2008) 012006. doi:10.1088/1742-6596/126/1/012006
71. Semmler M, Seitz J, Erbe F, Mayer P, Heyder J, Oberdörster G, Kreyling WG (2004) Long-term clearance kinetics of inhaled ultrafine insoluble iridium particles from the rat lung, including transient translocation into secondary organs. Inhal Toxicol 16:453–459
72. Sotiriou GA, Watson C, Murdaugh KM, Darrah TH, Pyrgiotakis G, Elder A, Brain JD, Demokritou P (2014) Engineering safer-by-design silica-coated ZnO nanorods with reduced DNA damage potential. Environ Sci Nano 1(2):144–153 PMC4060637
73. Surekha P, Kishore AS, Srinivas A, Selvam G, Goparaju A, Reddy PN, Murthy PB (2012) Repeated dose dermal toxicity study of nano zinc oxide with Sprague-Dawley rats. Cutan Ocul Toxicol 31(1):26–32
74. Tomer R, Ye L, Hsueh B, Deisseroth k (2014) Advanced clarity for rapid and high-resolution imaging of intact tissues. Nat Protoc 9:1682–1697. doi:10.1038/nprot.2014.123
75. Tosha T, Behera RK, Ng HL, Bhattasali O, Alber T, Theil EC (2012) Ferritin protein nanocage ion channels: gating by N-terminal extensions. J Biol Chem 287(16):13016–13025. doi:10.1074/jbc.M111.332734
76. Toumey C (2009) Plenty of room, plenty of history. Nat Nanotechnol 4:783–784
77. Turner S, Lazar S, Freitag B, Egoavil R, Verbeeck J, Put S, Strauven Y, VanTendeloo G (2011) High resolution mapping of surface reduction in ceria nanoparticles. Nanoscale 3(8):3385–3390. doi:10.1039/c1nr10510h
78. Utembe W, Potgieter K, Stefaniak AB, Gulumian M (2015) Dissolution and biodurability: important parameters needed for risk assessment of nanomaterials. Part. Fibre Toxicol 12(11):12
79. Walkey CD, Chan WC (2012) Understanding and controlling the interaction of nanomaterials with proteins in a physiological environment. Chem Soc Rev 41(7):2780–2799
80. Wang W, McCool G, Kapur N, Yuan G, Shan B, Nguyen M, Graham UM, Davis BH, Jacobs G, Cho K, Hao X (2012) Mixed-phase oxide catalyst based on Mn-mullite (Sm, Gd)Mn2O5 for NO oxidation in diesel exhaust. Science 337(6096):832–835

81. Williams DB, Carter CB (2009) ISBN-10: 0387765026 Transmission electron microscopy a textbook for materials science, 2nd edn. Springer, New York
82. Wong J, Lytle FW, Greegor RB, Maylotte DH, Lamson S, Glover B (1984) EXAFS and XANES studies of trace elements in coal, in EXAFS and Near Edge Structure III, Proc. in Physics 2. Hodgson KO, Hedman B, Penner-Hahn JE (eds), Springer Berlin, Heidelberg GmbH 1984, p 362–367
83. Wu J, Liu W, Xue C, Zhou S, Lan F, Bi L, Xu H, Yang X, Zeng FD (2009) Toxicity and penetration of TiO2 nanoparticles in hairless mice and porcine skin after subchronic dermal exposure. Toxicol Lett 191(1):1–8
84. Wu JS, Kim AM (2013) Imaging and elemental mapping of biological speciments with a dual-EDS dedicated scanning transmission electron microscope. Ultramicroscopy 128:24–31
85. Wu N, Xie Y, Nel A, Holian A (2013) Inter-laboratory comparison of in vitro nanotoxicological assays from the NIEHS NanoGo Consortium. Environ Health Perspect 121:683–690 PMC3672931
86. Yang H, Lozano JG, Pennycook SJ, Jones L, Hirsch PB, Nellist PD (2015) Imaging screw dislocations at atomic resolution by aberration-corrected electron optical sectioning. Nat Comm 6:7266. doi:10.1038/ncomms8266
87. Yokel RA, Au TC, MacPhail RC, Hardas SS, Butterfield DA, Sultana R, Goodman M, Tseng MT, Dan M, Haghnazar SS, Unrine JM, Graham UM (2012) Distribution, elimination, and biopersistence to 90 days of a systemically introduced 30 nm ceria-engineered nanomaterial in rats. Toxicol Sci 127(1):256–268
88. Zhang H, Ji Z, Xia T, Meng H, Low-Kam C, Liu R, Pokhrel S, Lin S, Wang X, Liao YP, Wang M, Li L, Rallo R, Damoiseaux R, Telesca D, Mädler L, Cohen Y, Zink JI, Nel AE (2012) Use of metal oxide nanoparticle band gap to develop a predictive paradigm for oxidative stress and acute pulmonary inflammation. ACS Nano 6(5):4349–4368

Part III
Modelling

Chapter 5
Literature Review of (Q)SAR Modelling of Nanomaterial Toxicity

Ceyda Oksel, Cai Y. Ma, Jing J. Liu, Terry Wilkins, and Xue Z. Wang

Abstract Despite the clear benefits that nanotechnology can bring to various sectors of industry, there are serious concerns about the potential health risks associated with engineered nanomaterials (ENMs), intensified by the limited understanding of what makes ENMs toxic and how to make them safe. As the use of ENMs for commercial purposes and the number of workers/end-users being exposed to these materials on a daily basis increases, the need for assessing the potential adverse effects of multifarious ENMs in a time- and cost-effective manner becomes more apparent. One strategy to alleviate the problem of testing a large number and variety of ENMs in terms of their toxicological properties is through the development of computational models that decode the relationships between the physicochemical features of ENMs and their toxicity. Such data-driven models can be used for hazard screening, early identification of potentially harmful ENMs and the toxicity-governing physicochemical properties, and accelerating the decision-making process by maximising the use of existing data. Moreover, these models can also support industrial, regulatory and public needs for designing inherently safer ENMs. This chapter is mainly concerned with the investigation of the applicability of (quantitative) structure-activity relationship ((Q)SAR) methods to modelling of ENMs' toxicity. It summarizes the key components required for successful application of data-driven toxicity prediction techniques to ENMs, the published studies in this field and the current limitations of this approach.

C. Oksel • C.Y. Ma • T. Wilkins
Institute of Particle Science and Engineering, School of Chemical and Process Engineering, University of Leeds, Leeds LS2 9JT, UK

J.J. Liu • X.Z. Wang (✉)
Institute of Particle Science and Engineering, School of Chemical and Process Engineering, University of Leeds, Leeds LS2 9JT, UK

School of Chemistry and Chemical Engineering, South China University of Technology, Guangzhou 510641, China
e-mail: x.z.wang@leeds.ac.uk; xuezhongwang@scut.edu.cn

L. Tran et al. (eds.), *Modelling the Toxicity of Nanoparticles*, Advances in Experimental Medicine and Biology, Vol. 947, DOI 10.1007/978-3-319-47754-1_5

Keywords Nanomaterial toxicity • Nanotoxicology • QSAR • NanoSAR • In silico toxicity prediction

5.1 Introduction

There has been much interest recently in assessing and managing the potential effects of ENMs on human health and the environment. There is now a significant amount of studies highlighting that although not all ENMs necessarily have side effects, certain types of ENMs can pose risks to human health and the environment [20, 53, 111] if not properly managed. It has been well reported in literature that a nano-sized form of some materials may exhibit significantly different toxicity despite their chemically identical structure [60, 64]. Although is evidently known that some ENMs exhibit adverse effects, their mode of toxic action and the factors affecting their toxicological responses are still not fully discovered. It is clear that large knowledge gaps still exist in areas that are essential for monitoring and minimizing the risks of potentially toxic ENMs [25, 31].

The current toxicity assessment approach primarily relies on animal-based testing that is not only time and cost demanding and but also ethically problematic. Considering the high number of ENMs requiring toxicity screening, the use of alternative approaches such as in silico tests relying on computational modelling methods are needed to predict health risks of a range of ENMs with less cost and time compared to animal testing. There are several computational techniques that have been developed and used in toxicology. The (quantitative) structure–activity relationship ((Q)SAR) analysis is one of the most promising computation approach for toxicity prediction of ENMs since they are capable of quantifying the relationship between relevant properties and biological activity of a certain class of materials. It is a data-driven method that attempts to make use of existing experimental data for in silico prediction of toxicological endpoints. The main assumption behind this approach is that toxicity depends on, and hence can be predicted by, physicochemical properties such as size, shape, surface characteristics and crystal structure. A schematic workflow of the nano-(Q)SAR technique is given in Fig. 5.1.

Although the need for the development of intelligent testing strategies based on in silico methods to assess the toxicity of ENMs has been emphasized by many scientists and regulators [35, 36, 98], scientific investigation of their applications as predictive tools for toxicological evaluation of ENMs has not received much attention. To address this research gap and devote systematic attention to this subject, this chapter is focused on investigating whether the computer-based structure-activity relationship methods are applicable to predict the toxicological effects of ENMs. The ultimate aim here is to contribute to moving the nanotoxicology research forward from individual assessments toward a more integrated hazard screening approach that can predict the toxicity potential of

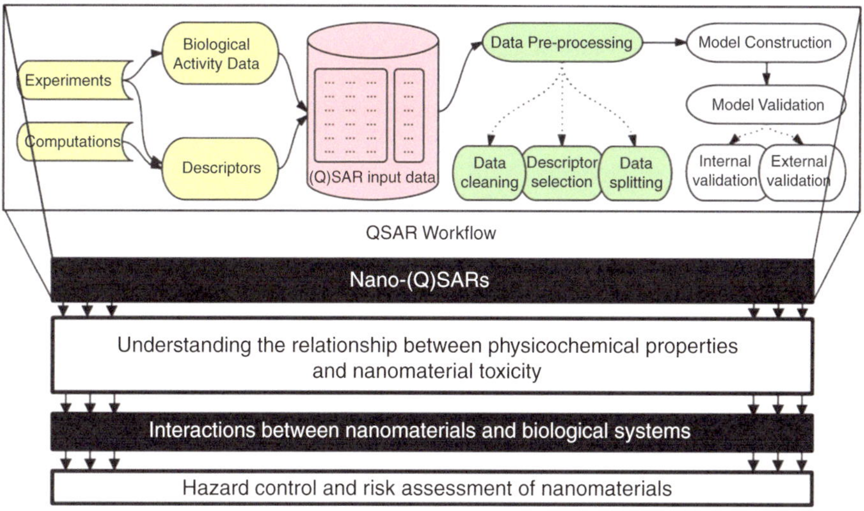

Fig. 5.1 (Q)SAR modelling of nanomaterial toxicity (Reprinted from *Particuology*, 21, Ceyda Oksel et al., (Q)SAR modelling of nanomaterial toxicity: A critical review, 1–19., Copyright (2015), with permission from Elsevier)

ENMs based on their structural and physical characteristics. The main aim is to give the reader a detailed understanding of the nano-(Q)SAR process, the concepts behind it, the appropriate tools to use, and the remaining knowledge gaps in this area.

5.2 Toxicity of Nanomaterials

Nanotechnology is a broadly applicable science with considerable potential for breakthroughs in a wide variety of fields. It has impact in almost all branches of engineering, resulting in a rapid increase in the number of ENMs being exploited commercially. However, the distinctive characteristics of ENMs not only make them a material of choice for various applications, but also affect their toxicity potential and present a challenge for the existing regulatory systems.

As numerous natural nano-sized materials have existed in the environment for centuries, and the nature has been filled with several nanofeatures, nanomaterials are not entirely new in many ways. However, unlike naturally occurring nanoparticles (NPs) or incidentally produced nano-sized byproducts, the intentionally designed and manufactured NPs are completely new and capable of exhibiting different biological effects [105]. The study of toxicity of these newly introduced ENMs falls into a new emerging discipline called nanotoxicology. Due to their small size, ENMs have generally been considered to be able to enter and damage living organisms [85].

In recent years, some particular ENMs have been shown to exhibit toxic responses. For example, carbon nanotubes (CNTs) are reported to induce oxidative stress [112] and pulmonary toxicity [115]. Another example of ENMs of significant toxicological concern is nanosilver. Despite initially being perceived to be non-hazardous material, recent studies have provided convincing evidence that nanosilver can exhibit more pronounced toxicity than larger-sized particles of the same substance [5, 32]. For an extensive review of the potential side effects of ENMs, the interested reader can refer to [4, 55, 60, 79, 111].

A toxicological endpoint is the measure of the toxic effect of a substance on human health or the environment, and it determines the harmfulness of a substance. The toxicity of compounds can be evaluated by conducting in vivo, in vitro, and in silico studies. For classical human health hazard assessment through in vivo testing, several toxicological endpoints are relevant, e.g. acute and chronic dermal, oral or inhalative toxicity as well as skin and eye irritation. Although in vitro assays are commonly preferred to in vivo assays as an initial test because of their time and cost effectiveness, there is also a well-recognised need in the nanoscience community to compare and validate in vitro findings with in vivo observations. In (Q)SAR analysis, it is the specific type of activity, such as cell viability or cytotoxicity, that is going to be modelled and predicted. (Q)SAR models can be built and used for the prediction of all toxicological endpoints as long as sufficient toxicity data is provided as input [99]. Ideally, the biological effects of various compounds with different sizes, structures, and complexities under relevant exposure conditions should be tested with standardized test methods for the successful development of nano-(Q)SAR models.

5.3 Descriptors of ENM Properties

Physicochemical descriptors are experimentally or theoretically derived parameters that are used to correlate the structural parameters with the endpoint of interest through a statistical method. In general, there are two ways of obtaining molecular descriptors representing physical and chemical properties of chemical: experimental measurements and theoretical calculations. Theoretical descriptors can be obtained from different theories and semi-empirical techniques. They provide various structural and compositional information and significant insight into correlations which are assumed to exist between physicochemical properties and biological activity. There are more than 5000 descriptors that have been derived to represent molecular structures but the majority of them are not directly applicable to ENMs. The main problem in using traditional descriptors for nanostructures is their complexity and non-uniformity which make the direct transformation of the nanostructures into a computer-readable form difficult. Experimentally measured parameters such as particle size, size distribution, surface area and surface charge can also be used as descriptors in computational studies. Due to the current limitations in representing nanostructures in a machine readable form, most of the existing nano-(Q)

SAR studies use experimental descriptors as an input when developing computational models of ENM toxicity.

There are several issues that are currently of concern to nano-(Q)SAR modellers. The main problem with experimental descriptors is the lack of agreement on how, when and where to characterize ENMs prior to toxicity testing. For theoretical descriptors, the main issue is the appropriate and useful representation of nanostructures enabling computational treatments. Another important issue is the development of novel descriptors for ENMs' structure. Although traditional QSAR analysis is almost standardized, their application to ENMs is still under development and involves several difficulties, given the complexity of nanostructures. In this section, physicochemical properties that are likely to influence the biological activity of ENMs are briefly explained and their measurement methods are summarised.

5.3.1 Parameters Influencing the Toxicity of ENMs and their Measurement

The first step in modelling ENM toxicity is identifying toxicity-related properties that can be used as potential determinants of adverse effects of ENMs. Because a complete and exact list of parameters influencing the toxicity of ENMs has not yet been established, detailed material characterization prior to toxicity testing is essential to determine the factors contributing to the biological activities of ENMs and their potential hazards. Although there is still no scientific consensus on the minimum set of relevant nanocharacteristics for toxicological evaluation, some particular physicochemical features are included in the majority of recommendations [95]. The size of ENMs is one of the most important characteristics that affects the properties and behavior of ENMs, and is hence included in the recommendation list of almost all nanotoxicologists. However, as mentioned by Oberdorster et al. [85], the size of the particles is not the only factor that causes changes in the biological activities of materials at the nanoscale. The following characteristics may also be linked to nanotoxicity: size distribution, agglomeration state, shape, crystal structure, chemical composition, surface area, surface chemistry, surface charge, and porosity. Powers et al. [95] investigated the important elements of NM characterization, and expanded the list reported by Oberdorster et al. [85] to include purity, solubility, and hydrophobicity. In a recent review on the minimum set of physicochemical properties required to characterize NMs, Pettitt and Lead [92] suggested that, in addition to the parameters that are most likely to have an effect on NM behavior such as size, surface properties, solubility, and aggregation characteristics, information about the production process and history of ENMs should also be provided to avoid incorrect interpretation of toxicity data. One of the most comprehensive lists of the important physicochemical characteristics for toxicological studies has been provided by the Organization for Economic Cooperation and Development (OECD) Working Group on Manufactured Nanomaterials, the OECD WPMN [88]. The WPMN suggested a list of physicochemical properties potentially needing to be addressed for

Table 5.1 Physicochemical properties and material characterization

Characterization (as on the shelf)		Characterization (in respective media)
Appearance (IA)	Dissociation constant (IA)	Composition/purity
Melting point (IA)	pH (IA)	Size, size distribution
Density (IA)	Agglomeration or aggregation	Agglomeration/aggregation
Size, size distribution	Crystalline phase	Zeta-potential
N-octanol-water partition coefficient (WR)	Crystallite and grain size	Biophysical properties (AA) (protein binding/corona characterization, residence times, adsorption enthalpy, conformation changes on binding)
Water solubility/dispersibility, hydrophilicity	Aspect ratio, shape	
Solubility/dispersibility in organic solvents, oleophilicity	Specific surface area	
Auto flammability (IA)	Zeta potential	Test item preparation protocol, conditioning, homogeneity and short term stability
Flammability (IA)	Surface chemistry (WA)	
Stability in solvents and identity of relevant degradation products	Stability and homogeneity (on the shelf, in water and organic solvents)	
Oxidizing properties (IA)	Dustiness	
Oxidation reduction potential	Porosity, pore and pour density	
Explosiveness (IA)	Photocatalytic activity	
Storage stability and reactivity towards container material	Catalytic activity	
Stability towards thermal, sunlight, metals	Radical formation potential	

WR where relevant, *IA* if applicable, *WA* where available, *AA* as appropriate

characterization relevant to (eco)toxicity, and devised a testing programme to investigate this. The physicochemical properties mentioned in this guidance are listed in Table 5.1. The term "composition" in Table 5.1 covers chemical identity and molecular structure, as well as degree of purity, impurities, and additives. Another term in this list that is often broadly defined is the "surface chemistry". Here, it is meant to identify various modifications of the surface (i.e., coatings) and the composition of the outer layer of the NMs. In OECD's list, there are also many properties, such as dustiness and *n*-octanol–water partition coefficient, that have not been specified as prerequisites for NM characterization by other researchers; within the OECD WPMN there is now agreement that the n-octanol-water partition coefficient is not relevant for NMs. Powers et al. [95] took dustiness as an example and argued that such a measurement for dry NM applications should first be standardized, because the presence of well-established analytical techniques for the measurement of intended properties is essential to express the results in comparable terms; dustiness is not an inherent property but depends on the sample tested. For a detailed description of the potential toxicity-related physicochemical properties shown in Table 5.1, please refer to OECD's guidance on testing ENMs [88].

5.3.1.1 Particle Size and Size Distribution

The size of ENMs is regarded as one of the most important properties determining the toxicity potential of ENMs. The surface area to volume ratio increases with decreasing particle size. The change in surface-to-volume ratio also affects the surface energy and hence the reactivity of the material. In addition to surface reactivity, the interaction of ENMs with living systems and the uptake and deposition of ENMs within the human body are also affected by particle size [96]. It is generally believed that the risk posed by materials containing nanosized particles increases with decreasing particle size [81]. Indeed, Gurr et al. [45] showed that the oxidative damage induced by TiO_2 particles is size-specific: the smaller the particle size, the greater the oxidative damage induced. Similarly, the toxicity of nanosilver is assumed to be dependent on the particle size. Park et al. [90] compared the cytotoxicity, inflammation, genotoxicity, and developmental toxicity induced by different-sized silver ENMs (20, 80, and 113 nm), and found that the smallest nanosilver particles exhibited higher toxicity than larger particles in the assays. More recently, in an interesting study, Xiu et al. [135] concluded that the toxicity of silver NPs are only indirectly associated with morphological features (i.e., these properties influence the release of silver ions which in turn has an effect on the toxicity). All such findings suggest that the size of particles is a possible factor that may directly or indirectly contribute to the toxicity of chemicals. However, in some cases, no relationship between the toxicity of particles and their sizes is observed [64, 70]. There are several techniques that can be used to measure the size of ENMs. Although not a comprehensive list, the most common particle size measurement techniques applicable to ENMs are given in Table 5.2.

The results of different particle size measurement techniques are usually not in agreement because the measurement principles behind each method are different. In general, it is possible to classify the particle size measurement methods applicable to ENMs into three categories: microscopy-based, light scattering-based, and separation techniques [107]. Electron microscopy techniques, which are based on scattered (SEM) or transmitted (TEM) electrons, provide very accurate information and give a clear view of individual and aggregated particles. Therefore, these methods can also be used for polydisperse particle samples. The scanning electron microscopy (SEM) technique provides information about the size, size distribution, particle shape, and morphology, but there is a risk of influencing particle properties during sample drying and contrasting [11]. SEM and TEM give two-dimensional information on the particles. Unlike electron microscopy techniques, a vacuum environment is not required to obtain atomic force microscopy (AFM) images, which allows the measurement of particle sizes under ambient conditions [47].

Table 5.2 Particle size measurement techniques

Method	Parameters measured	Sample required	Particle size range	Additional information
Electron microscopy	Particle size Size distribution Particle shape Agglomeration	Dry	0.3 nm– μm	(+) High resolution (−) Expensive and complex (−) Vacuum is needed [25]
Atomic force microscopy	Particle size Size distribution Morphology Surface structure Agglomeration	Wet/Dry	1 nm– μm	(+) 3D images (+) Works well in ambient air (−) Particles should be on the surface. [94]
Dynamic light scattering (DLS)	Particle size Size distribution Agglomeration Zeta potential	Wet	1 nm–6 μm	(+) Cheap and fast (−) Sample polydispersity may distort the results. [124]
NP tracking analysis (NPTA)	Particle size Size distribution Agglomeration	Wet	10 nm–2 μm	(+) Particle-by-particle basis (−) Dependence on the settings [52]
Centrifugal sedimentation	Particle size Size distribution	Wet	5 nm–10 μm	(+) Accurate and repeatable results (−) Takes long time for small particles to sediment [67]
BET surface area analysis	Particle size Surface area	Dry	5 nm–1 μm	(−) Size distribution is not provided. [25]
Laser diffraction	Particle size Size distribution	Wet/Dry	40 nm–3 mm	(+) Fast and flexible (−) Dependent on optical parameters [65]
Mobility analysis	Particle size Size distribution	Dry	2 nm–2 μm	(+) Commonly used for aerosols (−) Interpretation of results may require additional information. [85]
Acoustic methods	Particle size Size distribution Zeta potential	Wet	20 nm–10 μm	(+)Effective in concentrated suspensions (−) Difficult to interpret the data [94]

(+) represents advantageous, (−) means disadvantageous.

Dynamic light scattering (DLS) is based on the Brownian motion of suspended particles in solution and gives the hydrodynamic diameter of the particles measured, which is larger than results for dry-measurement diameters. The main advantages of DLS techniques are their simplicity and speed, while their main weaknesses are the high sensitivity to sample concentration and the inability to differentiate between large individual particles and aggregates [81], and furthermore, DLS cannot be successfully applied to polydisperse suspensions of particles as the intensity of the scattered light is proportional to diameter, D, to the power of six, D^6, meaning that large particles will overshadow smaller ones. Dynamic centrifugal sedimentation (DCS) and analytical ultracentrifugation use the difference in sedimentation rates of different sized particles to separate a sample. Tantra et al. [120] emphasized that one of the main disadvantages of DCS is the requirement to know the exact density of the particle including coatings and adsorbed analytes on the surface. A dry size measurement method is Brunauer–Emmett–Teller (BET) surface area analysis, which calculates the mean particle diameter from surface area measurement based on the assumption that the particles are nonporous and spherical. Additionally, there are several other size measurement methods, including laser diffraction, mobility analysis, acoustic methods, field-flow fractionation (FFF), and fluorescence correlation spectroscopy (FCS), each of which has its pros and cons. Domingos et al. [26] provided a good example of size measurement by multiple analysis methods including TEM, AFM, DLS, FCS, NP tracking analysis (NPTA), and flow field flow fractionation (FIFFF). They confirmed that the particle size measured by DLS is typically higher than those obtained using the other sizing methods. It was concluded that there is no ideal nanoscale measurement technique that is suitable for all sample types. Various factors, such as the nature of the substance to be measured, the constraints of cost and time, and the type of information required, play a decisive role in the choice of the sizing method. Additionally, the structural properties of ENMs, sample preparation, and polydispersity have significant effects on the results of different ENM size measurement techniques.

There are three important criteria that should be met for accurate measurement of particle size: a well-dispersed system, selection of a representative sample, and appropriate selection of the size measurement method considering the nature of the ENM and its intended use [96]. It should also be kept in mind that some methods require dispersion, such as DLS, NPTA, and DSC. The aggregation/agglomeration of particles in dispersions leads to an increase in the measured particle size, as does the formation of corona, when the hydrodynamic diameter is measured. The results from wet measurements may reflect well the biological situation in nano-toxicity studies, depending on the media, because ENMs will actually not be in a dry form when they are in contact with human cells/organs.

It is our view that the combination of a microscopic technique (e.g., TEM or AFM) and an ensemble technique (e.g., DLS) seems appropriate for monodisperse systems, because this can provide a complete picture of the size characteristics in the dry form and suspension. For polydisperse systems, the DLS technique has serious problems, hence it should be replaced or complemented with an alternative size measurement approach. In summary, it is usually useful to combine a single-particle size measurement technique with an ensemble method to obtain a rich dataset of particle sizes and the size distribution, especially when a priori knowledge on these parameters is unavailable for the test material. The results of seven studies by different researchers are given in Table 5.3, with the aim of comparing different ENM size measurement techniques. It should also be pointed out that, compared with the average value of the particle size, the size distribution provides a more realistic representation of particle size information, which is a critical attribute in nanotoxicology. However, measurement of particle size distributions usually provides a large amount of data (e.g., hundreds of size distribution components), which may cause problems in the (Q)SAR analysis (e.g., increased random correlations). Therefore, it is important to find a reasonable way to represent all components of the size distributions with a few variables that still retain all of the information present in the input data. Wang et al. [132] carried out principal component analysis on size distribution data consisting of a large number of particle size distribution measurements to reduce the number of descriptors to a manageable size. This study is a good example of how to handle large size distribution datasets prior to nano-(Q)SAR analysis. Instead of reporting mean particle size values, researchers should also take into account the variations in the size distribution as a whole, because the ENM samples consist of a range of particle sizes, not only a single type of particles.

5.3.1.2 Particle Shape

The shape of ENMs is another important feature influencing the biological activities of the particles. The hydrodynamic diameters of spherical and rectangular particles with the same mass, and hence their mobility in solution, vary because of shape effects. Moreover, shape characteristics greatly affect the deposition and absorption kinetics of NPs in a biological environment [81]. The importance of shape in toxicity has been proven for CNTs. Poland et al. [93] showed that long multiwalled CNTs (MWCNTs) are more toxic than short/tangled MWCNTs. The study undertaken by Powers et al. [96] revealed that the antibacterial activity of silver NPs is shape-dependent. In another study, Gratton et al. [41] demonstrated that rod-like (high aspect ratio) NPs are drawn or internalized more efficiently into cells than cylindrical NPs. Although there are several studies investigating and confirming the potential effect of NP shape on toxicity, it is still not possible to draw clear conclusions or define any particular shape inherently "toxic" with current knowledge. Further research is required on NPs with similar composition but different shape to investigate the role of NP shape in toxicity.

There are several nondimensional shape indexes that can be used to quantify the shape characteristics of particles, such as sphericity/circularity, aspect ratio/elongation, convexity, and fractal dimensions. The shape index of NPs is usually determined using microscopic methods such as SEM and TEM, which have the ability to simultaneously determine both particle size and shape. Additionally, the ratio of two particle sizes measured by different techniques, such as DLS and

Table 5.3 Nanoparticle mean size measurement results obtained by different size measurement methods

Particle size (nm)							
Ref.	Thiele, Poston, and Brown [122]				Lee et al. [69]	Akbari, Tavandashti, and Zandrahimi [1]	Borchert et al. [12]
Method	Ta	$TiSi_2$	Ni	C	SiO_2–7 nm	Al_2O_3	$CoPt_3$
BET	8	19	35	45	18	27	
TEM	7	13	24	31	19	24	4.86
DLS	316	157	1300		13		
Others						XRD:20; PCS:96	XRD:5; SAXS:4.97
Ref.	Hoo, Starostin, West, and Mecartney [54]				Supaka [117]		Boyd, Pichaimuthu, and Cuenat [14]
Method	PS-100	PS-20	PS-20&100	PS-20&101	CRM-60	CRM-100	Latex
DLS	114	23	109	245	73	105	110
AFM	99	16	15–95	16–98	58	58	98
Others					SEM:79	SEM:79	NTA:99

TEM/SEM, can be used as a simple expression of particle shape [56]. Because shape characteristics and the distribution of NPs may vary when they are in contact with organisms, shape measurements should also be made for "as-exposed" as well as "as-received" forms. Wang and Ma [130] defined the shape of a crystal according to the normal distance between each surface of the particle and its geometrical centre. They carried out principal component analysis (PCA) on the shape description dataset for data compression. The calculated surface–centre distances or the resultant principal component values can be directly used as shape indexes of NPs, especially nonspherical NPs, in nano-(Q)SAR. Moreover, these values can also be used as dynamic shape factors to investigate the time and size dependence of shape once this modelling methodology is applied to model the aggregation/agglomeration behaviour of NPs. If aggregation/agglomeration occurs, the normal distances for some faces may disappear with some new distances appearing. If breakage occurs, some new normal distances will be identified to represent the new faces. Such alternative approaches are useful for nano-(Q)SAR applications because they take into account the dynamic nature of NP shape.

5.3.1.3 Crystal Structure (Crystallinity)

ENMs with the same chemical composition may have different toxicological properties because of their different atomic arrangements and crystal structure. Jiang et al. [62] investigated the effect of crystallinity on NP activity by comparing the ROS generating capacity of TiO_2 NPs with similar size but different crystal phases (amorphous, anatase, rutile, and anatase/rutile mixtures). The study found that amorphous samples showed the highest level of ROS activity followed by pure anatase and anatase/rutile mixtures, while pure rutile produced the lowest level of ROS. Nanosilica, which occurs in multiple forms, is another ENM whose toxicity may vary depending on the nature of its crystal structure [82].

A widely used technique to obtain information about crystal phases, purity, crystal structure, crystallite size, lattice constants, and defects of NPs is X-ray diffraction (XRD). XRD is a useful tool to characterize nanostructures because it provides nondestructive evaluation of the structural characteristics without the need for exhaustive sample preparation [28]. Its noncontact and nondestructive features make XRD ideal for in situ measurements [113]. Measurement in a desired atmosphere is allowed in XRD. This makes XRD advantageous for toxicological characterization in which collection of crystal structure data in biologically relevant media becomes an important issue.

Additionally, high-resolution transmission electron microscopy (HR-TEM) and selected-area electron diffraction (SAED) can be used to obtain information about the crystal structure, especially when data acquisition from individual nanocrystals is required. We believe that conventional XRD is preferable over TEM for crystallographic investigation of nanostructures because of the sample-damaging and the user-dependent nature of TEM.

5.3.1.4 Surface Characteristics

Surface Functionalization

Surface chemistry is another factor that needs to be considered for the complete characterization of NPs, because it plays an important role in the surface interactions and aggregation behavior of NPs in liquid media. Therefore, if the surfaces of ENMs are intentionally functionalized, each chemical species and functional groups on the surface should be identified. The influence of surface coating on the toxicity of Ag-NPs has been investigated by many researchers [21, 83, 140]. The results from Nguyen et al. [83] showed that uncoated Ag-NPs are more toxic than coated Ag-NPs. However, the coating is not the only factor that reduces the toxicity of Ag-NPs. Changes in the aggregation state and particle size as a result of surface coating may also be important.

Information about how the ENM surface affects the interactions of NPs in a biological environment can be obtained from different techniques, such as electron spectroscopy (X-ray photoelectron spectroscopy (XPS) and Auger electron spectroscopy (AES)), scanning probe microscopy (AFM and scanning tunneling microscopy (STM)), ion-based methods (secondary ion mass spectrometry and low-energy ion scattering), and other spectroscopic techniques (e.g., IR, NMR, and Raman spectroscopy) [6]. The most important advantage of electron spectroscopy is its high surface sensitivity. XPS is one of the most commonly used techniques for surface analysis [125]. Both XPS and AES can be used to obtain information about the presence, relative surface enrichment, composition, and thickness of coatings.

Surface Charge

Surface charge is another important characteristic that may affect the toxicity of ENMs. The biological interactions of ENMs, and hence their biological activities, are highly surface-charge dependent. Park et al. [91] analyzed the effect of surface charge on toxicity using negatively and weakly negatively charged silica-NPs. They found that negatively charged silica-NPs have a higher level of cytotoxicity than weakly negatively charged silica-NPs. In another study, the core of silicon-NPs was covered with different organic monolayers to obtain different surface charges (positive, negative, and neutral) [10]. The study found that positively charged silicon-NPs are more toxic than neutral silicon-NPs, while negatively charged silicon-NPs induced almost no cytotoxicity.

Because it is challenging to directly measure the charge at the surface of particles, zeta potential measurement using dynamic or electrophoretic light scattering is usually used to quantify the surface charge. According to Xu [136], among the three techniques that can be used to determine the zeta potential (electrophoretic light scattering (ELS), and acoustic and electroacoustic methods), ELS is preferred for various applications because of its certainty, sensitivity, and versatility. However, classic ELS cannot successfully determine the zeta potential of turbid samples because the light cannot penetrate the sample. Preferably, the sample should be optically clean and nonturbid for accurate measurements. It was also noted in the same study that the accuracy of zeta potential measurements is greatly affected by environmental conditions, such as pH and ionic strength. The pH-dependence of the zeta potential should also be taken into account because changing the pH of a solution may greatly alter the distribution of surface charge.

The current understanding of the relationship between surface charge and toxicity is severely limited, mainly because of the incapability of existing in situ measurement techniques and the environment-dependence of zeta potential measurements [61]. Because the value of the zeta potential obtained may vary between different techniques and experiments [38], multiple tests should be conducted for the best possible accuracy and the results should be reported together with details on measurement conditions (e.g. pH value and sample concentration).

5.3.1.5 Aggregation State

Some NPs have the tendency to form large agglomerates both in the dry form and in suspension. If NPs form clusters, they may behave like larger particles because of their increased hydrodynamic size [20]. Because agglomeration could affect important physicochemical features, such as particle size and the size distribution, the biological effects of these changes should be identified to avoid incorrect estimation of the toxic potential of ENMs [61].

The aggregation state is often quantified by measuring the size distribution of existing agglomerates. It can be monitored and quantified by microscopic techniques such as TEM, SEM, and AFM. Additionally, DLS can also be used to investigate NP aggregation. However, characterization of the agglomerate size of NPs in suspensions is very challenging because the degree of aggregation can be influenced by external conditions (e.g., pH, temperature, and humidity). Ideally, in situ instruments that are capable of measuring the size, shape, and number of all agglomerates in the relevant medium are required to characterize the aggregation state. The particle size information used in early nanotoxicological studies usually refers to the primary size of individual NPs and ignores the effect of aggregation. Although accurate characterization of the aggregation state prior to nanotoxicity testing is seen as a prerequisite by several researchers [13, 61, 129], there is still no clear consensus on how to characterize aggregation. However, characterizing the aggregation shape using fractal dimensions, which provide an index of complexity by measuring the space-filling capacity of an object, may be the way forward [109].

5.3.2 NP-Specific Descriptors

Because some properties of ENMs are different from conventional materials, it is very likely that also the toxicity of ENMs could be different and associated to nanophenomena. Therefore, the development of nanospecific descriptors capable of describing the distinctive properties of NPs is one of the main research requirements in the area of computational nanotoxicology. In this section, the different approaches to develop novel NP-descriptors will be presented.

Glotzer and Solomon [39] proposed an approach to characterize NPs based on microscopic images. They defined eight orthogonal dimensions that can be used as NP-descriptors to compare the structural similarity of different NPs: surface coverage, aspect ratio, faceting, pattern quantization, branching, chemical ordering, shape gradient, and variation in roughness (Fig. 5.2). Although the development of new descriptors based on microscopic images is a promising idea, the numerical expression of these eight dimensions is still an unresolved problem.

The idea suggested by Glotzer and Solomon [39] has inspired other researchers to use microscopic images of NPs for the extraction of structural information. Puzyn et al. [97] proposed to quantify each pixel in SEM, TEM, and AFM images using RGB colour codes or gray-scale representation, and then produce a rectangular array of numbers (Fig. 5.3). They also emphasized that these numerical values of image pixels can be used as new descriptors for encoding the structural properties of NPs.

In another study, Xia et al. [133] developed a multidimensional biological surface adsorption index (BSAI) consisting of five quantitative nanodescriptors: lone-pair electrons, polarity/polarizability, hydrogen-bond donors, hydrogen-bond

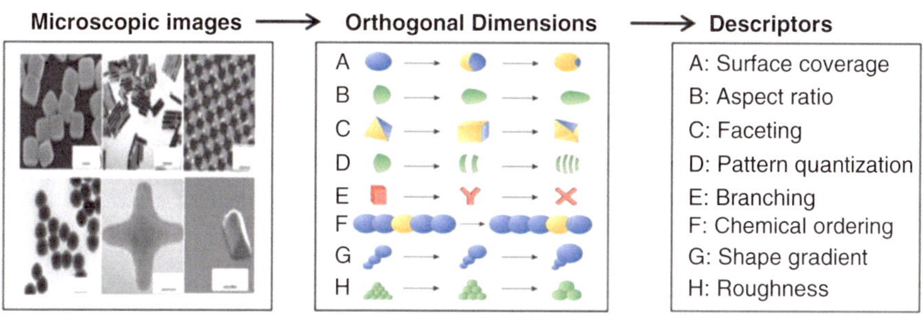

Fig. 5.2 Derivation of eight qualitative descriptors based on microscopic images (Reprinted from Nature materials, 6, Glotzer et al., Anisotropy of building blocks and their assembly into complex structures., 557–562., Copyright (2007), with permission from Elsevier)

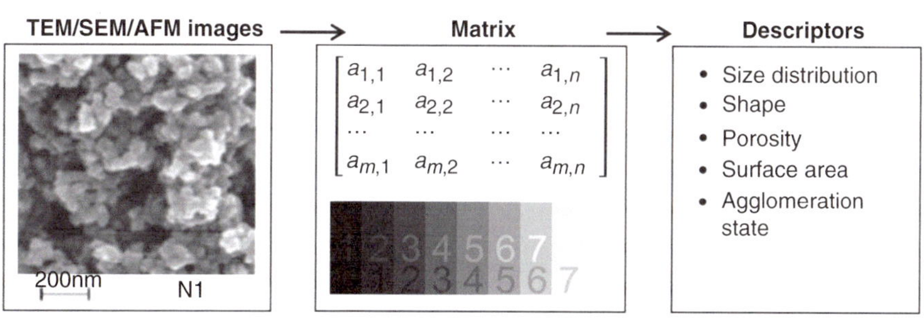

Fig. 5.3 Derivation of structural descriptors based on microscopic images (Reprinted from Small, 5, Puzyn et al., Toward the development of "nano-QSARs": advances and challenges., 2494–509., Copyright (2009), with permission from John Wiley and Sons)

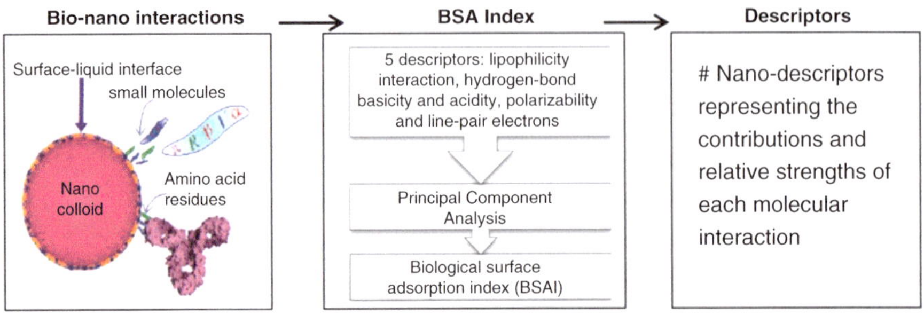

Fig. 5.4 Derivation of descriptors that represent the fundamental forces governing the adsorption process of NPs (Reprinted by permission from Macmillan Publishers Ltd.: Nat Nano, Xia, et al. An index for characterization of nanomaterials in biological systems. Nat Nano, 5, 671–675, copyright (2010))

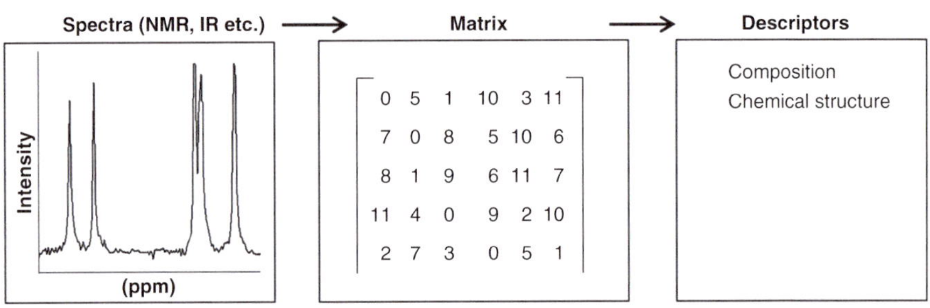

Fig. 5.5 Derivation of NP-descriptors based on the spectra of ENMs (Reprinted from Wiley Interdisciplinary Reviews: Nanomedicine and Nanobiotechnology, 3, Burello et al., QSAR modeling of nanomaterials., 298–306., Copyright (2011), with permission from John Wiley and Sons)

acceptors, and London dispersion. These five nanodescriptors represent the fundamental forces governing the adsorption process of NPs in a biological environment. In their follow-up study [134], they performed PCA on five-dimensional nanodescriptor datasets to reduce dimensionality, and obtained a two-dimensional representation of the molecular interaction forces in biological systems and hence facilitated characterization of the surface properties of ENMs (Fig. 5.4). After obtaining two-dimensional nanodescriptors via PCA, they managed to classify 16 different ENMs into separate clusters based on their surface adsorption properties.

Burello and Worth [19] proposed that different types of spectra (e.g., NMR, IR, Raman, and UV–Vis) can be used as nanodescriptors because they contain fingerprint-like information (Fig. 5.5). The first step is spectral measurement followed by conversion of the spectra into a numerical matrix. This data matrix can be seen as spectra-derived descriptors and used for (Q)SAR analysis. It is not entirely a new perspective because spectral information has already been used in a number of studies. The use of IR information for (Q)SAR analysis was shown to be promising by Benigni et al. [9]. They compared the IR spectra with several descriptors commonly used in (Q)SAR studies, and found that IR spectra contain unique information that cannot be obtained from molecular descriptors. Zhou et al. [141] used the spectra of multiwalled NTs for characterization, while Yang et al. [137] attempted to correlate XRD data with photocatalytic performance using the dye decolourization rate. We strongly believe that the use of spectra-derived descriptors in (Q)SAR modelling of ENMs is an interesting approach and deserves further investigation.

The final properties of materials are related not only to the chemical composition and structure of materials but also to the preparation, synthesis, and processing methods. Le et al. [68] suggested that molecular descriptors characterizing physicochemical properties of compounds could be combined with historical descriptors describing the sample preparation and synthesis techniques of materials to develop reliable and predictive models. Although historical descriptors can be useful for modelling traditional materials, their implementation to nano-(Q)SAR models can be very difficult because they probably have no ability to distinguish between ordinary and nanosized particles. The determination of three-dimensional descriptors that are suitable for nanostructures and NP representation is another promising approach and undoubtedly will be put into practice in the near future. In addition, the development of more sophisticated image analysis approaches (e.g., texture analysis-based methods) would facilitate the rapid extraction of morphological information (e.g., particle size, shape, surface area, and aggregation state) from microscopic images of NPs.

5.4 Nano-(Q)SAR and Modelling Techniques

A (Q)SAR is a mathematical model that attempts to relate the biological activities or properties of a series of chemicals to their physicochemical characteristics in a quantitative manner [99]. Although the first use of (Q)SAR models is attributed to Hansch [50], who brought physical organic chemistry and the study of chemical biological interactions together to propose the first (Q)SAR approach, the relationship between chemical structure and biological activity was reported in several earlier studies ([16, 103]., [89]). Hansch's (Q)SAR approach has found applications in many disciplines, such as drug design, and chemical and biological science. Moreover, numerous modification of Hansch's approach to QSAR modelling have been developed by many other researchers [66].

In (Q)SAR models, it is assumed that the observable biological activity is correlated with the structure of compounds, and this correlation can be expressed in a mathematical equation. The presumed relationship between the activity and structure is expressed with the following form of mathematical equation:

$$y = f\left(x_i\right), \tag{5.1}$$

where y is the biological activity of the chemical (i.e., toxicity) and $f(x_i)$ is a function of structural properties. A set of well-characterized compounds with known biological effects is required to obtain this mathematical equation. The structural features of compounds with known biological activities are represented by measured or calculated molecular descriptors. Then, a mathematical model relating the measured activity to the descriptor sets is obtained by regression analysis. The last step is evaluation of the reliability of the model and its applicability to other compounds. One of the most important steps, which is often omitted, is to define the model's boundaries and limitations to demonstrate how well it performs when applied to substances that are not used in building the model.

5.4.1 Nano-(Q)SAR Research

The papers focusing on the development of nano-(Q)SAR models are given in Table 5.4. Most of the nano-(Q)SAR studies focused on metal oxide (MO) ENMs because of their common commercial use and high production volume. One of the first attempts to show that computational (Q)SAR can give valuable information about nanotoxicity was reported by Liu and Hopfinger [71]. They used molecular dynamic simulations to investigate the effect of CNT insertion on the cellular membrane structure. Four potential toxicity sources were investigated through membrane interaction-(Q)SAR analysis. Although the result of this study was very informative and encouraging, a proven (Q)SAR model was not established because of the absence of experimental data.

Sayes and Ivanov [108] assessed the presence of ENM-induced cell damage based on the release of lactate dehydrogenase (LDH) from cells. Six different physical characteristics were measured for each of the selected MO ENMs (TiO_2 and ZnO): primary particle size, size in water and two buffered solutions, concentration, and zeta potential. First, they performed principal component and correlation analysis on the preprocessed dataset to reveal possible correlations between the physical properties and LDH release measurements. Although a strong correlation between some of the physical features were observed, such as particle size and concentration in water, no correlation was found between the measured physical properties and cellular cell damage in the principal component analysis. Their initial intention was to use the same dataset to develop a regression and classification model. However, they were unable to develop a statistically significant regression model using the TiO_2 and ZnO dataset. The results of classification analysis were better because they managed to produce a classifier with zero resubstitution error. A clear description of the experimental design, ENM preparation, cell culture conditions, and methodology were given in the paper. The inclusion of such knowledge in toxicological research is very important because it greatly improves the interpretability of collected data and enhances its comparability with other studies. The downside of the study is undoubtedly the small number of ENMs and physical descriptors used. It is unrealistic to build a (Q)SAR model with a few ENMs because it does not allow the splitting of the original datasets into training, validation, and test sets. The number of final descriptors used to develop a (Q)SAR model can be less than six, but it is desirable to have a much larger number of initial descriptors, especially in the

Table 5.4 Previously reported nano-(Q)SAR studies

References	NPs	Descriptors	Endpoints	(Q)SAR tool	Criteria met
Sayes and Ivanov [108]	24 NP susp., 2 MOs	Size measures, conc., zeta pot.	LDH	MLR, LDA	1,2,4
Fourches et al. [33]	44NPs, diverse core	Size, relaxivities, zeta potential	ATP, Red, Apop., Mito	SVM-classification	1,2,3,4
	109NPs, diverse modifier	150 MOE descriptors	Cellular uptake	KNN-regression	1,2,3,4
Puzyn et al. [100]	17 MO-NPs	12 theoretical descriptors	EC_{50}	MLR-GA	1,2,3,4
Chau and Yap [22]	105NPs, diverse modifier	679 theoretical descriptors	Cellular uptake	NB, LR,KNN,SVM	1,2,3,4
Zhang et al. [139]	24 MO-NPs	Size, crystallinity, band gap energy, conduction/valance band, dissolution, zeta pot.	MTS, ATP, LDH, DCF, MitoSox, Fluo4, JC1, PI	Regression tree	1,2,4
Epa et al. [29]	31NPs, diverse core	Indicator variables, size, relaxivities, zeta potential	ATP, Red, Apop., Mito	MLR, SLR, feature selection, ANN	1,2,4
	109NPs,diverse modifier	691 theoretical descriptors	Cellular uptake		
Wang et al. 132	18NPs, MOs and C-based	size, shape, area, porosity, free radicals, reactivity, metal conc. And charge	LDH, Apop., Nec., Proinflammatory, Hemolysis, MTT, DiOC6, morphology	PCA	1,2,4
Liu et al. [72]	44 iron oxide core NPs	Size, relaxivities, zeta potential	ATP, Red, Apop., Mito	NBC,LGR,LDA,NN	1,2,3,4
Liu et al. [73]	24 MO-NPs	30 molecular descriptors	MTS, ATP, LDH, DCF, MitoSox, Fluo4, JC1, PI	NBC, LR, LGR, LDA, SVM	1,2,3,4

(continued)

Table 5.4 (continued)

References	NPs	Descriptors	Endpoints	(Q)SAR tool	Criteria met
Singh and Gupta [116]	44 iron oxide core NPs	Size, relaxivities, zeta potential	ATP, Red, Apop., Mito	Ensemble learning (EL)-based techniques	1,2,3,4
	109NPs, diverse modifier	691 theoretical descriptors	Cellular uptake		
	17 MO-NPs	Oxygen percent, molar refractivity, polar surface area	Cytotoxicity (EC_{50})		
	80 MWCNTs	6 topo. And geo. Descriptors	Cell viability		
	48 fullerene derivatives	10 descriptors	The binding affinity		
Kar, Gajewicz, Puzyn, and Roy [63]	109 NPs, diverse modifier	307 theoretical descriptors	Cellular uptake	GFA, MLR, PLS	1,2,3,4

absence of specific knowledge regarding the relevance of particular properties to nanotoxicity.

In another study, two different experimental nanotoxicity datasets were used to derive a mathematical relationship between the toxicity of ENMs and their physicochemical properties [33]. The advantage of the data used in this study was the concurrent testing of ENMs under the same conditions. In the first case study, three distinct clusters of ENMs were identified based on their biological activity, and support vector machine (SVM) models with high accuracies were developed. In the second case study, a descriptor quantifying lipophilicity was the most significant predictor of biological activity because it accurately discriminated between ENMs with low and high values of $PaCa_2$ cellular uptake. Overall, it was shown that the (Q)SAR approach can provide useful information for toxicity prediction of new ENMs. The methodology used in this work fulfilled all the principles of the OECD for the validation of (Q)SAR models.

Puzyn et al. [100] were one of the first to derive a mathematical equation based on the dataset of cytotoxicity and molecular descriptors. Initially, a set of 12 structural descriptors were quantum-chemically calculated using the semiempirical PM6 method. Among the pool of descriptors, only one structural descriptor ($\Delta H_{\mathrm{Me+}}$) representing the enthalpy of formation of a gaseous cation with the same oxidation state as that in the MO structure was used to establish the following nano-(Q)SAR model:

$$\log\left(1/EC_{50}\right) = 2.59 - 0.50\Delta H_{Me+}. \tag{5.2}$$

A set of 17 MO-NPs used by Puzyn et al. [100] can be considered as small from a modelling perspective, but the development of such predictive nano-(Q)SAR models is helpful to encourage new investigations.

Another simple but statistically powerful nano-(Q)SAR model was developed by Epa et al. [29] based on the results of in vitro cell-based assays of ENMs. They used the same dataset as Fourches et al. [33] with minor changes. The difference was that new descriptors encoding the presence or absence of some particular features, such as coating, were added. They managed to build the following nano-(Q)SAR equation based on these dummy variables:

$$\begin{aligned}\text{Smooth muscle apoptosis} = 2.26(\pm 0.72) - 10.73(\pm 1.05) I_{\mathrm{Fe_2O_3}} - \\ 5.57(\pm 0.98) I_{\mathrm{dextran}} - 3.53(\pm 0.54) I_{\text{surface charge}},\end{aligned} \tag{5.3}$$

where $I_{\mathrm{Fe_2O_3}}$, I_{dextran}, and $I_{\text{surface charge}}$ stand for indicators (taking values of 1 or 0) for the core material, surface coating, and surface charge, respectively. This was the second quantitative model developed to predict the toxicity of nanostructures. Compared with Eq. (2), this mathematical expression was developed from a more diverse set of data.

Recently, the hypothesis that ENM toxicity is a function of some physicochemical properties was tested by Wang et al. [132]. A set of 18 ENMs including

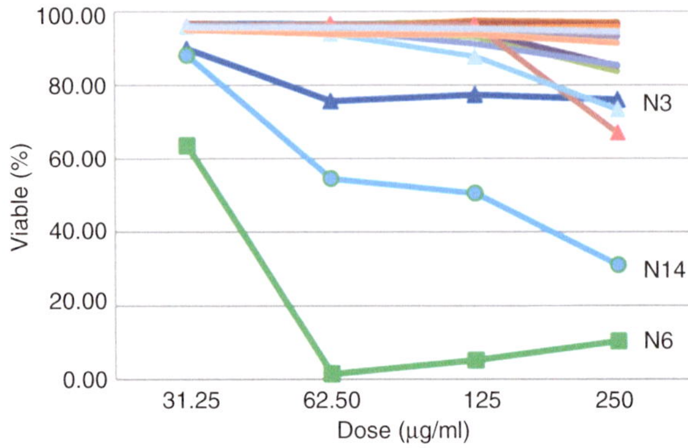

Fig. 5.6 Viability results for 18 NMs (Wang et al. [132], reprinted by permission of Taylor & Francis Ltd)

carbon-based materials and MOs were used in the study. Different types of cytotoxicity assays were performed, such as LDH, apoptosis, necrosis, hemolytic, and MTT, and several structural and compositional properties were measured. Initially, they applied PCA to the cytotoxicity data to combine the toxicity values measured at different doses into a single value that describes all the data points on the dose–response curve. It should be mentioned that, because toxicity is highly dose-dependent, the toxicological effects are usually evaluated at multiple concentrations in a series of tests, and the results are represented with a dose–response curve. Figure 5.6 shows examples of the dose–response curves obtained for the 18 ENMs. From this graph, the cell viability is lower in the cells treated with N3 (nanotubes), N14 (zinc oxide), and N6 (aminated beads) than the other ENMs. There are various methods to analyze and compare dose–response curves, such as area under the curve, slope of the curve, threshold values, min/max response, and the benchmark dose approach. In this study, Wang et al. [132] performed PCA to integrate the entire curve, and used the resulting principal components as an overall measure of cumulative response. They concluded that, compared with other approaches, PCA-based representation of the dose–response curves provides more reasonable results when ranking the ENMs according to their hazard potential. Because of the high toxicity level of four particular ENMs (zinc oxide, polystyrene latex amine, Japanese nanotubes, and nickel oxide), nano-(Q)SAR analysis focused on these four ENMs to investigate the potential factors behind their observed toxicity. It was concluded that the physicochemical characteristics leading to the toxicity of ENMs were different, and it was not possible to draw a general conclusion that was valid for all toxic ENMs screened in the study. However, the nano-(Q)SAR method was found to be useful to reveal that some of the measured properties, such as metal content, high aspect ratio, and particle charge, were correlated with the toxicity of different nanosized materials.

Liu et al. [72] developed a classification-based (Q)SAR model based on multiple toxicity assays, 44 iron oxide core NPs, and 4 simple descriptors (size, zeta potential, and relaxivities). They suggested that existing nano-(Q)SAR models did not take into account the acceptance level of false negative to false positive predictions. Unlike previously constructed nano-(Q)SAR models, they also investigated the decision boundaries of the nano-(Q)SARs subject to different acceptance levels of false negative/false positive predictions.

In another study, Liu et al. [73] attempted to relate the physicochemical properties of MO-NPs to their toxicity by developing a structure–activity relationship. A number of classification nano-(Q)SAR models were developed based on a large toxicity dataset of 24 MO-NPs. A set of 30 molecular descriptors were calculated for each NPs, and only two of them (conduction band energy and ionic index) were identified as important molecular descriptors on which the best performing nano-(Q) SAR model was built. Their conclusion was in a good agreement with the results of Burello and Worth [19], who found that the conduction band energy of oxide NPs is related to their toxicity. Similar findings have also been reported by Zhang et al. [139], who indicated that the oxidative stress induced by MO-NPs could be linked to their conduction and valance band energies.

More recently, Singh and Gupta [116] attempted to build classification and regression nano-(Q)SAR models using ensemble methods such as decision tree forest (DTF) and decision tree boost (DTB). Five different datasets were used to demonstrate and confirm the suitability of these techniques for the (Q)SAR modelling process by comparing the accuracy of the developed nano-(Q)SARs with past studies. It was concluded that the nano-(Q)SAR models constructed had high performance and statistical significance along with superior predictive ability to previous studies.

From our point of view, the common problem in the majority of published (Q) SAR studies is that it is not possible to generalize the results in the absence of explanatory information regarding the underlying reasons for the system behavior, thus making the usability of these studies limited for compounds outside the study. When the results of (Q)SAR analysis are only valid for the tested compounds, (Q) SAR becomes a data analysis tool with no predictive ability. To ensure the reliability of the established nano-(Q)SARs, researchers should also address model uncertainty arising from experimental error and lack of knowledge. Moreover, most of the existing nano-(Q)SAR studies used small datasets to establish a link between nanostructure and toxicity. Although small datasets can be useful to describe or explain the relationship between NP structure and activity, they may not be very useful for predictive purposes.

Table 5.4 summarises the previously reported nano-(Q)SAR studies and compares their methodologies with OECD principles: (1) a defined endpoint, (2) an unambiguous algorithm, (3) the applicability domain, and (4) model validation for stability and predictivity.

5.4.2 *Nano-(Q)SAR Modelling Techniques*

In principle, a variety of methods that have proven to be effective in classic (Q)SAR modelling, such as statistical methods, neural networks and decision trees, can be applied to nano-(Q)SAR. In practice, however, their direct use in ENM toxicity modelling has difficulties. The major obstacle originates from the availability of data, because some (Q)SAR algorithms require large datasets that are not currently available for ENMs. Considering the current scarcity of nanotoxicity data, it is reasonable to use modelling tools that can make effective use of smaller datasets. In addition, there is still insufficient knowledge about physicochemical descriptors that can predict the toxicity of ENMs. Therefore, current nano-(Q)SAR studies should focus on identifying toxicity-related physicochemical characteristics as well as predicting potential toxicity values. The ease of use (i.e., the ease of model building and interpretation of the results) is another important consideration, particularly in the nano-(Q)SAR world where the ability to interpret the resulting models is the key to understanding the correlation between different forms of biological activity and descriptors. Overall, the following factors have to be considered when selecting nano-(Q)SAR modelling techniques:

- Minimum data requirements. Effective use should be made of limited data without relying on the availability of large datasets.
- Transparency. Models should be transparent (rather than black-box), intuitive, and able to help identify the physicochemical descriptors that are related to the toxicity of ENMs
- Ease of model construction. The technique should be easy to use and easy to implement.
- Nonlinearity. The technique should be able to reveal nonlinear relationships/patterns in the dataset.
- Low overfitting risk. The technique should have low risk of overfitting, which may reduce the generalization of the model.
- Descriptor selection function. The technique should have the capability of feature selection to exclude redundant descriptors before model building.
- Ease of interpretation. The technique should be able to produce meaningful and interpretable outcomes and explain how the outcomes are produced.
- Low modeller dependency. The technique should have low sensitivity to changes in the model parameters.

Below, some (Q)SAR modelling methods are examined, including decision trees, statistical methods, support vector machines, neural networks, multidimensional visualisation, and knowledge-based expert systems. The focus is on discussing their suitability for nano-(Q)SAR modelling, rather than introducing the individual algorithms. Additionally, feature selection and model validation methods are briefly discussed.

5.4.2.1 Decision Trees (DTs)

Automatic generation of decision trees from data is a powerful machine learning technique that can be used as a classification or regression tool for categorical and numerical predictions of biological activity in (Q)SAR studies [76]. DTs can be constructed with small, large, or noisy datasets, and then used to detect nonlinear

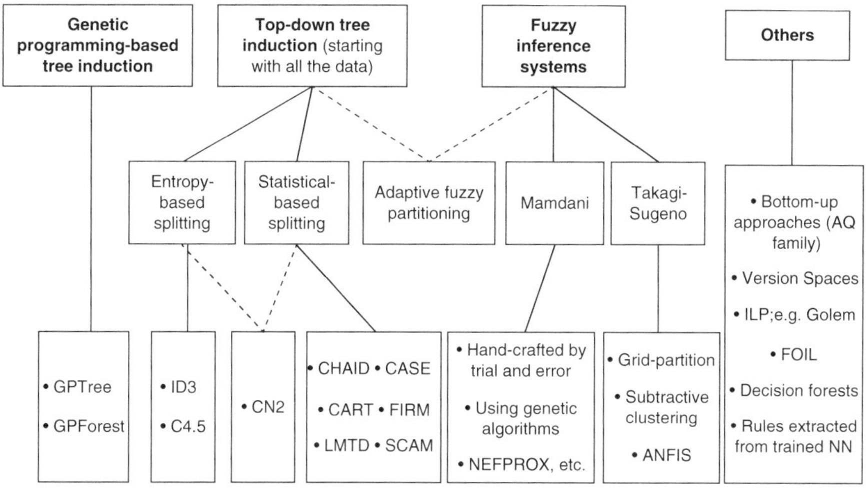

Fig. 5.7 Family tree of proposed inductive learning techniques showing a selection of specific implementations of each type (Reprinted from *Particuology*, 21, Ceyda Oksel et al., (Q)SAR modelling of nanomaterial toxicity: A critical review, 1–19., Copyright (2015), with permission from Elsevier)

relationships. They have a tree-like structure that splits data points into different classes based on decision rules to categorize and model input data. Various DT generation algorithms are available, and can be broadly classified as shown in Fig. 5.7. The most significant advantages of DT methods are their capability to automatically select the input variables (i.e., the physicochemical descriptors that contribute to the observed toxicity) and to remove descriptors that are not related to the endpoint of interest. In a previous study, Buontempo et al. [17] demonstrated the use of a genetic programming-based DT generation technique for in silico toxicity prediction. They developed a DT model containing five descriptors selected from a pool of more than a thousand descriptors that has good predictive performance for both training and test datasets. This "knowledge discovery" capability is no doubt valuable to identify the physicochemical descriptors that contribute to the toxic effects of ENMs. Such knowledge has even more benefits for eliminating or minimizing the risk of ENMs through engineering approaches (i.e., modification of physicochemical properties that influence the toxicological response through the active engineering of ENMs). Another benefit of DT analysis is its capability to avoid the (Q)SAR model being overbiased towards data in dense areas, which is a problem with some other techniques, such as linear regression and neural networks. Small data cases, i.e., data outside the dense data area, can also be modelled as branches of a decision tree. An additional advantage of DTs is the ease of their interpretability and transparency [77]. Investigation of DTs for modelling ENM toxicity requires more research, because, in addition to the abovementioned advantages, there are researchers who have voiced concerns about the generalization ability and predictive power of DTs [8]. DTs (and their extension known as "random forest") have been investigated for

(Q)SAR modelling in a number of studies [2, 3, 49, 76, 119]. Further research on DTs should focus on maximizing their advantages and overcoming their limitations. An interesting example is random decision forest, and several studies have shown its improved generalization ability over DTs [37, 77, 121].

5.4.2.2 Statistical Methods and Feature Selection

Several statistical methods, such as multiple linear regression (MLR), principal component regression (PCR), and partial least squares (PLS) regression, have been extensively studied in (Q)SAR analysis because of their ease of use and interpretation [138]. PLS is a linear regression method that handles data cases where the number of predictors is greater than the number of compounds. The PLS method works well when there are several noisy and intercorrelated descriptors, and also allows multiple responses to be simultaneously modelled . The usefulness of PLS in (Q)SAR studies, especially when the descriptors are highly correlated and numerous, has been proven by several researchers [23, 27, 30, 43, 75]. However, this method can only be used for the solution of linear regression problems. To overcome this problem, nonlinear versions of the PLS method have been developed based on different algorithms, such as kernel-based PLS [104], neural network PLS [101], and genetic algorithm-based PLS [51]. These extensions allow nonlinear relationships to be modelled in (Q)SAR studies, which is not otherwise possible with the simple PLS technique. Although MLR is one of the most common modelling techniques used to develop regression-based (Q)SAR models, there are three main factors limiting the use of MLR in nanotoxicity modelling: the linearity assumption, i.e., it cannot detect nonlinear causal relationship; the restriction on the ratio of compounds to predictors in the data, i.e., the lowest ratio of the number of ENMs to the number of descriptors should be 5:1; and the dependence of its performance on redundant variables, i.e., the presence of correlated input variables and input variables that are irrelevant to the output may lead to poor model performance [110]. Dimension reduction methods, such as PCA, can be useful for eliminating correlations between input variables (i.e., physicochemical descriptors) without removing information about irrelevant variables that may still affect the model performance. Overall, the main advantage of linear models, such as MLR and PLS, over nonlinear models is their transparency. Some information of the relative importance of the physicochemical descriptors can be directly obtained from a linear model by examining the weights, whereas some nonlinear models, such as neural networks, cannot give such direct information.

The feature selection process is different from the above mentioned dimension reduction technique, i.e., PCA, in that it selects only the inputs that have an effect on the outputs. The input variables that have little or no effect on the outputs are removed during the model building process. Among the various methods for automatic input feature selection, the genetic algorithm (GA) has shown excellent performance. The GA feature selection approach can be applied together with almost all (Q)SAR model building algorithms. The GA starts from a population of possible solutions (called individuals of chromosomes), which can be randomly generated. Each gene in the first generation of solutions consists of randomly selected descriptors. A (Q)SAR model can be built using the randomly selected descriptors in each chromosome. (Q)SAR models built based on the individuals in the initial population of solutions in this first generation are evaluated using a defined fitness function. Based on Darwin's theory of "survival of the fittest", individuals undergo operations such as mutation and crossover to generate the population of individuals in the next generation. In summary, a GA algorithm has the following essential steps:

1. Random generation of a set of solutions (the number of solutions can be set by the user) and code into a vector group with fixed length;
2. Generation of a new set of solutions by the method below, or generation of new solutions to substitute individuals in the current population;
    2.1 Selection of parent individuals based on the value of fitness function;
    2.2 Crossover to generate one or several subindividuals;
    2.3 Apply mutation operation to some individuals;
3. Repeat step (2) until one of the stopping criteria is met.

The stopping criteria are reaching the maximum number of generations or time limit, and satisfying the stop criterion for the fitness function. For more detail, please refer to (Reddy et al. [102], Goodarzi et al. [40], Ma and Wang [78]).

5.4.2.3 Support Vector Machines (SVMs)

There is increasing interest in the use of SVMs, which can handle both regression and classification problems, as an alternative to linear modelling methods such as MLR and PLS in (Q)SAR studies [24, 80]. SVMs can handle many issues that usually affect the performance of other (Q)SAR modelling techniques, such as nonlinear relationships, collinear descriptors, small datasets, and model overfitting [80]. SVMs have good potential for (Q)SAR analysis because of their accuracy and high generalization capability. On the other hand, the main disadvantages of SVMs are the high sensitivity of model performance to the selection of design parameters (e.g., kernel functions) and the complexity of direct interpretation of SVM decisions. SVMs have been used in numerous studies to construct classification [24] and regression [80, 84] based (Q)SAR models. As previously mentioned, GA-based feature selection can be integrated with SVM in (Q)SAR modelling, as shown in near-infrared chemometrics [78].

5.4.2.4 Artificial Neural Networks (ANNs)

ANNs are algorithms that imitate how the human brain works and computationally simulate human brain activity based on the neural structure of the brain. Although in some cases the poorly understood structure of this technique affects its practical reliability, successful applications of ANNs in the (Q)SAR world [48, 58, 128] keep interest in this method alive. ANNs offer several advantages to (Q)SAR developers, including the ability to deal with the nonlinear nature of structure–activity relationships and large descriptor datasets including unnecessary variables. However, ANNs also have several disadvantages, such as difficulty in interpreting the outcome, selecting the optimum complexity, risk of overfitting, and high sensitivity of the generalization power to changes in parameters and network topology. In some applications, ANN models are treated as a black-box because of their inability to give deep insight into the encoded relationship between the predictors and predicted outcomes [46]. Other studies have suggested that ANN systems should not still be seen as inexplicable models [118] because a number of methodologies facilitating the interpretation of model outcomes have been developed [7, 18, 44]. Furthermore, it should be pointed out that, like other modelling techniques, ANN can be used together with GA-based feature selection algorithms to remove redundant variables during the model building process. In addition, some researchers have investigated the use of the sensitivity analysis method for minimization of the input data dimension and extraction of information about the relative importance of inputs to an output [142].

5.4.2.5 Multidimensional Visualisation

Multi-dimensional data visualisation is an approach that allows visual exploration of high dimensional data sets in a lower-dimensional display. It significantly contributes to better understanding of the more complex statistical procedures and resulting models in relation to the dataset. It has many important applications and, in particular, can be considered as an important tool to summarise and visually explore the important characteristics of the dataset being analysed. Multidimensional visualisation techniques, such as parallel coordinates [15, 57, 131] and heat maps, are very effective tools for (Q)SAR analysis of toxicity data. They can visually display the causal relationships between ENM physicochemical descriptors and the toxicity endpoints, handle limited datasets, and allow interactive analysis with the aid of interactive functions and multiple colours built in to the software tools. To provide an example, in Fig. 5.8 the data generated by Shaw et al. [114] are scaled, displayed, and coloured using a parallel coordinates graph produced by C Visual Explorer (CVE) tool.

5.4.2.6 Knowledge-Based Expert Systems

(Q)SAR often refers to data-driven modelling. However, the usefulness of knowledge-based expert systems should not be underestimated, as evidenced by the success of the expert system DEREK of Lhasa Ltd. for toxicity predictions [42]. This expert system draws its knowledge from both literature and databases, and is considered to be one of the most powerful tools for the toxicity predictions of

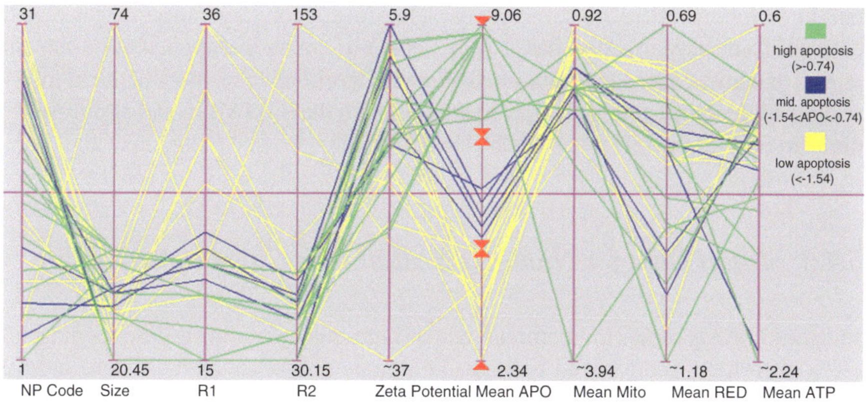

Fig. 5.8 CVE plot of the data collected by Shaw et al. [114] (descriptors: size, relaxivities (R1 and R2), and zeta potential; toxicity endpoints: apoptosis (APO), mitochondrial potential (Mito), reducing equivalents (RED), and ATP content (ATP)). The mean apoptosis data is divided into three categories: low (APO<−1.54), medium (−1.54<APO<−0.74), and high (APO>−0.74), and each category is highlighted in different colors (Reprinted from *Particuology*, 21, Ceyda Oksel et al., (Q)SAR modelling of nanomaterial toxicity: A critical review, 1–19., Copyright (2015), with permission from Elsevier)

molecules. Considering the gaps and variations in the available ENM toxicity data (i.e., incomplete characterization of physicochemical descriptors and different measures of toxicity), it is our belief that knowledge-based expert systems, ideally with some kind of "text data mining" capability that can continuously capture new knowledge appearing in the literature, might be one of the most effective approaches for nano-(Q)SAR.

5.4.2.7 Model Validation

Irrespective of the method used to construct the (Q)SAR models, the validity of the outcomes of the predictive models should be evaluated both internally and externally. Internal validation is the process of evaluating the prediction accuracy of (Q)SAR models based on the dataset used in the modelling process. The most common internal validation techniques used in (Q)SAR studies are least squares fit (R^2), chi-squared (χ^2), root-mean squared error (RMSE), leave-one-out or leave-many-out cross-validation, bootstrapping, and Y-randomization [127]. The use of external validation techniques in addition to internal validation methods is increasingly being recommended by researchers [126, 127] and authorities [86] for the assessment of (Q)SAR model reliability in the best and most trustworthy way. Moreover, it is always beneficial to use more than one validation metric to quantitatively measure the accuracy of the model prediction.

The definition of the applicability domain of the constructed and statistically validated model is the final, but one of the most important, steps in the (Q)SAR model building process. There are several approaches (e.g., geometry, range, distance, and probability density function based approaches) to define the applicability domain region of statistical models based on different algorithms. For more detailed information about the available approaches for defining the (Q)SAR model applicability domain, refer to the review papers of Jaworska et al. [59] and Sahigara et al. [106].

5.4.3 Input Data for Nano-(Q)SAR and Its Current Availability

In nano-(Q)SAR models, the importance of high-quality and well-described datasets is even more pronounced because the unique properties of ENMs are mostly associated with particular sizes and conditions [34]. Ideally, the input data required to build a reliable (Q)SAR model should be (1) obtained from a preferably single and standardized protocol, (2) examined in terms of accuracy and suitability for (Q) SAR analysis, and (3) large enough to allow rational division of the data into training and test sets. Because nano-(Q)SAR is a data-based method, the accuracy of the data determines the quality of the final model. Therefore, it is very important to create a comprehensive nanotoxicity database and make it broadly accessible.

In a recent study, Lubinski et al. [74] developed a framework to help modellers evaluate the quality of existing data for modelling (e.g., nano-(Q)SAR) purposes. In the first part of their study, they provided a set of criteria that are mostly related to the source and quantity of the data, experimental procedures, and international standards followed during the characterization process and documentation. In the second part, they assessed the quality of a collection of nanotoxicity data by scoring them according to the proposed criteria. The majority (201 out of 342 data points) of the dataset that was collected and scored was evaluated as useful with restrictions for developing (Q)SAR-like models.

In fact, there is now a great amount of data on nanotoxicity. However, the majority of the available data on ENM toxicity comes from studies focusing on a few ENMs, and hence is not useful for modelling purposes. At this point it should be noted that the data obtained by different research groups is often incomparable because of the differences in experimental procedures (e.g., sample preparation, dispersion protocols, assay types, cell types and exposure doses) and ENMs used (e.g., size, shape and surface modifications). Therefore, the data to be modelled should preferably come from the same study/project until standardized testing procedures and specific types of reference materials are available and accepted. Often, the physicochemical properties measured are not directly related to the toxicity of ENMs because characterization was carried out in the absence of a test medium.

Predefined data formats are necessary to facilitate the storage, maintenance, and exchange of ENM data between different researchers. There are a large number of freely available toxicity databases, most of which are more general in scope and not customized for particular purposes. Commercially available ENM-specific databases are still at the research stage and limited to a few applications. ISA-TAB-NANO introduced by Thomas et al. [123] is a standard NM data sharing format that facilitates the import/export of NM data and enables data exchange between different nanotechnology laboratories and researchers. The ISA-TAB-NANO specification uses four different spreadsheet-based file formats: investigation, study, assay, and material file format.

The OECD WPMN initially launched a database on Research into Safety of Manufactured Nanomaterials in 2009 [87]. However, it does not provide direct access to data because the overall outcomes and outputs section is usually filled in as "publications". Furthermore, as interest weaned, the systematic updating has been discontinued and the database put on hold.

NANOhub is a database for managing information about ENMs. It currently hosts several projects, but the access to data is usually restricted to only project participants. The experience of collecting data in NANOhub has been captured in OECD harmonised templates (OHTs) to report regulatory studies for some of the physicochemical endpoints for nanomaterials. These additional templates will also be integrated in the International Uniform Chemical Information Database (IUCLID) under REACH for registration. Another data sharing portal that provides access to ENM characterization and in vitro toxicity data is caNanoLab. The main aim of this data repository is to facilitate the sharing of knowledge on nanomedicine.

An alternative approach for collecting nanotoxicity data is to use text mining techniques to develop a customized knowledge repository system. The Nano Health and Environmental Commented Database (NHECD) is a text mining tool that allows automated extraction of information about the effects of ENMs on human health and the environment from scientific papers. However, the current performance of such NM databases using text mining algorithms is not very good because of the nonstandardized recording of ENM information and the difficulties in extracting numerical data from plots (i.e., a large amount of published data in nanotoxicity is available only in the form of plots) and. At this stage, it is important to ensure that all data is recorded in a universally agreed format to facilitate the extraction of ENM information from the literature. The existence of specifications for ENM information sharing is also very important from the viewpoint of (Q)SAR modelling, because the establishment of predictive (Q)SAR models requires close collaboration between different disciplines and research groups. The development of an agreed ontology for ENMs and nanosafety research (i.e., a formal representation of nanostructures, biological properties, experimental model systems, conditions, and protocols) will facilitate not only collection of nanotoxicity data, but also data mining and resource integration efforts.

5.5 Final Remarks

Despite the clear benefits that nanotechnology can bring to various sectors of industry, there are serious concerns about the potential health risks associated with ENMs, intensified by the limited understanding of what makes ENMs toxic and how to make them safe. As the use of ENMs for commercial purposes and the number of workers/end-users being exposed to these materials on a daily basis increases, the need for assessing the potential adverse effects of multifarious ENMs in a time- and cost-effective manner becomes more apparent. One strategy to alleviate the problem of testing a large number and variety of ENMs in terms of their toxicological properties is through the development of computational models that decode the relationships between the physicochemical features of ENMs and their toxicity. Such data-driven models can be used for hazard screening, early identification of potentially harmful ENMs and the toxicity-governing physicochemical properties, and accelerating the decision-making process by maximising the use of existing data. Moreover, these models can also support industrial, regulatory and public needs for designing inherently safer ENMs. Therefore, the idea of using time- and cost-saving computational approaches such as (Q)SAR in nanotoxicology has gained popularity in recent years and attracted the interest of regulators and researchers aiming at moving from animal-based individual toxicity assessments toward a more integrated hazard screening approach.

(Q)SAR models have been successfully used by engineers, and physical and medicinal chemists to predict hazardous properties of molecules for over 50 years. Although adaptation of the (Q)SAR approach to nanotoxicology has been encouraged by many investigators [19, 98], there are still several barriers that need to be overcome to establish predictive, reliable, and legally acceptable nano-(Q)SAR models. The current toxicity measurement methods used for bulk materials are not always fully adequate to examine ENMs and would, in any case, have to be used with due attention to the material tested. The WPMN launched a series of expert meetings to review the applicability of the OECD test guidelines to ENMs and to identify gaps in availability of test guidelines, resulting in a number of proposals to the OECD Test Guidelines Programme for updating existing guidelines and adding new ones with a view to better address the testing needs of ENMs.

As the available nanotoxicity data is far from ideal for modelling purposes, the choice of nano-(Q)SAR tools used should be made by considering the nature of the existing data (e.g. limited datasets, collinear input data) and desired outcomes (e.g. easily-interpretable models). Previous research on in silico analysis of ENMs toxicity has shown that although computerised (Q)SAR models are useful for modelling nanotoxicity endpoints, they have limited robustness and predictivity, and interpretation of the models they generate can be problematic. The main problem is caused due to the most commonly used (Q)SAR modelling methods working best with large data sets, but are not particularly good at feature selection, and cannot handle collinear input data. Ideally, new computational modelling tools or new ways of using existing tools are required to model the relatively sparse and sometimes lower quality data on the biological effects of ENMs.

Predictive models such as (Q)SAR have great potential to fill in data gaps on nanotoxicity and to be used as a priority-setting method for risk assessment of ENMs. Once all the potential risks are identified by means of toxicity screening methods including in silico models (e.g. (Q)SAR), the next step is the implementation of risk reduction measures for those risks that are outside the range of tolerable limits. While the past studies provided strong evidence that data-driven computational methods can provide useful information for hazard screening and risk assessment of ENMs, much research remains to be done on order to be able to develop optimal and regulatory acceptable nano-(Q)SAR models. Clearly, more comprehensive and high-quality datasets are necessary before obtaining optimal nano-(Q)SAR models. To improve the accuracy of computational models, quality issues associated with experimental data used to develop the model in the first place must be tackled. Moreover, the development of novel descriptors that are able to express the specificity of nano-characteristics would also be of interest. Another problem that complicates the development of predictive models is the heterogeneity of the ENM family. There is a need to generate homogeneous datasets that include specific types or individual classes of ENMs since different types of ENMs are likely to have different mechanisms of toxicity. Despite open questions and uncertainties addressed in this chapter, the results of published nano-(Q)SAR studies have provided quantitative insights leading to toxicity predictions of ENMs. To further prove the usefulness of nano-(Q)SAR approach for ENM toxicity prediction, there is a need for more case studies on high quality datasets associated with a set of ENMs with similar core composition but varying physicochemical properties (e.g. size, shape, surface charge etc.) examined under realistic and identical experimental conditions.

Acknowledgements The authors would like to acknowledge financial support from EU FP7 (Project: 236215, –Managing Risks of Nanomaterials (MARINA)) and the UK Department for Environment, Food & Rural Affairs (Project: 17857, Development and Evaluation of QSAR Tools for Hazard Assessment and Risk Management of Manufactured Nanoparticles) in support of the EU FP7 project entitled NANoREG: A common European approach to the regulatory testing of nanomaterials (FP7-NMP-2012-LARGE). We are also grateful to Kirsten Rasmussen for her comments and suggestions on an earlier version of this manuscript.

References

1. Akbari B, Tavandashti MP, Zandrahimi M (2011) Particle size characterization of nanoparticles–a practical approach. Iran J Mater Sci Eng 8:48–56
2. Andres C, Hutter MC (2006) CNS permeability of drugs predicted by a decision tree. QSAR Combin Sci 25:305–309
3. Arena VC, Sussman NB, Mazumdar S, Yu S, Macina OT (2004) The utility of structure–activity relationship (SAR) models for prediction and covariate selection in developmental toxicity: comparative analysis of logistic regression and decision tree models. SAR QSAR Environ Res 15:1–18
4. Arora S, Rajwade JM, Paknikar KM (2012) Nanotoxicology and in vitro studies: the need of the hour. Toxicol Appl Pharmacol 258:151–165
5. Asare N, Instanes C, Sandberg WJ, Refsnes M, Schwarze P, Kruszewski M, Brunborg G (2012) Cytotoxic and genotoxic effects of silver nanoparticles in testicular cells. Toxicology 291:65–72
6. Baer DR, Gaspar DJ, Nachimuthu P, Techane SD, Castner DG (2010) Application of surface chemical analysis tools for characterization of nanoparticles. Anal Bioanal Chem 396:983–1002
7. Baskin I, Ait A, Halberstam N, Palyulin V, Zefirov N (2002) An approach to the interpretation of backpropagation neural network models in QSAR studies. SAR QSAR Environ Res 13:35–41

8. Bengio Y, Delalleau O, Simard C (2010) Decision trees do not generalize to new variations. Comput Intell 26:449–467
9. Benigni R, Passerini L, Livingstone DJ, Johnson MA, Giuliani A (1999) Infrared spectra information and their correlation with QSAR descriptors. J Chem Inf Comput Sci 39:558–562
10. Bhattacharjee S, De Haan LH, Evers NM, Jiang X, Marcelis AT, Zuilhof H, Rietjens IM, Alink GM (2010) Role of surface charge and oxidative stress in cytotoxicity of organic monolayer-coated silicon nanoparticles towards macrophage NR8383 cells. Part Fibre Toxicol 7:25
11. Bootz A, Vogel V, Schubert D, Kreuter J (2004) Comparison of scanning electron microscopy, dynamic light scattering and analytical ultracentrifugation for the sizing of poly (butyl cyanoacrylate) nanoparticles. Eur J Pharm Biopharm 57:369–375
12. Borchert H, Shevchenko EV, Robert A, Mekis I, Kornowski A, Grübel G, Weller H (2005) Determination of nanocrystal sizes: a comparison of tem, saxs, and xrd studies of highly monodisperse CoPt3 particles. Langmuir 21:1931–1936
13. Boverhof DR, David RM (2010) Nanomaterial characterization: considerations and needs for hazard assessment and safety evaluation. Anal Bioanal Chem 396:953–961
14. Boyd RD, Pichaimuthu SK, Cuenat A (2011) New approach to inter-technique comparisons for nanoparticle size measurements; using atomic force microscopy, nanoparticle tracking analysis and dynamic light scattering. Colloids Surf A Physicochem Eng Asp 387:35–42
15. Brooks RW, Wilson JG (2011). Method and a system for operating a controllable multivariable process. US Patent No 7916140
16. Brown AC, Fraser TR (1868) V.—On the connection between chemical constitution and physiological action. Part. I.—On the physiological action of the salts of the ammonium bases, derived from strychnia, brucia, thebaia, codeia, morphia, and nicotia. Transact Royal Soc Edinburgh 25:151–203
17. Buontempo FV, Wang XZ, Mwense M, Horan N, Young A, Osborn D (2005) Genetic programming for the induction of decision trees to model ecotoxicity data. J Chem Inf Model 45:904–912
18. Burden FR, Winkler DA (1999) Robust QSAR models using Bayesian regularized neural networks. J Med Chem 42:3183–3187
19. Burello E, Worth A (2011) QSAR modeling of nanomaterials. Wiley Interdiscip Rev Nanomed Nanobiotechnol 3:298–306
20. Buzea C, Pacheco II, Robbie K (2007) Nanomaterials and nanoparticles: sources and toxicity. Biointerphases 2:Mr17–Mr71
21. Caballero-Díaz E, Pfeiffer C, Kastl L, Rivera-Gil P, Simonet B, Valcárcel M, Jiménez-Lamana J, Laborda F, Parak WJ (2013) The toxicity of silver nanoparticles depends on their uptake by cells and thus on their surface chemistry. Part Part Syst Charact 30(12): 1079–1085
22. Chau YT, Yap CW (2012) Quantitative nanostructure–activity relationship modelling of nanoparticles. RSC Adv 2:8489–8496
23. Cramer RD, Bunce JD, Patterson DE, Frank IE (1988) Crossvalidation, bootstrapping, and partial least squares compared with multiple regression in conventional QSAR studies. Quant Struct Act Relat 7:18–25
24. Czermiński R, Yasri A, Hartsough D (2001) Use of support vector machine in pattern classification: application to QSAR studies. Quant Struct Act Relat 20:227–240
25. Dhawan A, Sharma V (2010) Toxicity assessment of nanomaterials: methods and challenges. Anal Bioanal Chem 398:589–605
26. Domingos RF, Baalousha MA, Ju-Nam Y, Reid MM, Tufenkji N, Lead JR, Leppard GG, Wilkinson KJ (2009) Characterizing manufactured nanoparticles in the environment: multimethod determination of particle sizes. Environ Sci Technol 43:7277–7284
27. Dunn W, Wold S, Edlund U, Hellberg S, Gasteiger J (1984) Multivariate structure-activity relationships between data from a battery of biological tests and an ensemble of structure descriptors: the PLS method. Quant Struct Act Relat 3:131–137

28. Edelstein AS, Cammaratra R (1998) Nanomaterials: synthesis, properties and applications. CRC Press, New York
29. Epa VC, Burden FR, Tassa C, Weissleder R, Shaw S, Winkler DA (2012) Modeling biological activities of nanoparticles. Nano Lett 12:5808–5812
30. Eriksson L, Gottfries J, Johansson E, Wold S (2004) Time-resolved QSAR: an approach to PLS modelling of three-way biological data. Chemom Intel Lab Syst 73:73–84
31. Fadel TR, Steevens JA, Thomas TA, Linkov I (2015) The challenges of nanotechnology risk management. Nanotoday 10:6–10
32. Foldbjerg R, Dang DA, Autrup H (2011) Cytotoxicity and genotoxicity of silver nanoparticles in the human lung cancer cell line, A549. Arch Toxicol 85:743–750
33. Fourches D, Pu D, Tassa C, Weissleder R, Shaw SY, Mumper RJ, Tropsha A (2010) Quantitative nanostructure– activity relationship modeling. ACS Nano 4:5703–5712
34. Gajewicz A, Rasulev B, Dinadayalane TC, Urbaszek P, Puzyn T, Leszczynska D (2012) Advancing risk assessment of engineered nanomaterials: application of computational approaches. Adv Drug Deliv Rev 64(15):1663–1693
35. Gajewicz AEA (2012) Advancing risk assessment of engineered nanomaterials: application of computational approaches. Adv Drug Deliv Rev 64:1663–1693
36. Gallegos SA, Burello E, Worth A (2009) Review of computational approaches for predicting the physicochemical and biological properties of nanoparticles [online]. European Commission, Italy
37. Genuer R, Poggi J-M, Tuleau-Malot C (2010) Variable selection using random forests. Pattern Recogn Lett 31:2225–2236
38. Glawdel T, Ren C (2008) Zeta potential measurement. Encyclopedia of microfluidics and nanofluidics. Springer, New York
39. Glotzer SC, Solomon MJ (2007) Anisotropy of building blocks and their assembly into complex structures. Nat Mater 6:557–562
40. Goodarzi M, Saeys W, Deeb O, Pieters S, Vander Heyden Y (2013) Particle swarm optimization and genetic algorithm as feature selection techniques for the QSAR modeling of imidazo [1, 5-a] pyrido [3, 2-e] pyrazines, inhibitors of phosphodiesterase 10a. Chem Biol Drug Des 82:685–696
41. Gratton SE, Ropp PA, Pohlhaus PD, Luft JC, Madden VJ, Napier ME, Desimone JM (2008) The effect of particle design on cellular internalization pathways. Proc Natl Acad Sci 105:11613–11618
42. Greene N, Judson P, Langowski J, Marchant C (1999) Knowledge-based expert systems for toxicity and metabolism prediction: DEREK, StAR and METEOR. SAR QSAR Environ Res 10:299–314
43. Gu C, Goodarzi M, Yang X, Bian Y, Sun C, Jiang X (2012) Predictive insight into the relationship between ahr binding property and toxicity of polybrominated diphenyl ethers by PLS-derived QSAR. Toxicol Lett 208:269–274
44. Guha R, Stanton DT, Jurs PC (2005) Interpreting computational neural network quantitative structure-activity relationship models: a detailed interpretation of the weights and biases. J Chem Inf Model 45:1109–1121
45. Gurr J-R, Wang AS, Chen C-H, Jan K-Y (2005) Ultrafine titanium dioxide particles in the absence of photoactivation can induce oxidative damage to human bronchial epithelial cells. Toxicology 213:66–73
46. Guyon I, Elisseeff A (2003) An Introduction To Variable And Feature Selection. J Mach Learn Res 3:1157–1182
47. Gwaze P, Annegarn HJ, Huth J, Helas G (2007) Comparison of particle sizes determined with impactor, AFM and SEM. Atmos Res 86:93–104
48. Habibi-Yangjeh A, Danandeh-Jenagharad M, Nooshyar M (2006) Application of artificial neural networks for predicting the aqueous acidity of various phenols using QSAR. J Mol Model 12:338–347
49. Han L, Wang Y, Bryant SH (2008) Developing and validating predictive decision tree models from mining chemical structural fingerprints and high–throughput screening data in PubChem. BMC Bioinf 9:401

50. Hansch C (1969) Quantitative approach to biochemical structure-activity relationships. Acc Chem Res 2:232–239
51. Hasegawa K, Miyashita Y, Funatsu K (1997) GA strategy for variable selection in QSAR studies: GA-based PLS analysis of calcium channel antagonists⊥. J Chem Inf Comput Sci 37:306–310
52. Hassellöv M, Kaegi R (2009) Analysis and characterization of manufactured nanoparticles in aquatic environments. Wiley, London
53. Holgate ST (2010) Exposure, uptake, distribution and toxicity of nanomaterials in humans. J Biomed Nanotechnol 6:1–19
54. Hoo CM, Starostin N, West P, Mecartney ML (2008) A comparison of atomic force microscopy (AFM) and dynamic light scattering (DLS) methods to characterize nanoparticle size distributions. J Nanopart Res 10:89–96
55. Horie M, Fujita K (2011) Toxicity of metal oxides nanoparticles. Adv Mol Toxicol 5:145–178
56. Hosokawa M, Nogi K, Naito M, Yokoyama T (2007) Nanoparticle technology handbook. Elsevier, Amsterdam
57. Inselberg A (2009) Parallel coordinates: visual multidimensional geometry and its applications. Springer, New York
58. Jalali-Heravi M, Asadollahi-Baboli M, Shahbazikhah P (2008) QSAR study of heparanase inhibitors activity using artificial neural networks and Levenberg–Marquardt algorithm. Eur J Med Chem 43:548–556
59. Jaworska J, Aldenberg T, Nikolova N (2005) Review of methods for assessing the applicability domains of SARS and QSARS. Altern Lab Anim 33:445–459
60. Jeng HA, Swanson J (2006) Toxicity of metal oxide nanoparticles in mammalian cells. J Environ Sci Health A 41:2699–2711
61. Jiang J, Oberdörster G, Biswas P (2009) Characterization of size, surface charge, and agglomeration state of nanoparticle dispersions for toxicological studies. J Nanopart Res 11:77–89
62. Jiang J, Oberdörster G, Elder A, Gelein R, Mercer P, Biswas P (2008) Does nanoparticle activity depend upon size and crystal phase? Nanotoxicology 2:33–42
63. Kar S, Gajewicz A, Puzyn T, Roy K (2014) Nano-quantitative structure–activity relationship modeling using easily computable and interpretable descriptors for uptake of magnetofluorescent engineered nanoparticles in pancreatic cancer cells. Toxicol In Vitro 28:600–606
64. Karlsson HL, Gustafsson J, Cronholm P, Möller L (2009) Size-dependent toxicity of metal oxide particles—a comparison between nano-and micrometer size. Toxicol Lett 188:112–118
65. Kübart SA, Keck CM (2013) Laser diffractometry of nanoparticles: frequent pitfalls & overlooked opportunities. J Pharma Technol Drug Res 2:17
66. Kubinyi H (2008) QSAR, QSAR: Hansch analysis and related approaches. Wiley, New York
67. Laidlaw, I. & Steinmetz, M. 2005. Introduction to differential sedimentation. Analytical ultracentrifugation. The Royal Society of Chemistry, Cambridge, pp 270–290.
68. Le T, Epa VC, Burden FR, Winkler DA (2012) Quantitative structure–property relationship modeling of diverse materials properties. Chem Rev 112:2889–2919
69. Lee D-H, Cho G, Lim HM, Kim DS, Kim C, Lee S-H (2013) Comparisons of particle size measurement method for colloidal silica. J Ceram Process Res 14:274–278
70. Lin W, Xu Y, Huang C-C, Ma Y, Shannon KB, Chen D-R, Huang Y-W (2009) Toxicity of nano-and micro-sized ZnO particles in human lung epithelial cells. J Nanopart Res 11:25–39
71. Liu J, Hopfinger AJ (2008) Identification of possible sources of nanotoxicity from carbon nanotubes inserted into membrane bilayers using membrane interaction quantitative structure – activity relationship analysis. Chem Res Toxicol 21:459–466
72. Liu R, Rallo R, Weissleder R, Tassa C, Shaw S, Cohen Y (2013a) Nano-SAR development for bioactivity of nanoparticles with considerations of decision boundaries. Small 9(9–10): 1842–1852
73. Liu R, Zhang HY, Ji ZX, Rallo R, Xia T, Chang CH, Nel A, Cohen Y (2013b) Development of structure-activity relationship for metal oxide nanoparticles. Nanoscale 5(12):5644–5653

74. Lubinski L, Urbaszek P, Gajewicz A, Cronin M, Enoch S, Madden J, Leszczynska D, Leszczynski J, Puzyn T (2013) Evaluation criteria for the quality of published experimental data on nanomaterials and their usefulness for QSAR modelling. SAR QSAR Environ Res 24:995–1008
75. Luco JM (1999) Prediction of the brain-blood distribution of a large set of drugs from structurally derived descriptors using partial least-squares (PLS) modeling. J Chem Inf Comput Sci 39:396–404
76. Ma CY, Buontempo FV, Wang XZ (2008) Inductive data mining: automatic generation of decision trees from data for QSAR modelling and process historical data analysis. Comput Aided Chem Eng 25:581–586
77. Ma CY, Wang XZ (2009) Inductive data mining based on genetic programming: automatic generation of decision trees from data for process historical data analysis. Comput Chem Eng 33:1602–1616
78. Ma CY, Wang XZ (2011) Simultaneous characterization of multiple properties of solid and liquid phases in crystallization processes using NIR. Particuology 9:589–597
79. Magrez A, Kasas S, Salicio V, Pasquier N, Seo JW, Celio M, Catsicas S, Schwaller B, Forró L (2006) Cellular toxicity of carbon-based nanomaterials. Nano Lett 6:1121–1125
80. Mei H, Zhou Y, Liang G, Li Z (2005) Support vector machine applied in QSAR modelling. Chin Sci Bull 50:2291–2296
81. Monteiro-Riviere NA, Tran CL (2007) Nanotoxicology: characterization, dosing and health effects. CRC Press, Boca Raton
82. Napierska D, Thomassen L, Lison D, Martens JA, Hoet PH (2010) The nanosilica hazard: another variable entity. Part Fibre Toxicol 7:39
83. Nguyen KC, Seligy VL, Massarsky A, Moon TW, Rippstein P, Tan J, Tayabali AF (2013). Comparison of toxicity of uncoated and coated silver nanoparticles. J Phys Conf Series. IOP Publishing, 012025
84. Niu B, Su Q, Yuan X, Lu W, Ding J (2012) QSAR study on 5-lipoxygenase inhibitors based on support vector machine. Med Chem 8:1108–1116
85. Oberdorster G, Maynard A, Donaldson K, Castranova V, Fitzpatrick J, Ausman K, Carter J, Karn B, Kreyling W, Lai D, Olin S, Monteiro-Riviere N, Warheit D, Yang H, Group, A. R. F. T. I. R. F. R. S. I. N. T. S. W (2005) Principles for characterizing the potential human health effects from exposure to nanomaterials: elements of a screening strategy. Part Fibre Toxicol 2:8
86. OECD (2007) Guidance document on the validation of (quantitative)structure-activity relationships [(Q)SAR] models. OECD Environment Health and Safety Publications, P.03.03
87. OECD (2009) OECD database on research into the safety of manufactured nanomaterials. OECD Environment Directorate, Environment, Health and Safety Division, Paris/France
88. OECD (2010) Guidance manual for the testing of manufactured nanomaterials: OECD's sponsorship programme. Env/Jm/Mono(2009)20/Rev, Organization for Economic Co-operation and Development, Paris
89. Overton E (1901) Studien über die narkose zugleich ein Beitrag zur allgemeinen pharmakologie. Gustav Fischer, Jena
90. Park MV, Neigh AM, Vermeulen JP, De La Fonteyne LJ, Verharen HW, Briedé JJ, Van Loveren H, De Jong WH (2011) The effect of particle size on the cytotoxicity, inflammation, developmental toxicity and genotoxicity of silver nanoparticles. Biomaterials 32: 9810–9817
91. Park Y-H, Bae HC, Jang Y, Jeong SH, Lee HN, Ryu W-I, Yoo MG, Kim Y-R, Kim M-K, Lee JK (2013) Effect of the size and surface charge of silica nanoparticles on cutaneous toxicity. Mol Cell Toxicol 9:67–74
92. Pettitt ME, Lead JR (2013) Minimum physicochemical characterisation requirements for nanomaterial regulation. Environ Int 52:41–50
93. Poland CA, Duffin R, Kinloch I, Maynard A, Wallace WA, Seaton A, Stone V, Brown S, Macnee W, Donaldson K (2008) Carbon nanotubes introduced into the abdominal cavity of mice show asbestos-like pathogenicity in a Pilot study. Nat Nanotechnol 3:423–428

94. Powers KW, Brown SC, Krishna VB, Wasdo SC, Moudgil BM, Roberts SM (2006) Research strategies for safety evaluation of nanomaterials. Part VI. Characterization of Nanoscale Particles for Toxicological Evaluation. Toxicological Sciences, 90(2): 296–303
95. Powers KW, Carpinone PL, Siebein KN (2012) Characterization of nanomaterials for toxicological studies. Nanotoxicity. Springer, New York
96. Powers KW, Palazuelos M, Moudgil BM, Roberts SM (2007) Characterization of the size, shape, and state of dispersion of nanoparticles for toxicological studies. Nanotoxicology 1:42–51
97. Puzyn T, Leszczynska D, Leszczynski J (2009) Toward the development of "nano-QSARs": advances and challenges. Small 5:2494–2509
98. Puzyn T, Leszczynski J (2012) Towards efficient designing of safe nanomaterials: innovative merge of computational approaches and experimental techniques (no. 25). The Royal Society of Chemistry, London
99. Puzyn T, Leszczyński J, Cronin M (2010) Recent advances in QSAR studies: methods and applications. Springer, New York
100. Puzyn T, Rasulev B, Gajewicz A, Hu X, Dasari T, Michalkova A, Hwang H, Toropov A, Leszczynska D, Leszczynski J (2011) Using nano-QSAR to predict the cytotoxicity of metal oxide nanoparticles. Nat Nanotechnol 6:175–178
101. Qin SJ, Mcavoy TJ (1992) Nonlinear PLS modeling using neural networks. Comput Chem Eng 16:379–391
102. Reddy AS, Kumar S, Garg R (2010) Hybrid-genetic algorithm based descriptor optimization and QSAR models for predicting the biological activity of Tipranavir analogs for HIV protease inhibition. J Mol Graph Model 28:852–862
103. Richet C, Seances CR (1893) Comptes Rendus des Seances de la Societe de Biologie et de ses Filiales. Soc Biol Ses Fil 9:775–776
104. Rosipal R, Trejo LJ (2002) Kernel partial least squares regression in reproducing Kernel Hilbert space. J Mach Learn Res 2:97–123
105. Sadik OA (2013) Anthropogenic nanoparticles in the environment. Evnviron Sci Process Impacts 15:19–20
106. Sahigara F, Mansouri K, Ballabio D, Mauri A, Consonni V, Todeschini R (2012) Comparison of different approaches to define the applicability domain of QSAR models. Molecules 17:4791–4810
107. Savolainen K, Backman U, Brouwer D, Fadeel B, Fernandes T, Kuhlbusch T, Landsiedel R, Lynch I, Pylkkänen L (2013) Nanosafety in Europe 2015–2025: towards safe and sustainable nanomaterials and nanotechnology innovations. Helsinki, Finnish Institute Of Occupational Health
108. Sayes C, Ivanov I (2010) Comparative study of predictive computational models for nanoparticle-induced cytotoxicity. Risk Anal 30:1723–1734
109. Schaeublin NM, Braydich-Stolle LK, Maurer EI, Park K, Maccuspie RI, Afrooz AN, Vaia RA, Saleh NB, Hussain SM (2012) Does shape matter? Bioeffects of gold nanomaterials in a human skin cell model. Langmuir 28:3248–3258
110. Shahlaei M (2013) Descriptor selection methods in quantitative structure–activity relationship studies: a review study. Chem Rev 113:8093–8103
111. Sharifi S, Behzadi S, Laurent S, Forrest ML, Stroeve P, Mahmoudi M (2012) Toxicity of nanomaterials. Chem Soc Rev 41:2323–2343
112. Sharma CS, Sarkar S, Periyakaruppan A, Barr J, Wise K, Thomas R, Wilson BL, Ramesn GT (2007) Single-walled carbon nanotubes induces oxidative stress in rat lung epithelial cells. J Nanosci Nanotechnol 7:2466
113. Sharma R, Bisen D, Shukla U, Sharma B (2012) X-ray diffraction: a powerful method of characterizing nanomaterials. Recent Res Sci Technol 4(8):77–79
114. Shaw SY, Westly EC, Pittet MJ, Subramanian A, Schreiber SL, Weissleder R (2008) Perturbational profiling of nanomaterial biologic activity. Proc Natl Acad Sci 105: 7387–7392

115. Shvedova AA, Kisin ER, Mercer R, Murray AR, Johnson VJ, Potapovich AI (2005) Unusual inflammatory and fibrogenic pulmonary responses to single walled carbon nanotubes in mice. Am J Physiol Lung Cell Mol Physiol 289(5):L698–L708
116. Singh KP, Gupta S (2014) Nano-QSAR modeling for predicting biological activity of diverse nanomaterials. RSC Adv 4:13215–13230
117. Supaka N (2012) Measurement and compare particle size determined by DLS, AFM and SEM. J Microsc Soc Thai 5:38–41
118. Sussillo D, Barak O (2013) Opening the black box: low-dimensional dynamics in high-dimensional recurrent neural networks. Neural Comput 25:626–649
119. Sussman N, Arena V, Yu S, Mazumdar S, Thampatty B (2003) Decision tree SAR models for developmental toxicity based on an FDA/TERIS database. SAR QSAR Environ Res 14:83–96
120. Tantra R, Boyd R, Cackett A, Fry AT, Gohil DD, Goldberg S, Lee JLS, Minelli C, Peck R, Quincey P, Smith S, Snowden J, Spencer S, Tompkins J, Wang J, Yang L (2012) NPL report: final report on the physico-chemical characterisation of prospect engineered nanomaterials
121. Teixeira AL, Leal JP, Falcao AO (2013) Random forests for feature selection in QSPR models-an application for predicting standard enthalpy of formation of hydrocarbons. J Cheminf 5:1–15
122. Thiele G, Poston M, Brown R (2010) A case study in sizing nanoparticles [Online]. Micromeritics Instrument Corporation. Available: http://www.micromeritics.com/repository/files/a_case_study_in_sizing_nano_particles.pdf. Accessed Dec 2013
123. Thomas DG, Gaheen S, Harper SL, Fritts M, Klaessig F, Hahn-Dantona E, Paik D, Pan S, Stafford GA, Freund ET (2013) ISA-TAB-nano: a specification for sharing nanomaterial research data in spreadsheet-based format. BMC Biotechnol 13:2. doi:10.1186/1472-6750-13-2
124. Tomaszewska E, Soliwoda K, Kadziola K, Tkacz-Szczesna B, Celichowski G, Cichomski M, Szmaja W, Grobelny J (2013) Detection limits of DLS and UV-VIS spectroscopy in characterization of polydisperse nanoparticles colloids. J Nanomater 60–70
125. Tougaard S (2005) XPS for quantitative analysis of surface nano-structures. Microsc Microanal 11(S02):676–677
126. Tropsha A (2010) Best practices for QSAR model development, validation, and exploitation. Mol Inf 29:476–488
127. Veerasamy R, Rajak H, Jain A, Sivadasan S, Varghese CP, Agrawal RK (2011) Validation of QSAR models-strategies and importance. Int J Drug Design Discovery 3:511–519
128. Ventura C, Latino DA, Martins F (2013) Comparison of multiple linear regressions and neural networks based QSAR models for the design of new antitubercular compounds. Eur J Med Chem 70:831–845
129. Von Der Kammer F, Ferguson PL, Holden PA, Masion A, Rogers KR, Klaine SJ, Koelmans AA, Horne N, Unrine JM (2012) Analysis of engineered nanomaterials in complex matrices (environment and biota): general considerations and conceptual case studies. Environ Toxicol Chem 31:32–49
130. Wang XZ, Ma CY (2009) Morphological population balance model in principal component space. AICHE J 55:2370–2381
131. Wang XZ, Medasani S, Marhoon F, Albazzaz H (2004) Multidimensional visualization of principal component scores for process historical data analysis. Ind Eng Chem Res 43:7036–7048
132. Wang XZ, Yang Y, Li RF, Mcguinnes C, Adamson J, Megson IL, Donaldson K (2014) Principal component and causal analysis of structural and acute in vitro toxicity data for nanoparticles. Nanotoxicology 8(5):465–476
133. Xia X-R, Monteiro-Riviere NA, Riviere JE (2010) An index for characterization of nanomaterials in biological systems. Nat Nanotechnol 5:671–675
134. Xia XR, Monteiro-Riviere NA, Mathur S, Song X, Xiao L, Oldenberg SJ, Fadeel B, Riviere JE (2011) Mapping the surface adsorption forces of nanomaterials in biological systems. ACS Nano 5:9074–9081

135. Xiu Z-M, Zhang Q-B, Puppala HL, Colvin VL, Alvarez PJ (2012) Negligible particle-specific antibacterial activity of silver nanoparticles. Nano Lett 12:4271–4275
136. Xu R (2008) Progress in nanoparticles characterization: sizing and zeta potential measurement. Particuology 6:112–115
137. Yang Y, Guo Y, Hu C, Wang Y, Wang E (2004) Preparation of surface modifications of mesoporous titania with monosubstituted keggin units and their catalytic performance for organochlorine pesticide and dyes under UV irradiation. Appl Catal Gen 273:201–210
138. Yee LC, Wei YC (2012) Current modeling methods used in QSAR/QSPR. Statistical modelling of molecular descriptors in QSAR/QSPR, pp 1–31
139. Zhang H, Ji Z, Xia T, Meng H, Low-Kam C, Liu R, Pokhrel S, Lin S, Wang X, Liao Y-P (2012) Use of metal oxide nanoparticle band gap to develop a predictive paradigm for oxidative stress and acute pulmonary inflammation. ACS Nano 6:4349–4368
140. Zhao C-M, Wang W-X (2012) Importance of surface coatings and soluble silver in silver nanoparticles toxicity to *Daphnia magna*. Nanotoxicology 6:361–370
141. Zhou H, Mu Q, Gao N, Liu A, Xing Y, Gao S, Zhang Q, Qu G, Chen Y, Liu G (2008) A nanocombinatorial library strategy for the discovery of nanotubes with reduced protein-binding, cytotoxicity, and immune response. Nano Lett 8:859–865
142. Zurada JM, Malinowski A, Cloete I (1994) Sensitivity analysis for minimization of input data dimension for feedforward neural network. Circuits and systems. Iscas'94. IEEE international symposium on, pp 447–450

Chapter 6
Systems Biology to Support Nanomaterial Grouping

Christian Riebeling, Harald Jungnickel, Andreas Luch, and Andrea Haase

Abstract The assessment of potential health risks of engineered nanomaterials (ENMs) is a challenging task due to the high number and great variety of already existing and newly emerging ENMs. Reliable grouping or categorization of ENMs with respect to hazards could help to facilitate prioritization and decision making for regulatory purposes. The development of grouping criteria, however, requires a broad and comprehensive data basis. A promising platform addressing this challenge is the systems biology approach. The different areas of systems biology, most prominently transcriptomics, proteomics and metabolomics, each of which provide a wealth of data that can be used to reveal novel biomarkers and biological pathways involved in the mode-of-action of ENMs. Combining such data with classical toxicological data would enable a more comprehensive understanding and hence might lead to more powerful and reliable prediction models. Physico-chemical data provide crucial information on the ENMs and need to be integrated, too. Overall statistical analysis should reveal robust grouping and categorization criteria and may ultimately help to identify meaningful biomarkers and biological pathways that sufficiently characterize the corresponding ENM subgroups. This chapter aims to give an overview on the different systems biology technologies and their current applications in the field of nanotoxicology, as well as to identify the existing challenges.

Keywords Grouping • Nanomaterials • Transcriptomics • Proteomics • Metabolomics • Systems biology

---

C. Riebeling • H. Jungnickel • A. Luch • A. Haase (✉)
German Federal Institute for Risk Assessment, Department of Chemical and Product Safety, Berlin, Germany
e-mail: andrea.haase@bfr.bund.de

L. Tran et al. (eds.), *Modelling the Toxicity of Nanoparticles*, Advances in Experimental Medicine and Biology, Vol. 947, DOI 10.1007/978-3-319-47754-1_6

6.1 Introduction

Engineered nanomaterials (ENMs) are becoming a mainstream technology in modern product design due to their unique physico-chemical properties. An ever growing number of ENMs are used worldwide in very diverse products and applications such as construction, food packaging, cosmetics, textiles or medicines. ENMs can enhance mechanical properties for instance in concrete, facilitate cleaning of surfaces in paints, enhance gas barrier capabilities in beverage packaging, block ultraviolet radiation from human skin in sunscreens, or produce self-healing surfaces. With the widespread use of ENMs possible health risks for humans must be addressed properly. One of the obvious challenges is how to assess the multitude of already existing and newly emerging ENMs in a reasonable time frame in a reliable and relevant manner. Thus, in nanotoxicology there is an urgent need of powerful prediction tools, which can ultimately support decision making with respect to prioritization and facilitate ENMs grouping or categorization. Systems biology in combination with predictive statistical tools may become a central piece of the future nanotoxicological toolbox as it will allow for in-depth understanding of affected pathways and at the same time support and facilitate grouping on the basis of the mode-of-action of ENMs (Fig. 6.1).

A thorough understanding of affected pathways, so called "toxicity pathways", may then lead to the discovery of "adverse outcome pathways" (AOPs). The OECD [94] described the concept of AOPs in 2013, which integrate toxicological key events and describe in a cascade-like way the main steps from contact of the respective hazardous chemical to the ultimate adverse outcome. Importantly, AOPs do not only describe molecular and cellular events but also integrate data from tissues and whole organs [146]. Nowadays, an increasing number of toxicological endpoints can be described by AOPs as knowledge on underlying toxicity mechanisms is growing. Examples for well-defined AOPs are skin sensitization, cholestasis, liver fibrosis or liver steatosis. AOPs are not restricted to environmental chemicals, but may also be applied in nanotoxicology. However, so far examples in the field of nanotoxicology are still lacking. Thus, the use of systems biology and the description of AOPs for ENMs are of paramount interest in nanotoxicology. This would not only support the development of new and highly specific testing methods but also would allow grouping of ENMs. In a minimalistic view it would help to identify ENMs which require further testing, as well as ENMs where in-depth evaluation may not be needed [101]. Thus, the use of systems biology approaches in combination with modern statistical methods would also significantly reduce the number of animals for nanotoxicological testing and therefore implement the 3R paradigm (replacement, reduction and refinement) as first introduced in 1959 as a cornerstone of modern toxicology [116]. It also should be noted that the predictivity of animal models towards human adverse health effects is often limited [134]. It may be expected that knowledge on AOPs, however, will facilitate species-to-species extrapolation and thus the application of data obtained from animal models to humans, and also from animal models to in vitro approaches and vice versa [101].

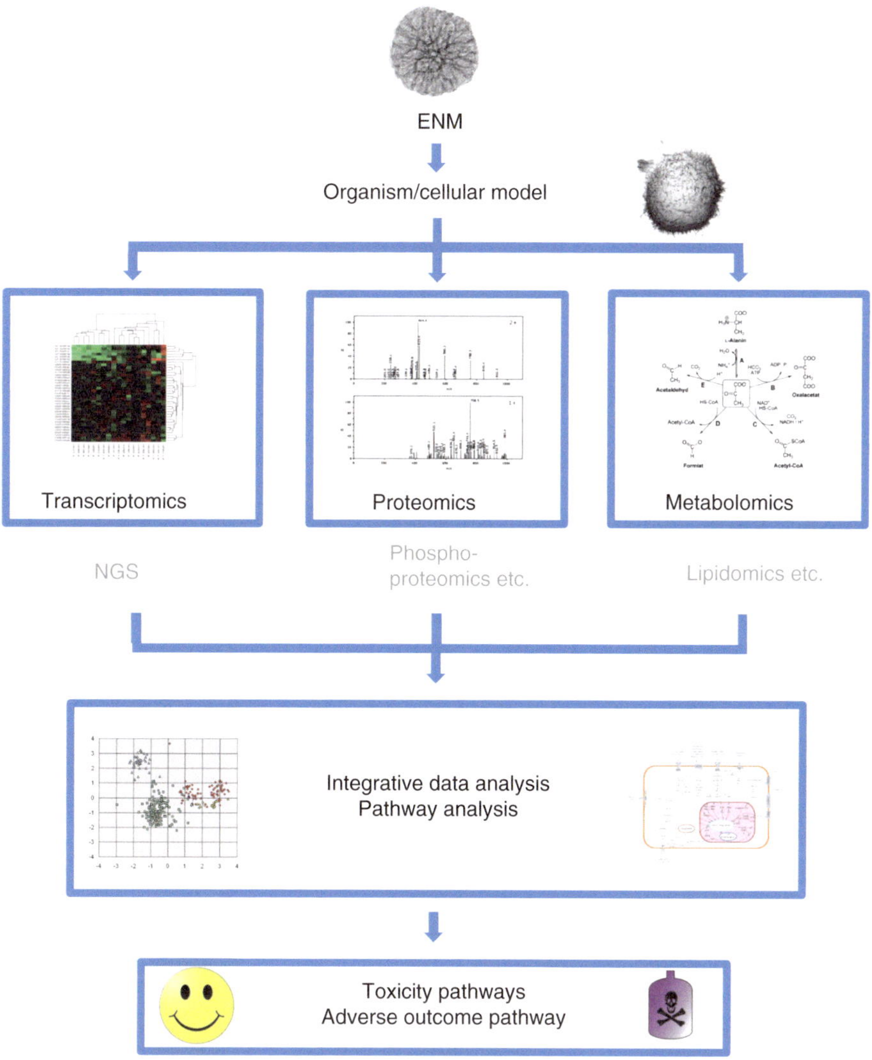

Fig. 6.1 Schematic work flow of systems toxicology (All images from Wikimedia, authors from top to bottom, left to right: Nandiyanto; Mikael Häggström; Dr. Timothy Triche, National Cancer Institute; Miguel Andrade; RonBeavis; Yikrazuul; Jheald; Boghog2; Pumbaa80; Nikn)

The goal in animal based experiments is to determine the highest possible dose at which a given substance or ENM does not cause any adverse effect, also referred to as "no observed adverse effect level" (NOAEL) together with the lowest dose that would cause a pathological effects after a defined treatment period, which is referred to as "lowest observed adverse effect level" (LOAEL). By applying empirically derived assessment factors to take into account the uncertainty from inter- and intra-species extrapolation hazard reference values like the "derived no effect level"

(DNEL) can be inferred, which are then employed for human health risk assessments. The use of NOAEL and DNEL values is applicable for toxicological effects with a threshold. In general, similar principles apply for ENMs despite the fact that some of the established guidelines may require additional adaptations for the assessment of ENMs.

Information about the molecular mechanisms of the observed adverse effects is currently considered a nice add-on and is no requirement for risk assessments. Traditionally, mechanistic insights are gained by hypothesis-driven research investigating perturbations after treatment with specific xenobiotics within one pathway at a time. Connections within and between pathways are inferred from additional experiments and from literature. This targeted approach has unravelled most of the vast molecular knowledge of biological systems available today.

Xenobiotics (and also ENMs) usually interact with multiple molecular targets in a biological system, often resulting in changes in cellular RNA molecules, proteins, lipids or metabolites. Often, affected proteins interact with other proteins and so on, creating networks of alterations that may even often overlap starting from different primary targets. The surface area of ENMs is large enough to provide the possibility for multiple interactions of a single nano-object with several biomolecules at the same time. The binding of molecules on the ENM surface may be used by the manufacturer to specifically alter the surface or may occur in a less specific and controlled way during application as soon as the ENM enters a biological environment [21], which results in the latter case in the formation of a biomolecule corona. Thus, different interactions are possible on a single nano-object. Due to interactions with ENMs biomolecules such as proteins might be denatured and/or presented out of context in the organism [37, 135]. In addition, on some ENMs surface reactions occur that generate secondary molecules that may interact with further undirected target molecules. For instance, reactive oxygen species may be generated due to surface catalysed reactions on ENMs that cause oxidative stress, which may lead to metabolite oxidation, DNA damage, and protein carbonylation [117]. Advances in life-sciences research in the fields of "omics" technologies provide tools to address the complexity of possible perturbations using descriptive approaches that record system-wide changes. The "omics" technologies collect very large data sets, often in a quantitative manner which mirror molecular responses in the genome, transcriptome, proteome and metabolome after exposure to environmental chemicals in living organisms. The four major approaches are genomics, transcriptomics, proteomics, and metabolomics. Examples of other more specialized fields are lipidomics and phosphoproteomics. Each of these technologies produces a static snapshot in one point of time, which represents only a small part of the whole organism even for the specific biomolecule class monitored. This limitation has to be considered in experimental design and data assessment such that for instance different time points are included. Genomics and transcriptomics are the predominantly used technologies in toxicological research. Both initiated the field of so-called toxicogenomics [90, 91]. Especially high-throughput micro-arrays are a valuable an often applied tool for the rapid screening and interpretation of toxicogenomic data, which subsequently can be used to define not only multiple modes of actions, but can ultimately in combination with other data be used to define adverse outcome pathways [4, 64, 109, 150]. Due to obvious similarities and common requirements for data analysis, data integration, and interpretation the use of "omics" technologies in toxicological research are summarised and commonly referred to as systems toxicology [148]. Systems toxicology usually applies methods of systems biology to toxicological problems and combines this knowledge with classical toxicological approaches. This allows for measurement of large networks of molecular and functional changes within one organism. The changes are recorded throughout multiple levels within a biological entity [122]. To this end, data from in vivo and in vitro experiments can be combined and also the combination of high-throughput data with "omics" proved useful [105]. Combining systems biology results with modern statistical approaches for data interpretation may support the elucidation of adverse outcome pathways. So far, systems biology is only beginning to be applied in nanotoxicology and currently only a few published studies are available. However, in particular in nanotoxicology the knowledge about affected pathways may help to design new test methods. Currently, there is high need of new test methods, which are suitable to screen many ENMs and which can support decision-making and prioritization. Furthermore, systems biology can substantially support the development of ENM grouping approaches based on the mode of action. In the next paragraphs we will give an overview of the different "omic" approaches and give examples how they have already been applied in nanotoxicology.

6.2 Transcriptomics

Gene-expression profiling, or transcriptomics, determines the changes in expression of mRNAs, rRNA, tRNA and other non-coding RNA molecules in a cell population [39], tissue/organ [14], or organism [136]. Transcriptomic analysis is arguably the best established and most widely used approach to investigate biological network responses [74]. Transcriptomics advanced with the development of oligonucleotide microarrays and the introduction of high-density array printing by Affymetrix. Limited by the known genomic sequence of the organism and using well-established bioinformatics tools to identify possible open reading frames, oligomer probe arrays can be designed covering the whole transcriptome. Clinical or environmental samples as well as samples derived from model systems can be investigated in high-throughput parallel analyses on these microarrays [82]. Current microarrays can cover each gene or its exons. Standard exon arrays are available for human, mouse, and rat. Especially the analysis of alternatively spliced RNA transcripts as well as the accuracy of the overall gene-expression has been greatly improved by the use of exon arrays [152]. However, with the establishment of next-generation sequencing (NGS) technologies it is now possible to achieve full cDNA sequencing and RNA sequencing of a cell or tissue in one analysis in a reasonable time frame. As NGS technologies are becoming more powerful and affordable they may replace microarray technologies in the near future [33, 97]. NGS approaches are expected to offer greater accuracy. In particular they provide transcript counts similar to quantitative PCR. Moreover, NGS methods are more flexible. They also allow for gene-expression studies in organisms for which microarrays are not available, such as many model systems used in environmental toxicology. NGS also allows for the assessment of the microbiome and its population responses to exposures [158].

6.3 Transcriptomics in Nanotoxicology

From all "omics" technologies transcriptomics is applied most often to study effects of ENMs due to the fact that approaches are well established in many laboratories. However, the amount of available transcriptome data for ENMs is still very low. Most available studies investigate only one ENM or a very small set of ENMs.

Transcriptomics has been applied in the nematode *Caenorhabditis elegans* to study the effects of gold nanoparticles (NPs), which were shown to induce transcriptional changes in the unfolded protein response [136]. Moreover, a comparison to previous data on silver NPs revealed diverse responses to the two ENMs [136].

Another study investigated carbon black Printex 90 NPs applied by instillation in mice. The authors found persistent elevated cytokine expression in dams and changes in liver mRNA of offspring at high doses [60]. A similar study revealed alteration of the hepatic cholesterol synthesis pathway in adult mice [14]. The latter study was combined with gene expression data of disease related studies for a subsequent bioinformatics analysis which revealed a similarity of carbon black induced effects and pulmonary injury and fibrosis [15].

Two types of multi-walled carbon nanotubes (MWCNTs) were instilled in mice showing overall similar transcriptional responses [107]. Larger MWCNTs exhibited an earlier stronger inflammatory response and stronger fibrosis [107].

Surface-modified TiO_2 NPs induced elevated cytokine transcript levels and also of several miRNAs after short term inhalation of mice [51].

No effects were found for coated and aged TiO_2 NPs on Caco-2 cells including in microarray analysis [36].

Comparing the effects of anatase and rutile NPs and bulk TiO_2 NPs in *Caenorhabditis elegans* revealed different expression patterns for the different materials [114]. However, nano- and bulk form of the materials exhibited similar profiles [114]. It was observed that anatase particles exerted a greater effect on metabolic pathways, whereas rutile particles had a greater effect on developmental processes [114].

Instilled TiO_2 NPs of different sizes and surface modifications in mice led to an overall similar inflammatory response by transcriptional analysis [52]. While this points to a common mechanism, closer analysis showed that the magnitude of the response was dependent on the ENM surface area [52].

Nephrotoxicity of nanoscale and microscale copper particles was addressed by a study in rats by gavage, demonstrating that a high dose of nanoscale but not microscale copper (by mass) induced strong transcriptional and necrotic responses

[75]. A lower dose induced a smaller, partially overlapping, set of transcripts which were suggested as possible low dose indicators of toxicity.

Differential effects of silver ions, citrate-coated, and PVP-coated silver NPs were uncovered in an ecotoxicological study in *Daphnia magna* [108]. Silver ions exhibited a clearly different expression profile to the NPs, and PVP-coated NPs elicited a response in DNA damage repair genes and an overall stronger response than citrate-coated NPs [108].

Similarly, comparing the effects of silver ions to citrate-coated silver NPs in *Oncorhynchus mykiss* showed only a small number of specifically regulated transcripts [40]. However, linear discriminant analysis was able to separate both forms of silver [40].

The gender difference in rats exposed to silver NPs was investigated in kidneys, showing a higher expression of genes involved in xenobiotic metabolism in males and of cellular signalling in females [28].

Different types of polystyrene NPs and carbon nanotubes were compared in a human endothelial cell line [39]. Inflammation, oxidative stress, and DNA damage were the most regulated processes [39]. The more cytotoxic particles induced more transcriptional changes while the presence of serum decreased overall cytotoxicity but had little effect on the top regulated transcripts [39].

6.4 Proteomics

The proteome encompasses the full complement of proteins in a cell [89], biological liquids [67], tissue/organ [49], or organism [69]. Proteomics is the systematic approach to characterizing ideally all proteins. However, the broad spectrum of physico-chemical properties of proteins dictates that only a part, albeit a large part, of all proteins is detected with current technologies. Thus, for some purposes a subset of proteins first needs to be selectively enriched. In proteomics an additional complexity arises due to the fact that not only changes in overall protein amounts are of interest but also the measurement of posttranslational protein modifications. Proteins are both acceptors and mediators of altered biological responses as a consequence of exposure to substances. Changes in protein levels may correspond directly to mRNA expression or may be due to post-transcriptional regulation such as regulated translation or regulated proteolytic turnover. In addition, altered protein function may be a consequence of posttranslational modifications. For instance, protein phosphorylation can be addressed by high-throughput phosphoproteomics to characterize molecular events proximal to disease-related signalling mechanisms [32, 88]. Moreover, oxidation events due to oxidative stress in response to toxic exposures, the most discussed possible effect of ENMs, can be described by redox proteomic analysis for instance by assessing carbonylated proteins [143].

With its unmatched sensitivity and throughput, mass spectrometry (MS) is the key technology of modern proteomics. It allows for the detection of peptides in biological samples in the sub-femtomolar range with a mass accuracy of less than 10 ppm. Such a high degree of accuracy is requisite to comparisons between samples of proteins derived from different exposed and control samples. For non-targeted comparisons of samples usually isotope tagging for relative and absolute quantification (iTRAQ) is used. Even higher accuracy can be achieved by complementation with a targeted method. Typically in such a combination the untargeted assessment is used to unravel possible alterations, which are then followed up in a targeted, more precise measurement. Often selected reaction monitoring (SRM) is used for this purpose, which allows for the precise quantification of predetermined proteins. Peptides are generated by a controlled enzymatic digestion of the proteome and quantified by MS. The selection of proteins and peptides for precise quantification is done either by prior, non-targeted approaches (e.g. iTRAQ) or by a careful review of available data in the scientific literature. Bioinformatics tools are employed to predict the cleavage pattern of the selected protein. From this list at least two proteolytic peptides are selected for SRM. Those ideally should be highly specific for the protein and distinguish it from all other proteins. SRM transitions for each of those peptides are selected and unique identification and accurate quantification have to be verified through optimization and validation. Multiplexed approaches are possible where hundreds of proteins are quantified in a single MS run [72]. For example, the Nrf2-mediated stress response of macrophages to oxidised LDL was investigated demonstrating up-regulation of a group of antioxidant proteins [65].

Often proteomics approaches require the enrichment of a protein subset. Most often this is done by cellular sub-fractionation (e.g. by differential centrifugation) or by antibody-based pulldown approaches. In particular, antibody-based enrichment is used for assessing posttranslational modifications of proteins. Posttranslational modifications of proteins are important cellular mechanisms for regulating and diversifying the cellular proteome. Identification and characterization of this layer of cellular regulation can provide deeper insight into the cellular physiology and affected pathways in response to toxic insults. Examples of posttranslational modifications include phosphorylation, glycosylation, ubiquitination, nitrosylation, methylation, acetylation, lipidation and proteolysis [159]. For enrichment, phosphorylation motif specific antibodies [83], di-glycine-lysine-specific antibodies for ubiquitinated peptides [147], and other antibodies are employed. To enrich phosphorylated peptides after digestion of the protein lysates TiO_2 resins are commonly used [131]. Another strategy is chemical labelling of proteins and immuno-detection after 2D gel electrophoresis. This approach is used in redox proteomics to detect protein carbonylation. In oxidative stress proteins become directly and indirectly oxidised generating various carbonyl groups. These protein carbonyl groups are conjugated to 2,4-dinitrophenylhydrazine and subsequently detected using anti-2,4-dinitrophenyl antibodies [153].

Proteomics and subsequent targeted proteomics yield important mechanistic details of toxicology pathways that were based on transcriptomic data.

6.5 Proteomics in Nanotoxicology

There are several studies using proteomics techniques to investigate toxicological effects of ENMs. However, similar to transcriptomics studies described above, most employ only one specific ENM and often they furthermore assess only a single concentration and/or one time point. While these studies provide important insights into ENM effects, the difference in ENM characteristics and in experimental settings makes it difficult or impossible to compare them in order to draw more general conclusions on ENM influences on the proteome and to understand how ENM properties influence toxicity.

Carbon nanotubes are one of the most investigated ENM. By comparing as-grown MWCNTs with thermally treated MWCNTs it was demonstrated that impurities were in large part responsible for the observed cytotoxicity [53]. However, stress response proteins were induced also by thermally treated MWCNTs [53]. A comparison of SWCNTs with graphene showed SWCNTs inducing proteins related to oxidative stress while graphene had little effect [156]. Lung tissue was investigated after a repeat-dose instillation of mice with SWCNTs, asbestos, and carbon black [130]. SWCNTs elicited the strongest response in regulated proteins with a similar profile to asbestos [130]. In a renal cell model, fullerenes, SWCNTs, and MWCNTs induced the most proteins in the lowest dose, suggesting that aggregation reduces the effect on cells [11]. Oxidized SWCNTs induced oxidative stress and interfered with intracellular metabolic routes, protein synthesis, and cytoskeletal systems in HepG2 cells [155]. Graphene oxide as a comparison had little effect on protein expression and was less cytotoxic [155]. Serum-free and surfactant-treated MWCNTs were compared for their effect on human aortic endothelial cells [144]. Different protein expression patterns were observed between the two suspensions with the eIF2 pathway as the only common pathway [144]. Lung tissue was investigated after repeat-dose instillation of rats with three different ENMs, Fe_3O_4, SiO_2, and SWCNTs [77]. Seventeen commonly regulated proteins were identified and the authors suggest all three ENMs induce lung damage [77]. The secreted proteins by a macrophage model in response to MWCNTs and asbestos were investigated by proteomics [98]. Long rigid MWCNTs and asbestos showed similarities while tangled MWCNTs exhibited only limited overlap with rigid MWCNTs. All materials showed release of lysosomal proteins while only for rigid MWCNTs apoptosis-related proteins were secreted [98]. Levels of proteins in lung bronchoalveolar lavage fluid of mice treated by oropharyngeal aspiration with uncoated or aluminium oxide coated MWCNTs were determined to uncover their effect on lung tissue [54]. Uncoated MWCNTs elicited a stronger response but in similar pathways to coated MWCNTs [54].

Another heavily researched ENM is nanosilver. A comparison of PVP-coated silver NPs versus $AgNO_3$ on plants revealed differing protein expression profiles, while redox regulation and sulphur metabolism were affected by both [142]. Similarly, silver NPs and $AgNO_3$ were tested in mussels showing different protein expression profiles yet similar affected pathways [45]. The authors suggest that toxicity of NPs was mediated by oxidative stress-induced cell signalling cascades [45]. Citrate capped 20 and 100 nm silver NPs were compared on a colon cell line [143]. While overall the same pathways were affected (e.g. DNA damage repair), more proteins were affected by the 20 nm NPs [143]. Reanalysis of the proteomic data revealed that proteins involved in cell death and mitochondrial activity were more affected by 20 nm NPs than by 100 nm NPs, while proteins involved in cell growth were affected similarly by both particle sizes [84]. Carbonylated proteins as a marker for oxidative stress were investigated in *Daphnia magna* after treatment with silver NPs or $AgNO_3$ [110]. Different profiles were found for the two treatments [110].

Recently, a large redoxproteomics study was published, which assessed protein carbonylation for a panel of 24 different ENMs [30]. The results reaffirmed that oxidative stress is a major affected pathway in response to cellular ENM exposure, and that protein carbonylation is a promising readout for this pathway.

6.6 Metabolomics

Metabolomics is the latest "omics" technology in the "omics" toolbox. Changes in metabolome are generally regarded to give an as-close-as-possible picture of the actual phenotype changes of the organism. The metabolome represents the ultimate change in the levels of chemical species usually resulting from molecular perturbations at the genomic and proteomic levels. Thereby, the metabolome ultimately represents the functional status of a cell.

This approach tries to quantify as many metabolites within the target organism as possible, e.g. sugars, lipids, steroids, amino acids, carnitines, nucleotides etc. To that end, hyphenated analytical techniques are applied, especially the combination of mass spectrometry with quantitative NMR is very common. The investigated metabolites encompass mostly products or substrates of enzyme-mediated processes [13]. It is also possible to detect and quantify internalized xenobiotics and their biotransformation products concomitant to the perturbed endogenous metabolome if the molecular size of the xenobiotic chemical is low enough. However, the latter requires some understanding of the kinetics of the xenobiotic toxicants and their metabolites as well as of related biomolecular adducts [111]. In a conventional approach, as many as possible metabolites are identified and changes in their abundance are quantified.

The biological matrix for metabolomics experiments can be very different and also highly complex. Generally a broad variety of different biological systems like cell cultures [12], 3D cell cultures, tissue samples [96] and whole organ cultures [76] including the emerging application area of microfluidic organ model systems [7] can be investigated for the assessment of the metabolome. Metabolomics experiments can also be performed with different body fluids like bronchoalveolar lavage fluid (BALF) [23] or serum. Even whole organisms can be assessed [112]. Due to the fact that the metabolome is very complex as it covers very different biomolecules it is not possible to use a single analytical technique to cover all metabolites within one biological system in one analysis. A variety of mass spectrometry methods (GC-MS, GCxGC-MS, LC-MS, Ion-mobility-MS) or quantitative NMR are used so far to assess quantitative changes in the metabolome after exposure to environmental chemicals. All these methods can also be used for the assessment of the metabolome in exposure assays to ENMs in nanotoxicology. Metabolomics has been also successfully used for the assessment of no observed adverse effect levels (NOAEL) in toxicology with similar or even higher sensitivity than common toxicological methods commonly used [140]. However, there are also pitfalls in the system, where metabolomics data show lower sensitivity than commonly used toxicological approaches. An overarching analysis found that 18 % of all investigated cases showed lower sensitivity than common toxicological approaches [140], indicating the need to combine not only results from other "omics" technologies, but also to combine "omics" approaches with established toxicological approaches in an integrated manner.

Of particular interest in toxicology are the assessment of metabolomic changes in animals and humans. This is closely linked to the identification of biomarker sets correlated with certain diseases like diabetes [123] or kidney disease [31]. For biomarker identification it is a paramount requisite that the sampling should be as easy as possible. Therefore, either metabolic profiles from blood [19], urine samples or breath samples, e.g. for the assessment of bronchoalveolar infections, [38] are investigated. In addition, tissue samples [29] or organ biopsy samples [44] are studied for the assessment of quantitative metabolite changes associated with various diseases or disease states. In particular, this is used for cancer diagnosis to distinguish non-malignant from malignant tissue samples or for stage determination of cancer.

6.7 Metabolomics in Nanotoxicology

So far, only a limited number of metabolomics studies investigated the influence of ENMs. Again, similarly to the situation in transcriptome or proteome research in most of the studies only one ENM is investigated.

Many studies focused on the effects of TiO_2 NPs. One study analysed metabolomics changes in human skin cells (HaCaT cells) after exposure to TiO_2 NPs and metabolite changes could be associated with oxidative stress and influenced mitochondrial activity [137]. Another study also tested TiO_2 NPs (anatase, 18 nm) in HGF cells [42], and subsequent metabolomics studies revealed an increase in prostaglandin levels within these cells after exposure together with an reduction of amino acid, urea cycle, polyamine, S- adenosylmethionine and glutathione biosynthesis. Metabolomics studies of mouse fibroblasts [12] showed a significant disturbance of the amino acid signature after exposure to colloidal nano-TiO_2 solutions and that these disturbances could be correlated to the observed cytotoxicity of the ENM. Urine and serum were investigated by metabonomics in rats exposed orally to TiO_2 NPs [18]. Disturbances in energy and amino acid metabolism and the gut microflora environment were found [18]. Intratracheally instilled TiO_2 in rats induced metabolite changes in serum that indicated slight liver and kidney injury which was corroborated by clinical chemistry [124].

Other intensely studied ENMs are nanosilver and nanosilica. Metabolomics was already used to assess metabolic profile changes in rats after oral gavage of silver ions and silver NPs [50]. The results showed that nanosilver increased not only uric acid levels, but also allantoin levels in rat urine.

Metabonomics was highlighted as a potential and robust non-destructive tool for monitoring the temporal effect of NPs in cell culture media [59]. The metabonomic assay revealed pronounced effects of SiO_2 NPs in lung alveolar A549 cells on glucose, lactate, histidine, phenylalanine, and tyrosine at early time points when cell viability was not impaired. Moreover, the data suggest that the different sizes of NPs induced different dose-dependent effects with different time courses [59]. The study also showed a dose-dependent increase of ROS formation. Different sizes of SiO_2 particles were intravenously injected in mice and liver tissues and serum analysed by integrated metabonomics analysis [80]. Disturbances in energy metabolism, amino acid metabolism, lipid metabolism, and nucleotide metabolism were reported that may be attributable to the observed hepatotoxicity. No major differences were found by the different NP sizes among the metabolite profiles. Surface area had a greater effect than particle number on toxicity [80]. Metabolite perturbations after intranasal SiO_2 NP application in rats implicated impairment in the tricarboxylic acid cycle and liver metabolism [99]. The authors suggested from their data that SiO_2 NPs may have a potential to induce hepatotoxicity in rats [99].

Other ENMs are less often studied. Metabolic responses to MnO NPs in biofluids (plasma and urine) and tissues (liver, spleen, kidney, lung and brain) from rats could be divided into four classes: MnO biodistribution-dependent, time-dependent, dose-dependent and complicated metabolic variations [73]. Particle size and surface chemistry of NPs were correlated to changes in the metabolic profile [73]. Single-walled carbon nanotubes after intratracheal instillation in rats induced changes in blood plasma and liver tissues indicating liver injury [76]. Changes in lipids and lipid associated molecules suggested a mechanism involving oxidative stress [76]. Iron oxide NPs were intravenously injected in rats, and metabonomics analysis performed on urine and plasma [35]. Subtle metabolic changes in response to NPs were found in a number of metabolic pathways including energy, lipid, glucose and amino acid metabolism [35]. The authors followed up their investigation by analysing tissues, including kidney, liver and spleen [34]. The metabonomics analysis demonstrated correlations between biofluids and tissues in their response to NPs [34]. Size and surface chemistry of the NPs affected their biological effects [34, 35]. Another study investigated the metabolic changes caused by antimicrobial effects of carboxyl-capped bismuth NPs in *Heliobacter pylori* colonies [86]. The results showed an increased release of acetate, formic acid, glutamate, valine, glycine, and uracil into the culture medium after NP treatment, indicating perturbations of various metabolic pathways like the Krebs cycle and nucleotide and amino acid metabolism. Thus, metabolomics in combination with other "omics" technologies could also give insights for new ENM applications, e.g. by discovering antimicrobial effects.

A few studies only investigate several ENMs at the same time. One study describes metabolic changes within 265 cellular metabolites after exposure of liver HepG2 cells to four TiO_2 and two CeO_2 materials [66]. The results showed that five out of the six investigated ENMs significantly reduced glutathione concentrations and associated metabolite levels within HepG2 cells. The study showed that 8 nm CeO_2 NPs significantly increased lipid levels including fatty acid concentrations within HepG2 cells, whilst all other investigated NPs did not show a similar effect. CeO_2, but not TiO_2, increased asymmetric dimethylarginine concentration and thus possible decreased iNOS activity and NO concentrations.

6.8 Lipidomics

The lipidome is an example of a specialized subset of the metabolome. It comprises the complete known lipid profile of a biological system [17]. Lipidomics is the systematic approach to characterize and quantify lipids in biological samples using analytical methods mostly based on MS.

Lipids are the fundamental constituents of all cellular membranes [47, 56], provide an important energy reserve [102], and exhibit intracellular as well as systemic signalling functions [141]. Exposure to environmental chemicals often induces considerable changes in the cellular and tissue lipid composition. Levels of specific lipids such as certain sphingolipids that are involved in lipid signalling can be indicative of a cells stress status [48]. Therefore, lipidomics as a powerful method to describe the overall lipid composition of biological matrices has great potential to identify and detect candidate biomarker signatures indicative of toxicity. Recent advances in MS-based techniques enable the identification and quantification of hundreds of molecular lipid species in a high throughput manner [145, 149]. It is possible to analyse large sample collections by automated methods in a 96-well format [61]. Multiple MS platforms can be employed to characterize the extracted lipid such as detecting the lipids by shotgun lipidomics or after separation by liquid chromatography to detect and quantify lipids of lower abundance.

Moreover, alterations of lipid homeostasis contribute to several pathophysiological conditions like diabetes, cardiovascular disease, Parkinson's disease, Alzheimer's disease or nonalcoholic fatty liver disease (NAFLD) to distinguish between the different disease states steatosis, nonalcoholic steatohepatitis (NASH), and cirrhosis [55, 106, 125]. Lipidomic studies, combined with other "omics" technologies in an integrated approach have the potential not only to get better understanding of the up- and downregulation of cellular signalling pathways [62] but may ultimately be one of the tools to assess adverse outcome pathways also for nanotoxicology.

6.9 Lipidomics in Nanotoxicology

Lipidomic studies were also already used to evaluate the nanotoxicity of single walled carbon nanotubes after inhalation and showed selective pulmonary peroxidation profiles [138]. Lipidomics in combination with proteomics was used to assess the influence of metal (silver) and metal oxide (CuO, TiO_2 and ZnO) NPs to primary mouse hepatocytes [128], thereby revealing particle specific effects. While silver NPs increased triacylglycerol levels and decreased sphingomyelin levels, CuO NPs decreased phosphatidylethanolamines and phosphatidylinositols and caused down-regulation of electron transferring protein subunit beta. TiO_2 caused the upregulation of ATP-synthase and electron transferring protein alpha. These investigations show the diversity of regulation mechanisms for a small subset of NPs and clearly indicate that a more general integrated "omics" approach is needed to fully assess nanotoxicity and possible involved adverse outcome pathways by combining resulting information from several "omics" technologies. Testing strategies for relevant "reference" subsets of nanoparticles are needed to establish and evaluate possible adverse outcome pathways and their subsequent "omics" perturbations not only when fully established but in a time-dependent manner [63]. That is how "omics" pattern gradually change from the time of exposure until adverse outcome effects like apoptosis are finally manifested. In that way, "omics" perturbations observed by exposure to newly emerging ENMs can not only be assessed against adverse outcome effects like apoptosis, but also against various intermitting time-dependent "omic" pattern changes. This would also ultimately give a tool to assess "positive" "omics" pattern changes after exposure to evaluate long-term effects caused by exposure to ENMs without acute manifestation of nanotoxicological effects.

6.10 High-Content Screening

Data of molecular changes gathered from "omics" technologies should ideally be corroborated by cellular or tissue-level observations measured under the same conditions. Histopathology is performed for the standard guidelines for regulatory assessment while "omics" technologies are employed to provide additional data. On the other hand, for small organisms and cells in culture, high-content screening (HCS) methods are available [85, 92]. These methods are based on the automated computer-aided visual detection of a panel of functional biomarkers in either a fixed specimen labelled with fluorescent reagents or directly on a living specimen during the time of the exposure. Mostly digital microscopy and flow cytometry are employed in HCS, which may provide precise temporal, spatial, and contextual information defining the biological status of the cells or organs and structure of small organisms. It should be noted that in terms of ENMs their possible interference with especially optical/visual techniques demands extra scrutiny for HCS methods [26]. A broad panel of biomarkers is available to quantify key cellular events such as apoptosis, autophagy, cell proliferation, cell viability, cytotoxicity,

DNA damage, mitochondrial health, mitotic index, oxidative stress, nascent protein synthesis, and phospholipidosis and steatosis. Some of these biomarkers can be used in multiplexed approaches and allow quantitative measurements of the abundance and localization of proteins and/or changes in the morphology of the cell.

6.11 Combinatorial Omics and Integrated Data Analysis

A small number of studies concerning ENMs have integrated more than one omics technique.

Cunningham et al. [22] combined high throughput omics biotechnologies with systems biology to screen for toxicity of single walled carbon nanotubes (SWCNTs) compared to nanosize TiO_2, quartzous SiO_2, carbon black (Printex 90), and carbonyl iron on human primary epidermal keratinocytes and bronchial epithelial cells. Expression arrays for mRNA and microRNA were used together with 2D protein gel electrophoresis and mass spectrometry detection. Expression profile comparison revealed similar profiles of SWCNTs and carbonyl iron at non-cytotoxic doses and of SWCNTs and quartzous SiO_2 at cytotoxic doses [22].

Silica-coated magnetic NPs containing Rhodamine B isothiocyanate MNPs@SiO_2(RITC) were investigated for gene expression and metabolic changes in human embryo kidney 293 cells [120]. Based on microarray gene chip and gas chromatography mass spectrometry analysis, glutamic acid was increased and expression of genes related to the glutamic acid metabolic pathway as well as organic acids related to the Krebs cycle were disturbed at a high dose of particles. Furthermore, a decreased capacity of ATP synthesis, increases in ROS concentration, and mitochondrial damage were observed in functional assays [120].

Proteomics and miRNA sequencing technologies were utilized to investigate effects of silver NPs on human dermal fibroblasts [58]. Of the 57 pathways found regulated in response to the ENM, four pathways were concurrently affected by differentially expressed miRNA, target mRNAs and target proteins: "Regulation of actin cytoskeleton", "Signalling of hepatocyte growth factor receptor", "Insulin signalling", and "MAPK signalling pathway". The results indicated that silver NPs might induce toxicity by affecting the cytoskeleton, ATP synthesis and apoptosis [58].

Exposure of three human cell lines to two high aspect ratio ENM types, TiO_2 nanobelts and multiwalled carbon nanotubes (MWCNT) was investigated by global transcriptome and proteome analyses [132]. Macrophage-like THP-1 cells, small airway epithelial HT29, and intestinal Caco-2 cells exhibited unique patterns of gene and protein expressions, with no differentially expressed genes or proteins overlapping across all three cell types. Exposure of 1 h induced similar expression patterns in response to both TiO_2 and MWCNT while being different for each cell type. This apparent general response to insult stood in contrast to the response after 24 h, which was unique to each ENM. In THP-1 cells TiO_2 exposure affected regulation of pathways associated with inflammation, apoptosis, cell cycle arrest, DNA replication stress and genomic instability, whereas MWCNTs elicited increased cell proliferation, DNA repair and anti-apoptotic pathways. The authors suggest that the differential regulation of the biological pathways might represent cellular responses to high (TiO_2) and low (MWCNT) ENM toxicity, respectively [132].

The mode of action of TiO_2 in the dark on *Escherichia coli* was investigated using transcriptomic and proteomic analysis [121]. Pathway enrichment was observed for the lipid A synthesis pathway, gluconeogenesis, the fatty acid β-oxidation pathway, and importantly for trehalose biosynthesis and several specific membrane transporters indicating osmotic stress. The study revealed that the bactericidal mechanism of TiO_2 in the dark comprises depolarization and loss of membrane integrity, resulting in cellular ion imbalance and depletion of the intracellular ATP content. At the molecular level it manifests as an osmotic stress response [121].

PVP-coated CeO_2 NPs were investigated in the alga *Chlamydomonas reinhardtii* [127]. While growth was unaffected, metabolomic and transcriptomic analysis revealed down-regulation of photosynthesis associated pathways at high concentrations. This response was ENM-specific as neither $CeNO_3$ nor PVP showed such an effect [127].

Overall, system toxicology attempts to combine all available data to reveal AOPs. AOPs are defined as a sequence of key events starting with a molecular initiating point and culminating in an adverse outcome of interest to risk assessment [6]. This provides a framework, which is different from the toxicant and species-specific mode of action concept. An AOP knowledge base (https://aopkb.org/) is made available by the OECD together with the US-EPA, the European commission, and the ERDC. The platform provides public access to a peer-reviewed wiki-based tool to develop AOPs (https://aopkb.org/aopwiki/). The OECD has also developed a handbook to guide in the development of AOPs (https://aopkb.org/common/AOP_Handbook.pdf). Not every technique is used for every toxicant, and it is believed that the wealth of data provided by "omics" technology allows for some extrapolation. However, the massive amounts of data also pose major challenges. Many of the techniques are in early development which means that data generation has the potential to still increase in large part because costs are decreasing.

Genomics, transcriptomics, proteomics, and metabolomics are involved in different ways in the definition of the phenotype. While the genome is rather static, epigenetics is a recent research field that involves regulation by DNA modifications as well as post-translational protein modification that has yet to acquire AOP relevant information. The transcriptome is much more dynamic and largely responsible for the regulation [74]. Proteomics and metabolomics have an even higher variability and therefore more directly participate in an observed change in phenotype. More immediate responses and rapid regulation of signalling pathways are for example mediated by post-translational modifications such as phosphorylation [95]. In addition, there are numerous examples for the regulatory influence of endogenous metabolites [2, 154]. The integration of data from the different "omics" techniques still represents a challenge as the techniques based on the measurement principles and molecular classes have different scales in terms of abundance, data accuracy and variance [5]. Methods and tools that manage, integrate, and process data are being developed [151]. Software tools are being developed that store and manage data and also provide details about the experimental setup, such as EMMA [27], and MIMAS [43]. A standard based on minimum information about microarray experiments (MIAME) [16], MAGE-TAB [113], has been adopted by public databases such as ArrayExpress (http://www.ebi.ac.uk/arrayexpress/) and Gene Expression Omnibus (http://www.ncbi.nlm.nih.gov/geo/). Similarly, for proteomics minimum information about proteomics experiments (MIAPE) has been developed as a standard [126]. Ideally, data would also be standardised so that exchange between platforms and techniques is facilitated. To this end, scoring methods are available that allow the direct combination of data, e.g. from proteomics and metabolomics [93]. With the development of genome-wide visualization and modelling platforms such as Cytoscape the situation has improved, and commercial vendors now provide the built-in inspection and analysis of data from different omics techniques, e.g. the Ingenuity Pathway Analysis (IPA) software [41]. The IPA software also provides analysis with respect to known molecular toxicological reactions [41]. Analysis starts with the identification of significantly modified individual genes, proteins or metabolites, and enrichments in certain pathways leads to the identification of affected signal transduction or other biosynthetic and metabolic pathways. The biologically relevant integration of many different marker molecules of multiple "omics" techniques in this higher level analysis makes it less susceptible to fluctuations in individual genes / proteins / metabolites. An increasing number of transcriptomics (e.g. [1, 3, 107], proteomics [70] and metabolomics studies [119] have been performed. Key to successful classification/grouping strategies is the identification of adverse outcome pathways, which needs to be as detailed and accurate as possible by integrating various omics data. These can then add to the definition of AOPs [146]. Definition of AOPs for ENMs is seen as an important step towards the classification of their effects and grouping of ENMs.

Different algorithms are required for the bioinformatics data analysis of signalling pathways [24, 139]. The various algorithms allow for different perspectives on the data for their evaluation. Statistical analyses are highly susceptible to the quality of the underlying data and this still presents a challenge for increasing the reliability of the conclusions reached [104]. Data integration over the different platforms still represents a formidable challenge [46]. The more so as for ENMs additionally physico-chemical parameters must be brought together with classical toxicity data, transcriptome, proteome, metabolome and possibly heterogeneous data from other sources (publications, other projects) of which the structure varies widely. Tools for data integration are being developed, but even more work is needed for heterogeneous data sets [151]. Once a data matrix is created, the data can be examined for correlations by means of principal component analysis (principal component analysis, PCA), hierarchical cluster analysis (Hierarchical Cluster Analysis, HCA) and other statistical analysis methods such as partial-least-squares (PLS), and orthogonal projection to latent structures discriminant analysis (OPLS-DA), or Random Forest.

For instance, it was possible to correlate oxidative stress to the conduction band energy levels of metal oxide NPs in a large data set of physico-chemical conditions and in vitro experiments [157]. Twenty four metal oxide NPs were investigated by different in vitro cytotoxicity assays not addressing specific mechanisms, and in addition using an automated multi-parametric HTS assay. Changes in ROS production (DCF and MitoSox red fluorescence), intracellular calcium flux (Fluo-4 fluorescence), mitochondrial membrane potential (JC-1 fluorescence), and surface membrane permeability (PI uptake) were quantitatively assesses in two cell lines cells. A selection of materials was also tested for acute pro-inflammatory effects by oropharyngeally instillation in mice.

Induction of ROS production and pro-inflammatory effects were strongly correlated to overlap conduction band energy levels with the cellular redox potential. Both cellular assays exhibited good correlation with the generation of acute neutrophilic inflammation and cytokine responses in vivo. This analysis is based primarily on the use of high-throughput methods and the interpretation of the resulting large amounts of data [87].

Another type of data can be obtained from reporter gene library expression data and select panel quantitative PCR. A comparative study investigated the genotoxic effects of anatase TiO_2, carbon black, single wall carbon nanotube (SWCNT) and fullerene in *Escherichia coli*, *Saccharomyces cerevisiae*, and human A549 cells [71]. Through integration of data from the different assays, it was demonstrated that anatase TiO_2 and carbon black induce oxidative stress which contributes to DNA damage in eukaryotic cells [71]. On the other hand, single wall carbon nanotube (SWCNT) and fullerene appear to induce DNA double strand breaks in a different way [71]. Gene expression profiles also indicate different types of DNA repair mechanisms involved for the different materials [71].

Supervised machine learning can also be used, as demonstrated by a decision tree developed on the toxicity of cobalt ferrite NPs [57]. In addition to the grouping based on the aforementioned band gap of metal oxide NPs, the size of the particle surface has been associated with oxidative stress responses to ENM [115]. Due to the huge variety of possible nano-objects it may be necessary to additionally perform an expert-assisted weight of evidence analysis in most cases [78, 160].

Direct interpretation of results obtained from in vitro studies in the context of potential in vivo exposures is not possible in most cases. To date, most in vitro models do not yield information on pharmacokinetics, i.e. the processes regarding absorption, distribution, metabolism and excretion. However, these processes govern the exposure of the target tissue in the intact organism, making it a crucial difference between the situation in vitro and in vivo. Moreover, this issue is not limited to the in vivo-in vitro comparison, many differences in toxicity from test animals to humans originate in differences in pharmacokinetics [59]. For this reason, data on the mechanisms of action as well as data on pharmacokinetic behaviour are required for a comprehensive prediction of the biological activity of compounds [8, 9].

Quantitative in vitro *to* in vivo extrapolation (QIVIVE) models the environmental exposures to a chemical that could produce target tissue exposures in humans equivalent to those associated with effects in an in vitro toxicity test. Typically, in vitro toxicity tests yield an EC50, a Benchmark concentration, or an interaction threshold identified by a biologically based dose-response model for the toxicity pathway of concern that can be used in such calculations.

Cellular assays can reveal specific molecular and cellular perturbations, and can be used to characterize dose-dependent transitions that may result in organ/system insult. Using these data together with in vitro and in silico approaches including quantitative structure activity relationship (QSAR) modelling, physiologically based pharmacokinetic (PBPK) modelling, and information on metabolism, transport, binding, and other model parameters from cell- and/or cell derived material-based assays, QIVIVE can provide an estimate of the likelihood of harmful effects in vivo from expected environmental exposures. Blaauboer et al. recommended a scheme for the incorporation of in vitro assay data, QSAR and QSPR information, in vitro metabolism data, and pharmacokinetic modelling in the estimation of human toxicity [10]. In this scheme, a chemical-specific pharmacokinetic model is parameterised using the available in vitro data on the absorption, tissue distribution, metabolism, and excretion of a chemical. While this scheme holds true also for ENMs, much less data are available and novel parameters concerning physico-chemical properties have to be taken into account. For chemicals, currently available quantitative structure-property relationship (QSPR) techniques can be used in many cases to estimate chemical properties and kinetics when the specific data for that chemical are lacking. For example, tissue partitioning of a chemical can be estimated using simple empirical correlations from its water solubility, vapour pressure, and octanol/water partitioning co-efficient [25, 100, 118]. QSPR techniques are currently being developed for ENMs, and require the input of systematic and high quality data [20, 81, 133]. The complexity of the possible changes to ENMs in the body, such as (partial) solubility, protein corona formation and evolution, and aggregation has to be reflected in a pharmacokinetic model. Pharmacokinetic models are not only useful in estimating expected equivalent doses associated with toxicity by in vivo exposure from concentrations at which toxicity is observed in an in vitro toxicity assay. Modelling of the in vitro toxicity assay can also provide important information on the temporal profile of cellular exposure to free chemical that can be used in the design of the most appropriate in vitro experimental protocol [129].

Estimation of the metabolic clearance is arguably the greatest challenge in parameterizing even the simplest pharmacokinetic model. Currently, the most extensive data in this respect are on drug pharmacokinetics. ENMs pose an extra challenge in that they are often composed of more than one material, might release chemicals depending on the different compartments, or even dissolve and reform in the body [103]. For soluble chemicals, e.g. released by an ENM, it would be necessary to perform in vitro assays of the dose-response (capacity and affinity) for metabolic clearance [79]. A qualitative classification system has been developed based on physico-chemical properties to predict whether a chemical was likely to be cleared by metabolism (including the CYP isozyme involved) or by urinary excretion [68]. As data accumulates for a greater number of chemicals across a wider range of chemical classes, it may be possible to predict both qualitative and quantitative clearance using QSAR approaches over a broader domain of applicability.

6.12 Conclusion

Systems biology is increasingly used in toxicology as we currently observe a paradigm change and there is increasing interest in understanding underlying toxicity mechanisms and defining AOPs. It is advisable to combine "omics" technology with classical toxicological endpoints. If possible, different "omics" techniques should be used to assess the full complexity of changes and also to derive more reliable information on affected pathways. However, data integration over the different platforms still represents a formidable challenge as the techniques are based on different measurement principles and different molecular classes have different scales in terms of abundance, data accuracy and variance. The more so this holds true for ENMs, where additional factors account for an even larger variability. Currently, knowledge is only beginning to emerge how different physico-chemical parameters truly affect toxicity and which influence batch-to-batch variations play. Thus, the material characterization and the sample preparation (e.g. preparation of ENM dispersions or also mode of ENM presentation to the cells) deserve much more attention when assessing ENMs.

Additionally one should take into account different possible uptake routes for ENMs (ingestion, dermal, inhalation or injection). Another important issue is the choice of the cell model for in vitro studies or the strain & species for animal studies. Large differences in responses may be expected in different cell lines as well as in different strains of a given species.

Ultimately only the combination of "omics" technologies with high power statistical integrative data interpretation methodologies will unravel important and relevant information with respect to toxicity. In part, concepts already exist how omics data can be used for risk assessment, e.g. for quantitative assessment of the metabolome. Thus systems biology is getting more and more established. It may be expected that current limitations, e.g. in data integration and data analysis, might be overcome soon. Systems biology, by providing very large data sets offers the unique advantage of getting information on underlying molecular mechanisms and identifying affected signalling pathways, often referred to as toxicity pathways. This in turn may allow the development of AOPs. For ENMs such mechanistic based knowledge is highly needed in order to develop grouping approaches. It is well accepted that traditional risk assessment paradigms, e.g. assessing each ENM variant in a case-by-case basis, will not be sufficient to deal with the large amount of ENMs in a reasonable time frame. Systems biology can support the development of grouping approaches. However, prerequisite is the development of better standardized approaches starting for instance with the definition of benchmark materials which allow for comparison between different studies. The largest bottleneck is that currently most studies assess only one ENM at a time or a very limited number of ENMs only. This renders it very difficult to compare outcomes of different studies.

However, intensive research efforts are ongoing. Many large currently funded European projects focus on the use of systems biology for a larger set of ENMs. First possible grouping approaches for ENMs are already discussed in scientific literature. By integrating omics based data one may expect a huge progress.

References

1. Adam N, Vergauwen L, Blust R, Knapen D (2015) Gene transcription patterns and energy reserves in *Daphnia magna* show no nanoparticle specific toxicity when exposed to ZnO and CuO nanoparticles. Environ Res 138C:82–92
2. Alisi A, Leoni S, Piacentani A, Conti Devirgiliis L (2003) Retinoic acid modulates the cell-cycle in fetal rat hepatocytes and HepG2 cells by regulating cyclin-cdk activities. Liver Int Off J Int Assoc Study Liver 23:179–186
3. Bajak E, Fabbri M, Ponti J, Gioria S, Ojea-Jimenez I, Collotta A, Mariani V, Gilliland D, Rossi F, Gribaldo L (2015) Changes in Caco-2 cells transcriptome profiles upon exposure to gold nanoparticles. Toxicol Lett 233:187–199
4. Bandara LR, Kennedy S (2002) Toxicoproteomics – a new preclinical tool. Drug Discov Today 7:411–418
5. Baumann S, Kalkhof S, Hackermuller J, Otto W, Tomm JM, Wissenbach DK, RK U, von Bergen M (2013) Requirements and perspectives for integrating metabolomics with other omics data. Curr Metabolomics 1:15–27
6. Becker RA, Ankley GT, Edwards SW, Kennedy SW, Linkov I, Meek B, Sachana M, Segner H, Van Der Burg B, Villeneuve DL, Watanabe H, Barton-Maclaren TS (2015) Increasing Scientific Confidence in Adverse Outcome Pathways: Application of Tailored Bradford-Hill Considerations for Evaluating Weight of Evidence. Regul Toxicol Pharmacol 72:514–537
7. Bhatia SN, Ingber DE (2014) Microfluidic organs-on-chips. Nat Biotechnol 32:760–772
8. Blaauboer BJ (2002) The necessity of biokinetic information in the interpretation of in vitro toxicity data. Altern Lab Anim ATLA 30(Suppl 2):85–91
9. Blaauboer BJ (2003) The integration of data on physico-chemical properties, in vitro-derived toxicity data and physiologically based kinetic and dynamic as modelling a tool in hazard and risk assessment. A commentary. Toxicol Lett 138:161–171
10. Blaauboer BJ, Clewell HJ, Clothier R, Crespi C, Gerson B, Hawksworth G, Kedderis GL, Rozman K, Willhite C (2001) In vitro methods for assessing acute toxicity: biokinetic determinations. In: Report of the International Workshop on in vitro methods for assessing acute systemic toxicity: results of an International Workshop Organized by ICCVAM and NICEATM (NIH Publication 01–4499). NIEHS, Research Triangle Park. pp 47–60
11. Blazer-Yost BL, Banga A, Amos A, Chernoff E, Lai X, Li C, Mitra S, Witzmann FA (2011) Effect of carbon nanoparticles on renal epithelial cell structure, barrier function, and protein expression. Nanotoxicology 5:354–371
12. Bo Y, Jin C, Liu Y, Yu W, Kang H (2014) Metabolomic analysis on the toxicological effects of TiO(2) nanoparticles in mouse fibroblast cells: from the perspective of perturbations in amino acid metabolism. Toxicol Mech Methods 24:461–469
13. Bouhifd M, Hartung T, Hogberg HT, Kleensang A, Zhao L (2013) Review: toxicometabolomics. J Appl Toxicol JAT 33:1365–1383
14. Bourdon JA, Halappanavar S, Saber AT, Jacobsen NR, Williams A, Wallin H, Vogel U, Yauk CL (2012) Hepatic and pulmonary toxicogenomic profiles in mice intratracheally instilled with carbon black nanoparticles reveal pulmonary inflammation, acute phase response, and alterations in lipid homeostasis. Toxicol Sci Off J Soc Toxicol 127:474–484

15. Bourdon JA, Williams A, Kuo B, Moffat I, White PA, Halappanavar S, Vogel U, Wallin H, Yauk CL (2013) Gene expression profiling to identify potentially relevant disease outcomes and support human health risk assessment for carbon black nanoparticle exposure. Toxicology 303:83–93
16. Brazma A, Hingamp P, Quackenbush J, Sherlock G, Spellman P, Stoeckert C, Aach J, Ansorge W, Ball CA, Causton HC, Gaasterland T, Glenisson P, Holstege FC, Kim IF, Markowitz V, Matese JC, Parkinson H, Robinson A, Sarkans U, Schulze-Kremer S, Stewart J, Taylor R, Vilo J, Vingron M (2001) Minimum information about a microarray experiment (MIAME)-toward standards for microarray data. Nat Genet 29:365–371
17. Brugger B (2014) Lipidomics: analysis of the lipid composition of cells and subcellular organelles by electrospray ionization mass spectrometry. Annu Rev Biochem 83:79–98
18. Bu Q, Yan G, Deng P, Peng F, Lin H, Xu Y, Cao Z, Zhou T, Xue A, Wang Y, Cen X, Zhao YL (2010) NMR-based metabonomic study of the sub-acute toxicity of titanium dioxide nanoparticles in rats after oral administration. Nanotechnology 21:125105
19. Buesen R, Landsiedel R, Sauer UG, Wohlleben W, Groeters S, Strauss V, Kamp H, van Ravenzwaay B (2014) Effects of SiO(2), ZrO(2), and BaSO(4) nanomaterials with or without surface functionalization upon 28-day oral exposure to rats. Arch Toxicol 88:1881–1906
20. Carbo-Dorca R, Besalu E (2011) Construction of coherent nano quantitative structure-properties relationships (nano-QSPR) models and catastrophe theory. SAR QSAR Environ Res 22:661–665
21. Cheng LC, Jiang X, Wang J, Chen C, Liu RS (2013) Nano-bio effects: interaction of nanomaterials with cells. Nanoscale 5:3547–3569
22. Cunningham MJ, Shah M, Lema C, Magnuson SR, Falduto MT, Balzano L, Resasco DE (2007) An OMICs approach for assessing the safety of single-walled carbon nanotubes in human skin and lung cells. In: 2007 NSTI nanotechnology conference and trade show – NSTI Nanotech 2007, Technical Proceedings, vol 2. pp 651–654
23. Dailey LA, Hernandez-Prieto R, Casas-Ferreira AM, Jones MC, Riffo-Vasquez Y, Rodriguez-Gonzalo E, Spina D, Jones SA, Smith NW, Forbes B, Page C, Legido-Quigley C (2014) Adenosine monophosphate is elevated in the bronchoalveolar lavage fluid of mice with acute respiratory toxicity induced by nanoparticles with high surface hydrophobicity. Nanotoxicology 9:106–115
24. de Ridder D, de Ridder J, Reinders MJ (2013) Pattern recognition in bioinformatics. Brief Bioinform 14:633–647
25. DeJongh J, Verhaar HJ, Hermens JL (1997) A quantitative property-property relationship (QPPR) approach to estimate in vitro tissue-blood partition coefficients of organic chemicals in rats and humans. Arch Toxicol 72:17–25
26. Dhawan A, Sharma V (2010) Toxicity assessment of nanomaterials: methods and challenges. Anal Bioanal Chem 398:589–605
27. Dondrup M, Albaum SP, Griebel T, Henckel K, Junemann S, Kahlke T, Kleindt CK, Kuster H, Linke B, Mertens D, Mittard-Runte V, Neuweger H, Runte KJ, Tauch A, Tille F, Puhler A, Goesmann A (2009) EMMA 2 – a MAGE-compliant system for the collaborative analysis and integration of microarray data. BMC Bioinformatics 10:50
28. Dong MS, Choi JY, Sung JH, Kim JS, Song KS, Ryu HR, Lee JH, Bang IS, An K, Park HM, Song NW, Yu IJ (2013) Gene expression profiling of kidneys from Sprague-Dawley rats following 12-week inhalation exposure to silver nanoparticles. Toxicol Mech Methods 23:437–448
29. Dowling P, Hughes DJ, Larkin AM, Meiller J, Henry M, Meleady P, Lynch V, Pardini B, Naccarati A, Levy M, Vodicka P, Neary P, Clynes M (2015) Elevated levels of 14-3-3 proteins, serotonin, gamma enolase and pyruvate kinase identified in clinical samples from patients diagnosed with colorectal cancer. Clin Chim Acta 441:133–141
30. Driessen MD, Mues S, Vennemann A, Hellack B, Bannuscher A, Vimalakanthan V, Riebeling C, Ossig R, Wiemann M, Schnekenburger J, Kuhlbusch TA, Renard B, Luch A, Haase A (2015) Proteomic analysis of protein carbonylation: a useful tool to unravel nanoparticle toxicity mechanisms. Part Fibre Toxicol 12:36

31. Duranton F, Lundin U, Gayrard N, Mischak H, Aparicio M, Mourad G, Daures JP, Weinberger KM, Argiles A (2014) Plasma and urinary amino acid metabolomic profiling in patients with different levels of kidney function. Clin J Am Soc Nephrol 9:37–45
32. Edelmann MJ, Shack LA, Naske CD, Walters KB, Nanduri B (2014) SILAC-based quantitative proteomic analysis of human lung cell response to copper oxide nanoparticles. PLoS One 9:e114390
33. Feliu N, Kohonen P, Ji J, Zhang Y, Karlsson HL, Palmberg L, Nystrom A, Fadeel B (2015) Next-generation sequencing reveals low-dose effects of cationic dendrimers in primary human bronchial epithelial cells. ACS Nano 9:146–163
34. Feng J, Liu H, Bhakoo KK, Lu L, Chen Z (2011) A metabonomic analysis of organ specific response to USPIO administration. Biomaterials 32:6558–6569
35. Feng J, Liu H, Zhang L, Bhakoo K, Lu L (2010) An insight into the metabolic responses of ultra-small superparamagnetic particles of iron oxide using metabonomic analysis of biofluids. Nanotechnology 21:395101
36. Fisichella M, Berenguer F, Steinmetz G, Auffan M, Rose J, Prat O (2012) Intestinal toxicity evaluation of TiO2 degraded surface-treated nanoparticles: a combined physico-chemical and toxicogenomics approach in caco-2 cells. Part Fibre Toxicol 9:18
37. Fleischer CC, Payne CK (2014) Secondary structure of corona proteins determines the cell surface receptors used by nanoparticles. J Phys Chem B 118:14017–14026
38. Fowler SJ, Basanta-Sanchez M, Xu Y, Goodacre R, Dark PM (2015) Surveillance for lower airway pathogens in mechanically ventilated patients by metabolomic analysis of exhaled breath: a case-control study. Thorax 70:320–325
39. Frohlich E, Meindl C, Wagner K, Leitinger G, Roblegg E (2014) Use of whole genome expression analysis in the toxicity screening of nanoparticles. Toxicol Appl Pharmacol 280:272–284
40. Gagne F, Andre C, Skirrow R, Gelinas M, Auclair J, van Aggelen G, Turcotte P, Gagnon C (2012) Toxicity of silver nanoparticles to rainbow trout: a toxicogenomic approach. Chemosphere 89:615–622
41. Ganter B, Zidek N, Hewitt PR, Muller D, Vladimirova A (2008) Pathway analysis tools and toxicogenomics reference databases for risk assessment. Pharmacogenomics 9:35–54
42. Garcia-Contreras R, Sugimoto M, Umemura N, Kaneko M, Hatakeyama Y, Soga T, Tomita M, Scougall-Vilchis RJ, Contreras-Bulnes R, Nakajima H, Sakagami H (2015) Alteration of metabolomic profiles by titanium dioxide nanoparticles in human gingivitis model. Biomaterials 57:33–40
43. Gattiker A, Hermida L, Liechti R, Xenarios I, Collin O, Rougemont J, Primig M (2009) MIMAS 3.0 is a Multiomics Information Management and Annotation System. BMC Bioinformatics 10:151
44. Giskeodegard GF, Bertilsson H, Selnaes KM, Wright AJ, Bathen TF, Viset T, Halgunset J, Angelsen A, Gribbestad IS, Tessem MB (2013) Spermine and citrate as metabolic biomarkers for assessing prostate cancer aggressiveness. PLoS One 8:e62375
45. Gomes T, Pereira CG, Cardoso C, Bebianno MJ (2013) Differential protein expression in mussels *Mytilus galloprovincialis* exposed to nano and ionic Ag. Aquat Toxicol 136:79–90
46. Gomez-Cabrero D, Abugessaisa I, Maier D, Teschendorff A, Merkenschlager M, Gisel A, Ballestar E, Bongcam-Rudloff E, Conesa A, Tegner J (2014) Data integration in the era of omics: current and future challenges. BMC Syst Biol 8(Suppl 2):I1
47. Goni FM (2014) The basic structure and dynamics of cell membranes: an update of the Singer-Nicolson model. Biochim Biophys Acta 1838:1467–1476
48. Grosch S, Schiffmann S, Geisslinger G (2012) Chain length-specific properties of ceramides. Prog Lipid Res 51:50–62
49. Guo L, Panderi I, Yan DD, Szulak K, Li Y, Chen YT, Ma H, Niesen DB, Seeram N, Ahmed A, Yan B, Pantazatos D, Lu W (2013) A comparative study of hollow copper sulfide nanoparticles and hollow gold nanospheres on degradability and toxicity. ACS Nano 7:8780–8793
50. Hadrup N, Lam HR, Loeschner K, Mortensen A, Larsen EH, Frandsen H (2012) Nanoparticulate silver increases uric acid and allantoin excretion in rats, as identified by metabolomics. J Appl Toxicol JAT 32:929–933

51. Halappanavar S, Jackson P, Williams A, Jensen KA, Hougaard KS, Vogel U, Yauk CL, Wallin H (2011) Pulmonary response to surface-coated nanotitanium dioxide particles includes induction of acute phase response genes, inflammatory cascades, and changes in microRNAs: a toxicogenomic study. Environ Mol Mutagen 52:425–439
52. Halappanavar S, Saber AT, Decan N, Jensen KA, Wu D, Jacobsen NR, Guo C, Rogowski J, Koponen IK, Levin M, Madsen AM, Atluri R, Snitka V, Birkedal RK, Rickerby D, Williams A, Wallin H, Yauk CL, Vogel U (2015) Transcriptional profiling identifies physicochemical properties of nanomaterials that are determinants of the in vivo pulmonary response. Environ Mol Mutagen 56:245–264
53. Haniu H, Matsuda Y, Takeuchi K, Kim YA, Hayashi T, Endo M (2010) Proteomics-based safety evaluation of multi-walled carbon nanotubes. Toxicol Appl Pharmacol 242:256–262
54. Hilton GM, Taylor AJ, McClure CD, Parsons GN, Bonner JC, Bereman MS (2015) Toxicoproteomic analysis of pulmonary carbon nanotube exposure using LC-MS/MS. Toxicology 329:80–87
55. Hla T, Dannenberg AJ (2012) Sphingolipid signaling in metabolic disorders. Cell Metab 16:420–434
56. Holthuis JC, Menon AK (2014) Lipid landscapes and pipelines in membrane homeostasis. Nature 510:48–57
57. Horev-Azaria L, Baldi G, Beno D, Bonacchi D, Golla-Schindler U, Kirkpatrick JC, Kolle S, Landsiedel R, Maimon O, Marche PN, Ponti J, Romano R, Rossi F, Sommer D, Uboldi C, Unger RE, Villiers C, Korenstein R (2013) Predictive toxicology of cobalt ferrite nanoparticles: comparative in-vitro study of different cellular models using methods of knowledge discovery from data. Part Fibre Toxicol 10:32
58. Huang Y, Lü X, Ma J (2014) Toxicity of silver nanoparticles to human dermal fibroblasts on MicroRNA level. J Biomed Nanotechnol 10:3304–3317
59. Irfan A, Cauchi M, Edmands W, Gooderham NJ, Njuguna J, Zhu H (2014) Assessment of temporal dose-toxicity relationship of fumed silica nanoparticle in human lung A549 cells by conventional cytotoxicity and (1)H-NMR-based extracellular metabonomic assays. Toxicol Sci Off J Soc Toxicol 138:354–364
60. Jackson P, Hougaard KS, Vogel U, Wu D, Casavant L, Williams A, Wade M, Yauk CL, Wallin H, Halappanavar S (2012) Exposure of pregnant mice to carbon black by intratracheal instillation: toxicogenomic effects in dams and offspring. Mutat Res 745:73–83
61. Jung HR, Sylvanne T, Koistinen KM, Tarasov K, Kauhanen D, Ekroos K (2011) High throughput quantitative molecular lipidomics. Biochim Biophys Acta 1811:925–934
62. Jungnickel H, Potratz S, Baumann S, Tarnow P, von Bergen M, Luch A (2014) Identification of lipidomic biomarkers for coexposure to subtoxic doses of benzo[a]pyrene and cadmium: the toxicological cascade biomarker approach. Environ Sci Technol 48:10423–10431
63. Kalkhof S, Dautel F, Loguercio S, Baumann S, Trump S, Jungnickel H, Otto W, Rudzok S, Potratz S, Luch A, Lehmann I, Beyer A, von Bergen M (2015) Pathway and time-resolved benzo[a]pyrene toxicity on Hepa1c1c7 cells at toxic and subtoxic exposure. J Proteome Res 14:164–182
64. Kim KB, Um SY, Chung MW, Jung SC, Oh JS, Kim SH, Na HS, Lee BM, Choi KH (2010) Toxicometabolomics approach to urinary biomarkers for mercuric chloride (HgCl(2))-induced nephrotoxicity using proton nuclear magnetic resonance ((1)H NMR) in rats. Toxicol Appl Pharmacol 249:114–126
65. Kinter CS, Lundie JM, Patel H, Rindler PM, Szweda LI, Kinter M (2012) A quantitative proteomic profile of the Nrf2-mediated antioxidant response of macrophages to oxidized LDL determined by multiplexed selected reaction monitoring. PLoS One 7:e50016
66. Kitchin KT, Grulke E, Robinette BL, Castellon BT (2014) Metabolomic effects in HepG2 cells exposed to four TiO2 and two CeO2 nanomaterials. Environ Sci Nano 1:466–477
67. Kreyling WG, Fertsch-Gapp S, Schaffler M, Johnston BD, Haberl N, Pfeiffer C, Diendorf J, Schleh C, Hirn S, Semmler-Behnke M, Epple M, Parak WJ (2014) In vitro and in vivo interactions of selected nanoparticles with rodent serum proteins and their consequences in biokinetics. Beilstein J Nanotechnol 5:1699–1711

68. Kusama M, Toshimoto K, Maeda K, Hirai Y, Imai S, Chiba K, Akiyama Y, Sugiyama Y (2010) In silico classification of major clearance pathways of drugs with their physiochemical parameters. Drug Metab Dispos Biol Fate Chem 38:1362–1370
69. Kuznetsova GP, Larina OV, Petushkova NA, Kisrieva YS, Samenkova NF, Trifonova OP, Karuzina II, Ipatova OM, Zolotaryov KV, Romashova YA, Lisitsa AV (2014) Effects of fullerene C60 on proteomic profile of *Danio rerio* fish embryos. Bull Exp Biol Med 156:694–698
70. Lai ZW, Yan Y, Caruso F, Nice EC (2012) Emerging techniques in proteomics for probing nano-bio interactions. ACS Nano 6:10438–10448
71. Lan J, Gou N, Gao C, He M, Gu AZ (2014) Comparative and mechanistic genotoxicity assessment of nanomaterials via a quantitative toxicogenomics approach across multiple species. Environ Sci Technol 48:12937–12945
72. Lange V, Picotti P, Domon B, Aebersold R (2008) Selected reaction monitoring for quantitative proteomics: a tutorial. Mol Syst Biol 4:222
73. Li J, Zhao Z, Feng J, Gao J, Chen Z (2013) Understanding the metabolic fate and assessing the biosafety of MnO nanoparticles by metabonomic analysis. Nanotechnology 24:455102
74. Li JJ, Biggin MD (2015) Gene expression. Statistics requantitates the central dogma. Science 347:1066–1067
75. Liao M, Liu H (2012) Gene expression profiling of nephrotoxicity from copper nanoparticles in rats after repeated oral administration. Environ Toxicol Pharmacol 34:67–80
76. Lin BC, Zhang HS, Lin ZQ, Fang YJ, Tian L, Yang HL, Yan J, Liu HL, Zhang W, Xi ZG (2013) Studies of single-walled carbon nanotubes-induced hepatotoxicity by NMR-based metabonomics of rat blood plasma and liver extracts. Nanoscale Res Lett 8:236
77. Lin ZQ, Ma L, X ZG, Zhang, HS, Lin BC (2013) A comparative study of lung toxicity in rats induced by three types of nanomaterials. Nanoscale Res Lett 8:521.
78. Linkov I, Massey O, Keisler J, Rusyn I, Hartung T (2015) From "weight of evidence" to quantitative data integration using multicriteria decision analysis and Bayesian methods. ALTEX 32:3–8
79. Lipscomb JC, Meek ME, Krishnan K, Kedderis GL, Clewell H, Haber L (2004) Incorporation of pharmacokinetic and pharmacodynamic data into risk assessments. Toxicol Mech Methods 14:145–158
80. Lu X, Tian Y, Zhao Q, Jin T, Xiao S, Fan X (2011) Integrated metabonomics analysis of the size-response relationship of silica nanoparticles-induced toxicity in mice. Nanotechnology 22:055101
81. Lubinski L, Urbaszek P, Gajewicz A, Cronin MT, Enoch SJ, Madden JC, Leszczynska D, Leszczynski J, Puzyn T (2013) Evaluation criteria for the quality of published experimental data on nanomaterials and their usefulness for QSAR modelling. SAR QSAR Environ Res 24:995–1008
82. Malone JH, Oliver B (2011) Microarrays, deep sequencing and the true measure of the transcriptome. BMC Biol 9:34
83. Mandell JW (2003) Phosphorylation state-specific antibodies: applications in investigative and diagnostic pathology. Am J Pathol 163:1687–1698
84. Miethling-Graff R, Rumpker R, Richter M, Verano-Braga T, Kjeldsen F, Brewer J, Hoyland J, Rubahn HG, Erdmann H (2014) Exposure to silver nanoparticles induces size- and dose-dependent oxidative stress and cytotoxicity in human colon carcinoma cells. Toxicol In Vitro Int J Publ Assoc BIBRA 28:1280–1289
85. Mikut R, Dickmeis T, Driever W, Geurts P, Hamprecht FA, Kausler BX, Ledesma-Carbayo MJ, Maree R, Mikula K, Pantazis P, Ronneberger O, Santos A, Stotzka R, Strahle U, Peyrieras N (2013) Automated processing of zebrafish imaging data: a survey. Zebrafish 10:401–421
86. Nazari P, Dowlatabadi-Bazaz R, Mofid MR, Pourmand MR, Daryani NE, Faramarzi MA, Sepehrizadeh Z, Shahverdi AR (2014) The antimicrobial effects and metabolomic footprinting of carboxyl-capped bismuth nanoparticles against Helicobacter pylori. Appl Biochem Biotechnol 172:570–579
87. Nel AE (2013) Implementation of alternative test strategies for the safety assessment of engineered nanomaterials. J Intern Med 274:561–577

88. Newman RH, Zhang J, Zhu H (2014) Toward a systems-level view of dynamic phosphorylation networks. Front Genet 5:263
89. Ng CT, Yung LY, Swa HL, Poh RW, Gunaratne J, Bay BH (2015) Altered protein expression profile associated with phenotypic changes in lung fibroblasts co-cultured with gold nanoparticle-treated small airway epithelial cells. Biomaterials 39:31–38
90. NRC (2007) Applications of toxicogenomic technologies to predictive toxicology and risk assessment. Washington, DC
91. Nuwaysir EF, Bittner M, Trent J, Barrett JC, Afshari CA (1999) Microarrays and toxicology: the advent of toxicogenomics. Mol Carcinog 24:153–159
92. O'Brien PJ (2014) High-content analysis in toxicology: screening substances for human toxicity potential, elucidating subcellular mechanisms and in vivo use as translational safety biomarkers. Basic Clin Pharmacol Toxicol 115:4–17
93. Oberbach A, Bluher M, Wirth H, Till H, Kovacs P, Kullnick Y, Schlichting N, Tomm JM, Rolle-Kampczyk U, Murugaiyan J, Binder H, Dietrich A, von Bergen M (2011) Combined proteomic and metabolomic profiling of serum reveals association of the complement system with obesity and identifies novel markers of body fat mass changes. J Proteome Res 10:4769–4788
94. OECD (2013) Guidance document on developing and assessing adverse outcome pathways. In: Series on testing and assessment. Environment Directorate of the OECD, Paris
95. Olsen JV, Blagoev B, Gnad F, Macek B, Kumar C, Mortensen P, Mann M (2006) Global, in vivo, and site-specific phosphorylation dynamics in signaling networks. Cell 127:635–648
96. Overmyer KA, Thonusin C, Qi NR, Burant CF, Evans CR (2015) Impact of anesthesia and euthanasia on metabolomics of mammalian tissues: studies in a C57BL/6 J mouse model. PLoS One 10:e0117232
97. Ozsolak F, Milos PM (2011) RNA sequencing: advances, challenges and opportunities. Nat Rev Genet 12:87–98
98. Palomaki J, Sund J, Vippola M, Kinaret P, Greco D, Savolainen K, Puustinen A, Alenius H (2015) A secretomics analysis reveals major differences in the macrophage responses towards different types of carbon nanotubes. Nanotoxicology 9:719–728
99. Parveen A, Rizvi SH, Gupta A, Singh R, Ahmad I, Mahdi F, Mahdi AA (2012) NMR-based metabonomics study of sub-acute hepatotoxicity induced by silica nanoparticles in rats after intranasal exposure. Cell Mol Biol 58:196–203
100. Paterson S, Mackay D (1989) Correlation of tissue, blood, and air partition coefficients of volatile organic chemicals. Br J Ind Med 46:321–328
101. Patlewicz G, Simon TW, Rowlands JC, Budinsky RA, Becker RA (2015) Proposing a scientific confidence framework to help support the application of adverse outcome pathways for regulatory purposes. Regul Toxicol Pharmacol 71:463–477
102. Peirce V, Carobbio S, Vidal-Puig A (2014) The different shades of fat. Nature 510:76–83
103. Peters R, Kramer E, Oomen AG, Rivera ZE, Oegema G, Tromp PC, Fokkink R, Rietveld A, Marvin HJ, Weigel S, Peijnenburg AA, Bouwmeester H (2012) Presence of nano-sized silica during in vitro digestion of foods containing silica as a food additive. ACS Nano 6:2441–2451
104. Pettit S, des Etages SA, Mylecraine L, Snyder R, Fostel J, Dunn RT 2nd, Haymes K, Duval M, Stevens J, Afshari C, Vickers A (2010) Current and future applications of toxicogenomics: results summary of a survey from the HESI Genomics State of Science Subcommittee. Environ Health Perspect 118:992–997
105. Plant NJ (2015) An introduction to systems toxicology. Toxicol Res Uk 4:9–22
106. Platt FM (2014) Sphingolipid lysosomal storage disorders. Nature 510:68–75
107. Poulsen SS, Saber AT, Williams A, Andersen O, Kobler C, Atluri R, Pozzebon ME, Mucelli SP, Simion M, Rickerby D, Mortensen A, Jackson P, Kyjovska ZO, Molhave K, Jacobsen NR, Jensen KA, Yauk CL, Wallin H, Halappanavar S, Vogel U (2014) MWCNTs of different physicochemical properties cause similar inflammatory responses, but differences in transcriptional and histological markers of fibrosis in mouse lungs. Toxicol Appl Pharmacol 284:16–32

108. Poynton HC, Lazorchak JM, Impellitteri CA, Blalock BJ, Rogers K, Allen HJ, Loguinov A, Heckman JL, Govindasmawy S (2012) Toxicogenomic responses of nanotoxicity in *Daphnia magna* exposed to silver nitrate and coated silver nanoparticles. Environ Sci Technol 46:6288–6296
109. Rabilloud T, Lescuyer P (2015) Proteomics in mechanistic toxicology: History, concepts, achievements, caveats, and potential. Proteomics 15:1051–1074
110. Rainville LC, Carolan D, Varela AC, Doyle H, Sheehan D (2014) Proteomic evaluation of citrate-coated silver nanoparticles toxicity in *Daphnia magna*. Analyst 139:1678–1686
111. Rappaport SM, Li H, Grigoryan H, Funk WE, Williams ER (2012) Adductomics: characterizing exposures to reactive electrophiles. Toxicol Lett 213:83–90
112. Ratnasekhar C, Sonane M, Satish A, Mudiam MK (2015) Metabolomics reveals the perturbations in the metabolome of *Caenorhabditis elegans* exposed to titanium dioxide nanoparticles. Nanotoxicology 9:1–11
113. Rayner TF, Rocca-Serra P, Spellman PT, Causton HC, Farne A, Holloway E, Irizarry RA, Liu J, Maier DS, Miller M, Petersen K, Quackenbush J, Sherlock G, Stoeckert CJ Jr, White J, Whetzel PL, Wymore F, Parkinson H, Sarkans U, Ball CA, Brazma A (2006) A simple spreadsheet-based, MIAME-supportive format for microarray data: MAGE-TAB. BMC Bioinformatics 7:489
114. Rocheleau S, Arbour M, Elias M, Sunahara GI, Masson L (2014) Toxicogenomic effects of nano- and bulk-TiO particles in the soil nematode *Caenorhabditis elegans*. Nanotoxicology 9(4):502–512
115. Rushton EK, Jiang J, Leonard SS, Eberly S, Castranova V, Biswas P, Elder A, Han X, Gelein R, Finkelstein J, Oberdorster G (2010) Concept of assessing nanoparticle hazards considering nanoparticle dosemetric and chemical/biological response metrics. J Toxicol Environ Health Part A 73:445–461
116. Russell WMS, Burch RL (1959) The principles of humane experimental technique. Methuen, London, 238 pp
117. Sarkar A, Ghosh M, Sil PC (2014) Nanotoxicity: oxidative stress mediated toxicity of metal and metal oxide nanoparticles. J Nanosci Nanotechnol 14:730–743
118. Schmitt W (2008) General approach for the calculation of tissue to plasma partition coefficients. Toxicol In Vitro Int J Publ Assoc BIBRA 22:457–467
119. Schnackenberg LK, Sun J, Beger RD (2012) Metabolomics techniques in nanotoxicology studies. Methods Mol Biol 926:141–156
120. Shim W, Paik MJ, Nguyen DT, Lee JK, Lee Y, Kim JH, Shin EH, Kang JS, Jung HS, Choi S, Park S, Shim JS, Lee G (2012) Analysis of changes in gene expression and metabolic profiles induced by silica-coated magnetic nanoparticles. ACS Nano 6:7665–7680
121. Sohm B, Immel F, Bauda P, Pagnout C (2015) Insight into the primary mode of action of TiO2 nanoparticles on *Escherichia coli* in the dark. Proteomics 15:98–113
122. Sturla SJ, Boobis AR, FitzGerald RE, Hoeng J, Kavlock RJ, Schirmer K, Whelan M, Wilks MF, Peitsch MC (2014) Systems toxicology: from basic research to risk assessment. Chem Res Toxicol 27:314–329
123. Suhre K, Meisinger C, Doring A, Altmaier E, Belcredi P, Gieger C, Chang D, Milburn MV, Gall WE, Weinberger KM, Mewes HW, Hrabe de Angelis M, Wichmann HE, Kronenberg F, Adamski J, Illig T (2010) Metabolic footprint of diabetes: a multiplatform metabolomics study in an epidemiological setting. PLoS One 5:e13953
124. Tang M, Zhang T, Xue Y, Wang S, Huang M, Yang Y, Lu M, Lei H, Kong L, Wang Y, Pu Y (2011) Metabonomic studies of biochemical changes in the serum of rats by intratracheally instilled TiO2 nanoparticles. J Nanosci Nanotechnol 11:3065–3074
125. Taskinen MR, Boren J (2015) New insights into the pathophysiology of dyslipidemia in type 2 diabetes. Atherosclerosis 239:483–495
126. Taylor CF, Paton NW, Lilley KS, Binz PA, Julian RK Jr, Jones AR, Zhu W, Apweiler R, Aebersold R, Deutsch EW, Dunn MJ, Heck AJ, Leitner A, Macht M, Mann M, Martens L, Neubert TA, Patterson SD, Ping P, Seymour SL, Souda P, Tsugita A, Vandekerckhove J, Vondriska TM, Whitelegge JP, Wilkins MR, Xenarios I, Yates JR 3rd, Hermjakob H (2007)

The minimum information about a proteomics experiment (MIAPE). Nat Biotechnol 25:887–893
127. Taylor NS, Merrifield R, Williams TD, Chipman JK, Lead JR, Viant MR (2015) Molecular toxicity of cerium oxide nanoparticles to the freshwater alga *Chlamydomonas reinhardtii* is associated with supra-environmental exposure concentrations. Nanotoxicology 1–10. doi: 10.3109/17435390.2014.948941
128. Tedesco S, Bayat N, Danielsson G, Buque X, Aspichueta P, Fresnedo O, Cristobal S (2015) Proteomic and lipidomic analysis of primary mouse hepatocytes exposed to metal and metal oxide nanoparticles. JIOMICS 5:44–57
129. Teeguarden JG, Barton HA (2004) Computational modeling of serum-binding proteins and clearance in extrapolations across life stages and species for endocrine active compounds. Risk Anal Off Publ Soc Risk Anal 24:751–770
130. Teeguarden JG, Webb-Robertson BJ, Waters KM, Murray AR, Kisin ER, Varnum SM, Jacobs JM, Pounds JG, Zanger RC, Shvedova AA (2011) Comparative proteomics and pulmonary toxicity of instilled single-walled carbon nanotubes, crocidolite asbestos, and ultrafine carbon black in mice. Toxicol Sci Off J Soc Toxicol 120:123–135
131. Thingholm TE, Jensen ON, Larsen MR (2009) Analytical strategies for phosphoproteomics. Proteomics 9:1451–1468
132. Tilton SC, Karin NJ, Tolic A, Xie Y, Lai X, Hamilton RF Jr, Waters KM, Holian A, Witzmann FA, Orr G (2014) Three human cell types respond to multi-walled carbon nanotubes and titanium dioxide nanobelts with cell-specific transcriptomic and proteomic expression patterns. Nanotoxicology 8:533–548
133. Toropova AP, Toropov AA, Benfenati E, Korenstein R, Leszczynska D, Leszczynski J (2015) Optimal nano-descriptors as translators of eclectic data into prediction of the cell membrane damage by means of nano metal-oxides. Environ Sci Pollut Res Int 22:745–757
134. Tralau T, Riebeling C, Pirow R, Oelgeschlager M, Seiler A, Liebsch M, Luch A (2012) Wind of change challenges toxicological regulators. Environ Health Perspect 120:1489–1494
135. Treuel L, Jiang X, Nienhaus GU (2013) New views on cellular uptake and trafficking of manufactured nanoparticles. J R Soc Interface R Soc 10:20120939
136. Tsyusko OV, Unrine JM, Spurgeon D, Blalock E, Starnes D, Tseng M, Joice G, Bertsch PM (2012) Toxicogenomic responses of the model organism *Caenorhabditis elegans* to gold nanoparticles. Environ Sci Technol 46:4115–4124
137. Tucci P, Porta G, Agostini M, Dinsdale D, Iavicoli I, Cain K, Finazzi-Agro A, Melino G, Willis A (2013) Metabolic effects of TiO2 nanoparticles, a common component of sunscreens and cosmetics, on human keratinocytes. Cell Death Dis 4:e549
138. Tyurina YY, Kisin ER, Murray A, Tyurin VA, Kapralova VI, Sparvero LJ, Amoscato AA, Samhan-Arias AK, Swedin L, Lahesmaa R, Fadeel B, Shvedova AA, Kagan VE (2011) Global phospholipidomics analysis reveals selective pulmonary peroxidation profiles upon inhalation of single-walled carbon nanotubes. ACS Nano 5:7342–7353
139. Valerio LG Jr, Choudhuri S (2012) Chemoinformatics and chemical genomics: potential utility of in silico methods. J Appl Toxicol JAT 32:880–889
140. van Ravenzwaay B, Montoya GA, Fabian E, Herold M, Krennrich G, Looser R, Mellert W, Peter E, Strauss V, Walk T, Kamp H (2014) The sensitivity of metabolomics versus classical regulatory toxicology from a NOAEL perspective. Toxicol Lett 227:20–28
141. Vance DE (2008) In: Vance DE, editor. Biochemistry of lipids, lipoproteins and membranes Elektronische Ressource, 5th edn. Elsevier, Amsterdam. 1st ed
142. Vannini C, Domingo G, Onelli E, Prinsi B, Marsoni M, Espen L, Bracale M (2013) Morphological and proteomic responses of *Eruca sativa* exposed to silver nanoparticles or silver nitrate. PLoS One 8:e68752
143. Verano-Braga T, Miethling-Graff R, Wojdyla K, Rogowska-Wrzesinska A, Brewer JR, Erdmann H, Kjeldsen F (2014) Insights into the cellular response triggered by silver nanoparticles using quantitative proteomics. ACS Nano 8:2161–2175

144. Vidanapathirana AK, Lai X, Hilderbrand SC, Pitzer JE, Podila R, Sumner SJ, Fennell TR, Wingard CJ, Witzmann FA, Brown JM (2012) Multi-walled carbon nanotube directed gene and protein expression in cultured human aortic endothelial cells is influenced by suspension medium. Toxicology 302:114–122
145. Vihervaara T, Suoniemi M, Laaksonen R (2014) Lipidomics in drug discovery. Drug Discov Today 19:164–170
146. Vinken M (2013) The adverse outcome pathway concept: a pragmatic tool in toxicology. Toxicology 312:158–165
147. Wagner SA, Beli P, Weinert BT, Nielsen ML, Cox J, Mann M, Choudhary C (2011) A proteome-wide, quantitative survey of in vivo ubiquitylation sites reveals widespread regulatory roles. Mol Cell Proteomics 10(M111):013284
148. Waters M, Boorman G, Bushel P, Cunningham M, Irwin R, Merrick A, Olden K, Paules R, Selkirk J, Stasiewicz S, Weis B, Van Houten B, Walker N, Tennant R (2003) Systems toxicology and the Chemical Effects in Biological Systems (CEBS) knowledge base. EHP Toxicogenomics J Natl Inst Environ Health Sci 111:15–28
149. Wenk MR (2010) Lipidomics: new tools and applications. Cell 143:888–895
150. Wilson VS, Keshava N, Hester S, Segal D, Chiu W, Thompson CM, Euling SY (2013) Utilizing toxicogenomic data to understand chemical mechanism of action in risk assessment. Toxicol Appl Pharmacol 271:299–308
151. Wruck W, Peuker M, Regenbrecht CR (2014) Data management strategies for multinational large-scale systems biology projects. Brief Bioinform 15:65–78
152. Xiao X, Lee JH (2010) Systems analysis of alternative splicing and its regulation. Wiley Interdiscip Rev Syst Biol Med 2:550–565
153. Yan LJ, Forster MJ (2011) Chemical probes for analysis of carbonylated proteins: a review. J Chromatogr B Analyt Technol Biomed Life Sci 879:1308–1315
154. Yang M, Soga T, Pollard PJ (2013) Oncometabolites: linking altered metabolism with cancer. J Clin Invest 123:3652–3658
155. Yuan J, Gao H, Sui J, Duan H, Chen WN, Ching CB (2012) Cytotoxicity evaluation of oxidized single-walled carbon nanotubes and graphene oxide on human hepatoma HepG2 cells: an iTRAQ-coupled 2D LC-MS/MS proteome analysis. Toxicol Sci Off J Soc Toxicol 126:149–161
156. Yuan JF, Gao HC, Ching CB (2011) Comparative protein profile of human hepatoma HepG2 cells treated with graphene and single-walled carbon nanotubes: An iTRAQ-coupled 2D LC-MS/MS proteome analysis. Toxicol Lett 207:213–221
157. Zhang H, Ji Z, Xia T, Meng H, Low-Kam C, Liu R, Pokhrel S, Lin S, Wang X, Liao YP, Wang M, Li L, Rallo R, Damoiseaux R, Telesca D, Madler L, Cohen Y, Zink JI, Nel AE (2012) Use of metal oxide nanoparticle band gap to develop a predictive paradigm for oxidative stress and acute pulmonary inflammation. ACS Nano 6:4349–4368
158. Zhang Y, Zhao F, Deng Y, Zhao Y, Ren H (2015) Metagenomic and metabolomic analysis of the toxic effects of trichloroacetamide-induced gut microbiome and urine metabolome perturbations in mice. J Proteome Res 14:1752–1761
159. Zhao Y, Jensen ON (2009) Modification-specific proteomics: strategies for characterization of post-translational modifications using enrichment techniques. Proteomics 9:4632–4641
160. Zuin S, Micheletti C, Critto A, Pojana G, Johnston H, Stone V, Tran L, Marcomini A (2011) Weight of evidence approach for the relative hazard ranking of nanomaterials. Nanotoxicology 5:445–458

Chapter 7
Multiscale Modelling of Bionano Interface

Hender Lopez, Erik G. Brandt, Alexander Mirzoev, Dmitry Zhurkin, Alexander Lyubartsev, and Vladimir Lobaskin

Abstract We present a framework for coarse-grained modelling of the interface between foreign nanoparticles (NP) and biological fluids and membranes. Our model includes united-atom presentations of membrane lipids and globular proteins in implicit solvent, which are based on all-atom structures of the corresponding molecules and parameterised using experimental data or atomistic simulation results. The NPs are modelled by homogeneous spheres that interact with the beads of biomolecules via a central force that depends on the NP size. The proposed methodology is used to predict the adsorption energies for human blood plasma proteins on NPs of different sizes as well as the preferred orientation of the molecules upon adsorption. Our approach allows one to rank the proteins by their binding affinity to the NP, which can be used for predicting the composition of the NP-protein corona for the corresponding material. We also show how the model can be used for studying NP interaction with a lipid bilayer membrane and thus can provide a mechanistic insight for modelling NP toxicity.

Keywords Nanoparticle • Toxicity • Coarse-grained molecular dynamics • Protein corona • Cell membrane

7.1 Introduction

Over the last decade, in vitro and in vivo experiments have produced significant amount of veritable information that can be integrated into theoretical models with the aim of predicting possible health and environmental effects of engineered

H. Lopez • V. Lobaskin (✉)
School of Physics, Complex and Adaptive Systems Lab, University College Dublin, Belfield, Dublin 4, Ireland
e-mail: vladimir.lobaskin@ucd.ie

E.G. Brandt • A. Mirzoev • D. Zhurkin • A. Lyubartsev
Department of Materials and Environmental Chemistry, Stockholm University, SE-10691 Stockholm, Sweden

L. Tran et al. (eds.), *Modelling the Toxicity of Nanoparticles*, Advances in Experimental Medicine and Biology, Vol. 947, DOI 10.1007/978-3-319-47754-1_7

nanoparticles (NP) [1]. However, even the most systematic studies leave the question of precise toxicity mechanisms associated with NPs wide open [2–4]. An important finding arising from these studies is that the toxic effects can emerge either from membrane damage or from interaction of NPs, once they are inside the cell, with the internal cell machinery. Therefore, an evaluation of possible risks should include an assessment of NP ability to penetrate, modify, or destroy the cell membrane and bind to key biomolecules [4]. Being selectively permeable, membranes participate in control of the transport of vital substances into and out of cells. Whereas some biomolecules may penetrate or fuse with cell membranes without overt membrane disruption, no synthetic material of comparable size has shown this property [5]. Among the factors determining the outcome of NP-membrane interaction the surface properties of nanomaterials play a critical role, which can implicate the membrane or plasma proteins in conditioning NP prior to cell penetration.

The detailed understanding of the crucial stages of NP-cell membrane interaction can be achieved with computer simulation. Molecular dynamics is now a well-recognized tool for studying intermolecular interactions, self-assembly, and structure of biomolecules or their complexes. The reliability and predictive character of molecular modelling has improved significantly during the last few years, with development of new, carefully parameterized force fields, simulation algorithms, and greatly increased computer power [6]. The role of computer simulation is now well recognized in many fields including drug design and toxicology [7–9]. In the same way, one can attempt to predict the detrimental effect of NPs from physical considerations. Establishing a qualitative and quantitative connection between physicochemical properties of NPs and their effect on biological functioning of membranes can help to identify the possible pathways leading to toxicity and give a mechanistic interpretation of toxicological data. To achieve this goal, one has to understand the processes occurring at the bionano interface or on the initial stages of contact between the foreign nanomaterial and the organism such as formation of NP-biomolecule complexes, NP-cell membrane interaction, and NP uptake into the cell.

Understanding the corona formation and NP uptake requires one to address the lengthscales at the range of up to 100 nanometres, which is currently beyond the reach of direct atomistic modelling. Though lipid membranes have been very intensively studied by molecular simulations during last decade [10], in general, modelling NP translocation through a lipid membrane is a significant challenge. Depending on the size of the NP and any associated proteins (corona) tens of thousands, or more, of lipid and other molecules may be needed to model a representative fraction of the membrane. For small (under 5 nm) NPs, cytotoxicity effects such as membrane disruption and poration can be addressed at the atomistic scale and at this scale significant insights have already been gained using molecular simulation using atomistic or coarse-grained (CG) force fields [11–14]. To assess interactions of larger NPs with membranes mesoscopic simulations based on greatly reduced number of degrees of freedom are required. To build a quantitative mesoscale model, information on NP-biomolecule association should be transferred from atomistic simulations to the larger scale using coarse-graining.

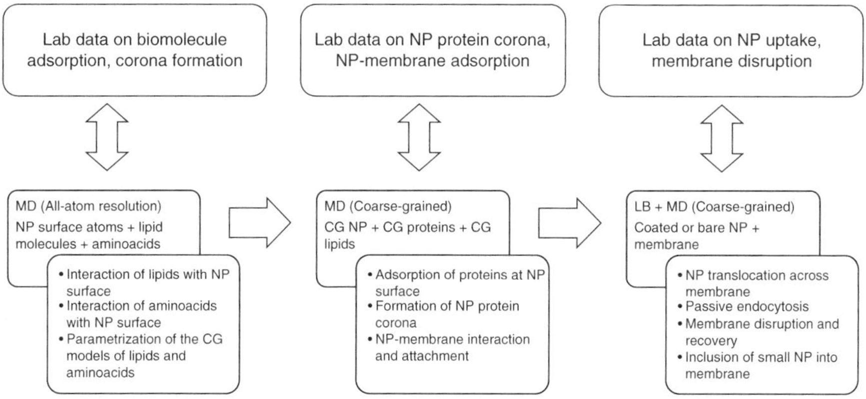

Fig. 7.1 Scheme of the multiscale simulation approach for modelling NP uptake

Many of today's CG models use empirical parameterization of effective interaction potentials. There exist several basic approaches for systematically constructing effective CG potentials from the results of atomistic simulations. One common approach is based on reproduction of forces for specific snapshots of the system (the force matching approach [15, 16] and the other, is based on fitting of structural properties, for which the radial distribution functions are typically used (the inverse Monte Carlo (IMC), or Newton Inversion method) [17, 18]. The IMC method was previously used to build CG models of various molecular systems including ion-DNA solutions [19]. In the same spirit, CG models of plasma and membrane proteins have been developed [20–23] using the united-atom scheme, i.e. replacing the common groups of atoms by single beads, and thus drastically reducing the number of degrees of freedom. The solvent is usually removed from the CG model and is integrated into effective interaction potentials between the CG beads, which on itself provides a big gain in efficiency. A systematic coarse-graining based on the all-atom presentations will preserve the shape and size of the relevant molecules and thus molecular specificity. In this approach we sacrifice a number of internal degrees of freedom, such as protein conformations, which can be justified a posteriori. Although neglecting the protein internal degrees of freedom is a necessarily shaky approximation, this could be the most beneficial one as we can get around the dynamic bottlenecks related to slow protein unfolding.

Similar to molecules, one can use IMC and other coarse-graining methods to model effective interactions between NPs [9, 24, 25]. Thus, the construction of the mesoscale modelling tool involves the following steps, with each consecutive stage based on a systematic coarse-graining of the more detailed description and validated by experimental data (Fig. 7.1):

In the following sections, we describe a minimum set of such CG tools that allow one to simulate the interaction of the NP-protein corona complex with a lipid bilayer. The remainder of the paper is constructed as follows. First, we describe a CG model to calculate the adsorption energies and the most favorable adsorption orientations of proteins onto a hydrophobic NP. The proposed method is then used to calculate the adsorption energies of the two common proteins in human blood onto NPs with negative or positive surface charge or neutral surface. We also report on the effect of the NP radius on the adsorption energies and validate the proposed methodology against full atomistic simulations. Then, in Sect. 7.3 we describe a methodology, in which full atomistic simulations of a lipid bilayer and various lipid-Cholesterol mixtures are used for the extraction of CG pair potentials. In Sect. 7.4, we present a CG simulation of the interaction a bare NP and of a NP-protein complex with a lipid bilayer. Finally, in Sect. 7.5 we summarise the main results.

7.2 Nanoparticle-Protein Interaction

It is now well accepted that foreign surfaces are modified by the adsorption of biomolecules such as proteins or lipids in a biological environment, and that cellular responses to materials in a biological medium might reflect the adsorbed biomolecule layer, rather than the material itself [26]. Recently, the concept of the NP-protein corona has been introduced to describe the proteins in association with NPs in biological fluids [27–30]. The composition of NP corona is flexible and is determined by many affinity constants and concentrations of the components of the blood plasma. One can speculate that in many practically relevant situations, the protein corona is the surface that is exposed to the cell membrane and is the entity the cell protective mechanisms have to deal with. Thus, for most cases it is more likely that the biologically relevant unit is not the particle itself, but a nanoobject of specified size, shape, and with certain protein corona structure. Naked particle surfaces will have a much greater (non-specific) affinity for the cell surface than a particle hiding behind a corona of "bystander" proteins – that is proteins for which no suitable cellular recognition machinery exists. The evidence suggests that, in comparison to typical cell-membrane-biology event timescales, the particle corona is likely to be a defining property of the particle in its interactions with the cell surface, whether it activates cellular machinery or not. Similar observations and outcomes exist for particles inside the cell, in key locations, though we cannot discuss details here [27–30]. We assume that the actual content of the corona is determined by (i) the NP exposure to the protein solution (blood plasma), (ii) a competition between the adsorbed proteins and the glycoproteins/membrane lipids. We will model the protein and lipid interaction with the NP surface at the CG united-atom level for selected set of proteins (see Table 7.1) and lipids. These simulations will provide interaction energies and will be used to predict the kinetics of protein/lipid corona formation. The data on aminoacid interaction with NP will allow us to compute binding affinities of arbitrary proteins of known structure within an additive model implying that the total protein-NP interaction energy is computed as a sum of NP interactions with aminoacids in contact with NP surface. From the typical protein concentrations and adsorption energies one can also predict the average content of the corona using ideal adsorbed solution theory [31]. It is important to understand

Table 7.1 Proteins, PDB ID used for the coarse-graining and the abbreviations used in the text, and their size and abundance in human blood plasma

Protein	PDB ID	Abbreviation	Weight fraction in plasma, %	Molar mass in, kDa
Human Serum Albumin	1N5U	HSA	5.0	67
Fibrinogen	3GHG	Fib	0.4	340

that at this stage we would be able neither to scan all the plasma proteins nor to take into account any change of protein conformation or bonding between the adsorbed proteins. However, as the effect of the corona is still largely unknown, we can only hope to capture the most important contributions of the plasma protein to the NP dispersion stability and their interaction with the cell membrane.

Due to complexity of blood plasma, we can only model it at a simplified level. It seems reasonable to include the elements, which are more likely to affect the NP interactions and aggregation, and mediate their interaction with the membrane. The plasma can then be modelled as a solution of biomolecules in an implicit solvent with a dielectric constant of water and the Debye length corresponding to physiological ionic strength, van der Waals interactions set to corresponding triplets NP-protein-water, or protein-water-protein, and appropriate surface charges on the molecules. In this work, we study the adsorption of two of the most abundant proteins in blood plasma, Human Serum Albumin (HSA) and Fibrinogen (Fib). In Table 7.1, we summarise their relative content in blood and their molar mass. Although this two proteins represent important components of the blood plasma because of their abundance, recent observations [27–29, 32] demonstrate that the protein corona can include hundreds of different plasma proteins. As of now, it is mostly not known what proteins dominate the content of the corona or play the most crucial role in the NP coating and uptake, although some progress has been made [33] and there is hope that such information will become available in the coming years.

7.2.1 *Adsorption of Proteins onto Nanoparticles*

The starting point for development of a CG model for the interaction of NPs with proteins is to decide how much detail from the molecular structure of the protein one needs to keep. There is an active and extensive research activity on the different CG models that can be used to simulate proteins under different conditions (for more detailed reviews see [20, 22, 34]). The aim of this work is to propose a set of tools that could be used to simulate the interaction of one or more proteins (and in some cases quite big proteins) with a NP, for relatively long timescales. To meet this goal with a reasonable computational effort the number of beads representing the protein should be kept as small as possible but the proposed model should also preserve enough structural information about the molecule. For these reasons we

Table 7.2 Normalized hydrophobicities ε_i (taken from Table II in [21]) and σ_i for each amino acid (taken from [38])

Residue	LYS	GYU	ASP	ASN	SER	ARG	GLU	PRO	THR	GLY
ε_i, ε	0.00	0.05	0.06	0.10	0.11	0.13	0.13	0.14	0.16	0.17
σ_i, nm	0.64	0.59	0.56	0.57	0.52	0.66	0.60	0.56	0.56	0.45
Residue	HIS	ALA	TYR	CYS	TRP	VAL	MET	ILE	PHE	LEU
ε_i, ε	0.25	0.26	0.49	0.54	0.64	0.65	0.67	0.84	0.97	1.00
σ_i, nm	0.61	0.50	0.65	0.55	0.68	0.59	0.62	0.62	0.64	0.62

The most hydrophilic residue has a ε_i of 0, while the most hydrophobic has a value of 1. For residue-residue interactions, we use the Lorentz-Berthelot mixing rules $\sigma_{i,j} = \left(\sigma_i + \sigma_j\right)/2$, $\varepsilon_{i,j} = \sqrt{\varepsilon_i \varepsilon_j}$

propose a one-bead-per-residue model and consider the structure of the protein as a rigid body. We have studied the predictions of this model in more detail in Ref. [35]. The crystal structures of the proteins are obtained from the literature and one bead is per amino-acid is placed at the position of the α-carbon. At the end of this section, we will test the validity of this first approximation. The second approximation is what level of detail will be needed to represent the NP. In this work, we will consider spherical homogeneous NPs so a single-bead representation is justified.

In our model, the total NP surface-protein interaction potential (U) is a function of distance from the surface to the centre of mass (COM) of the protein, d_{COM} and of protein orientation. It is given by a sum of two contributions:

$$U = \sum_{i=1}^{N} \left(U_i^{\mathrm{VdW}} + U_i^{\mathrm{el}}\right) \tag{7.1}$$

where N is the total number of residues in the protein, U_i^{VdW} is the van der Waals interaction of residue i with the surface and U_i^{el} is the electrostatics interaction of residue i with the surface.

For van der Waals contribution to the potential energy we propose a modified version of the residue-residue interaction potential as suggested in [21]. The model is based on the widely used residue-residue interaction energies proposed by Miyazawa and Jernigan [36], but instead of having a 20 × 20 interaction matrix this is reduced to a table of normalized hydrophobicities, ε_i, one for each amino acid (see Table 7.2 in [21]). A hydrophobicity index 0 is assigned to the most hydrophilic residue (LYS) and an index 1 to the most hydrophobic one (LEU). We should stress that any other hydrophobicity scale can also be used, it just has to be transformed such that the indices have to be between 0 and 1, where 0 is assigned to the most hydrophilic residue while 1 to the most hydrophobic one. In this work, we consider a generic surface which chemical reactivity that can be modeled as another residue with a hydrophobicity index ε_s.

To model interaction of biomolecules with particles of different sizes we use the following model for the nanomaterial. We assume that the interaction between a residue i and a bead of the NP s being at a distance r from each other is given by a modified 12-6 Lennard-Jones potential:

$$U_{s,i}(r)=\begin{cases}4\varepsilon_{e,n}\left[\left(\frac{\sigma_{s,i}}{r}\right)^{12}-\left(\frac{\sigma_{s,i}}{r}\right)^{6}\right]+\varepsilon_{e,n}\left(1-\varepsilon_{s,i}\right) & r<r_{c,i}\\ 4\varepsilon_{e,n}\varepsilon_{s,i}\left[\left(\frac{\sigma_{s,i}}{r}\right)^{12}-\left(\frac{\sigma_{s,i}}{r}\right)^{6}\right] & r_{c,i}\le r\le r_{\text{cut}}\\ 0 & r>r_{\text{cut}}\end{cases} \tag{7.2}$$

$\varepsilon_{e,n}$ is a parameter that scales the interaction energy, $\varepsilon_{s,i}$ is the combined hydrophobicity index of residue i and the nanomaterial and is given by $\varepsilon_{i,s}=\sqrt{\varepsilon_i\,\varepsilon_s}$, $\sigma_{s,i}$ is the average van der Waals radius of residue i and the nanomaterial bead, $\sigma_{s,i}=(\sigma_s+\sigma_i)/2$, $r_{c,i}$ is the position of the minimum of the pair potential.

An integration of the 12-6 potential over the volume of the nanomaterial as defined in [21] gives a 9-3 Lennard-Jones-type potential. For a flat surface, the interaction can be expressed in terms of d, the distance between the residue centre of mass the closest element of the surface. An integration over a semi-space gives:

$$U_{si}^{\text{vdW}}(d)=\begin{cases}\varepsilon_{es}\rho\sigma_{s,i}^{3}\left[\left(\frac{\sigma_{s,i}}{d}\right)^{9}-\frac{15}{2}\left(\frac{\sigma_{s,i}}{d}\right)^{3}+\left(\frac{125}{2}\right)^{\frac{1}{2}}\left(1-\varepsilon_{s,i}\right)\right] & d<d_{c,i}\\ \varepsilon_{es}\varepsilon_{s,i}\rho\sigma_{s,i}^{3}\left[\left(\frac{\sigma_{s,i}}{d}\right)^{9}-\frac{15}{2}\left(\frac{\sigma_{s,i}}{d}\right)^{3}\right] & d_{c,i}\le d\le d_{\text{cut}}\\ 0 & d>d_{\text{cut}}\end{cases} \tag{7.3}$$

where $\varepsilon_{es}=4\pi/45\,\varepsilon_{e,n}$, ρ is the number density of beads in the nanomaterial, d is the distance from the residue i to the surface, $d_{c,i}=(2/5)^{1/6}\sigma_{s,i}$. Although the density ρ seems to be an important parameter scaling the interaction, it is not independent and therefore is not crucial for our method. From fitting the adsorption energy to experimental or MD simulation data, we can find the composite quantity $\varepsilon_{es}\rho$, which is sufficient for further calculations. For a NP of radius R, a similar integration over the particle volume gives:

$$U_{si}^{\mathrm{vdW}}(r)=\begin{cases}4\varepsilon_{es}\rho\sigma_{s,i}^{3}\left[\frac{\left(15r^{6}R^{3}+63r^{4}R^{5}+45r^{2}R^{7}+5R^{9}\right)\sigma_{s,i}^{9}}{\left(r^{2}-R^{2}\right)^{9}}-\frac{15R^{3}\sigma_{s,i}^{3}}{\left(r^{2}-R^{2}\right)^{3}}\right]-U_{c}^{\mathrm{vdW}}\left(1-\varepsilon_{s,i}\right) & r<r_{c,i}\\ 4\varepsilon_{es}\varepsilon_{s,i}\rho\sigma_{s,i}^{3}\left[\frac{\left(15r^{6}R^{3}+63r^{4}R^{5}+45r^{2}R^{7}+5R^{9}\right)\sigma_{s,i}^{9}}{\left(r^{2}-R^{2}\right)^{9}}-\frac{15R^{3}\sigma_{s,i}^{3}}{\left(r^{2}-R^{2}\right)^{3}}\right] & r_{c,i}\leq r\leq r_{\mathrm{cut}}\\ 0 & r>r_{\mathrm{cut}}\end{cases} \tag{7.4}$$

where r is the distance from residue i to the centre of the NP. The distance $r_{c,i}$ corresponds to the minimum of the potential and U_c^{vdW} is the value of the function $U_{s,i}^{\mathrm{vdW}}(r_{c,i})$ as defined in the range $r_{c,i} \leq r \leq r_{\mathrm{cut}}$. We do not show the general expression for the position of the minimum as it is too bulky. The minimum is located at $r_{c,i} - R \approx (2/5)^{1/6}\,\sigma_{s,i}$ at $R \gg \sigma_{s,i}$ and is displaced to shorter distances at smaller R. The variation, however, is not very large, at $R \to \infty$, $r_{c,i} - R \approx 0.858374\sigma_{s,i}$, at $R = 200\sigma_{s,i}$ it is $0.858375\sigma_{s,i}$, at $R = 20\sigma_{s,i}$ it is $0.858469\sigma_{s,i}$, at $R=2\sigma_{s,i}$ it is $0.865242\sigma_{s,i}$.

Note that the potential in Eq. (7.2) will only give a repulsive interaction between a highly hydrophilic surface and any residue (*i.e.* defining $\varepsilon_s = 0$, gives $\varepsilon_{s,i} = 0$ for all residues). On the other hand, assigning a non-zero value for ε_s will only change the magnitude of the interaction between any residue and the surface but not the shape of the potential. In this way, the proposed potential is limited to model only hydrophobic surfaces. Because of this limitation, we set the value of $\varepsilon_s = 1$ for all simulations. Alternatively, a potential that includes desolvatation penalties, as the 12-10-6 Lennard-Jones potential proposed in [37, 38] for residue-residue interactions or the modified version proposed in [23] used to model residue-surface interactions, can be used to generate a more general interaction potential. The main drawback of the use of these more refine formulas for the potential is that the parameterization is more challenging, and the applicability of a set of parameters could be very narrow.

The electrostatic interactions in Eq. (7.1) is modeled by adding point charges on the NP surface. This charges interact with the charged residues via a Debye-Hückel potential. The electrostatic interaction energy between a residue i and all the charges on the surface is given by:

$$U_i^{\mathrm{el}} = \sum_{j}^{N_e} \lambda_B k_B T q_i q_j \frac{\exp\left(-r_{ij}/\lambda D\right)}{r_{ij}} \tag{7.5}$$

where r_{ij} is the distance between the residue i and the point charge on the surface j, $\lambda_B = e^2/(4\pi\varepsilon_0\varepsilon_r k_B T)$ is the Bjerrum length, k_B is the Boltzmann constant, T the temperature, ε_0 the dielectric permittivity of vacuum, ε_r the relative dielectric permittivity of water, q_i the charge of residue *i*, q_j the charge of the point charge *j* on the surface, N_e the total number of point charges on the surface and λ_D is the Debye length (defined through $\lambda_D^{-2} = 8\pi\lambda_B c_0$, with c_0 is the background electrolyte concentration). In practice, the points charges are evenly distributed on the spherical surface of the NP using a Golden Section spiral algorithm and all points will have the same charge q_j given by $q_j = 4\pi\sigma R^2/N_e$, where σ is the surface charge density of the NP and *R* is the radius of the NP.

7.2.2 Orientational Sampling and the Calculation of the Adsorption Energy

In this work, we are not considering conformational changes during the adsorption process and assume that proteins are rigid. Although the adsorption process might conduce to conformational changes, this events happen at longer times than orientational changes on the surface [39]. Taking this into account, the adsorption energies calculated here will give a valuable insight into the long-time evolution of the of the NP-protein corona content.

In our CG model, each residue of a protein is represented by a single bead located at the α-carbon position. The native structures are obtained from the Protein Data Bank, and in Table 7.1 we report the proteins studied in this work, the PDB ID from which the CG model were built and the abbreviation that will be used in the rest of the text. The chosen proteins are some of the most abundant in human blood and will have a major influence in the formation of the NP protein-corona.

To identify the most favourable orientation of adsorbed protein globule (the one with the minimum adsorption enthalpy) we will follow the method suggested in [40], which is not as efficient as *e.g.* a genetic algorithm, but can provide additional information about the adsorption process. Briefly, a configuration space search is performed, where a systematic rotation of the protein allows us to build an adsorption map. There are three degrees of freedom (DOF) that have to be scanned. Figure 7.2 shows that any point on the surface of the protein can be defined by a position vector from the COM of the protein. This vector is characterised by two angles: ϕ and θ and by rotating the molecule an angle $-\phi$ around the *z* direction and then by an angle $-\theta + 180°$ around the *y* axes will make the position vector point towards the surface. The third DOF is the distance from the COM to the closest point of the surface, d_{COM}. Here, we sample ϕ from 0 to 350° in steps of 10° and θ from 0 to 170° in steps of 10° (note that $\phi = 0°$ is equivalent to $\phi = 360°$, and that $\theta = 0°$ is equivalent to $\theta = 180°$). Instead of obtaining the "real" adsorption free energy by calculating the potential of mean force for all orientations, we only calculate the potential energy *U* (given by Eq. (7.1)), which is the sum of all the interactions between the surface and the protein. As the adsorption energies are expected to be at least five times k_BT and as the proteins are assumed to be rigid, neglecting thermal fluctuations

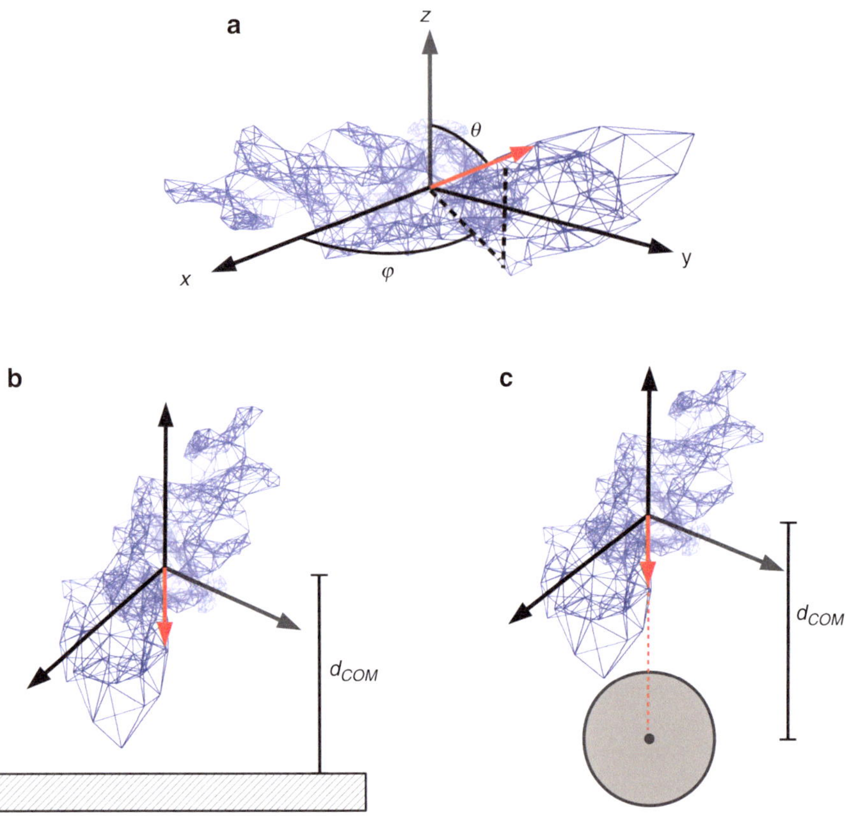

Fig. 7.2 Definition of the protein orientation. (**a**): Any point on the surface of the protein can be defined by a position vector from the COM to that point and depends on two angles ϕ and θ. The remaining degree of freedom is the distance of the COM, d_{COM} to (**b**) the surface for a slab or (**c**) to the center of the NP

is clearly justified. For each configuration (ϕ_i, θ_j), the total potential energy is calculated as a function of distance of the COM, U (d_{COM}, ϕ_i, θ_j), to the surface for the case of a slab (Fig. 7.2b) or to the center of the NP for the case of a NP (Fig. 7.2c). Following a similar approach as in [41], and denoting the reaction coordinate $d_{COM} = z$, the adsorption energy for any particular configuration in the case of a protein adsorbing on a flat surface is given by:

$$E\left(\phi_i,\theta_j\right) = -k_B T \ln\left[\frac{1}{a\left(\phi_i,\theta_j\right)}\int_0^{a\left(\phi_i,\theta_j\right)} \exp\left(-U\left(\phi_i,\theta_j,z\right)/k_B T\right)dz\right] \quad (7.6)$$

where $a(\phi_i, \theta_j)$ is the maximum interaction distance from the COM of the protein to the surface for the given orientation. For the case of a NP-protein interaction, the mean interaction energy for any particular orientation is given by:

$$E\left(\phi_i,\theta_j\right) = -k_B T \times \ln\left[\frac{3}{\left(R+a\left(\phi_i,\theta_j\right)\right)^3 - R^3}\int_R^{R+a\left(\phi_i,\theta_j\right)} z^2 \exp\left(\frac{-U\left(z,\phi_i,\theta_j\right)}{k_B T}\right) dz\right] \tag{7.7}$$

Then the total mean adsorption energy of the system for both cases (slab and NP), E_{ad}, can be estimated by averaging over all adsorbed states with Boltzmann weighting [40]:

$$E_{\text{ad}} = \frac{\sum_i \sum_j P_{ij} E\left(\phi_i,\theta_j\right)}{\sum_i \sum_j P_{ij}} \tag{7.8}$$

where $P_{ij} = \sin(\theta_j)\exp[-E(\phi_i, \theta_j)/k_B T]$ is the Boltzmann weighting factor.

7.2.3 *Details of the Simulations, Parameterisation and Validation*

All simulation were performed using ESPResSo MD package [42] and the cutoff for the interaction potential in Eq. (7.2) was set to $r_{\text{cut}} = 6$ nm. For all calculations the simulation box was taken big enough to fit the NP and the protein. The method described here only involves the calculation of the total energy of the system given by Eq. 7.1, therefore a coupling to a thermostat is not required. After the CG model were built from the PDB files, the obtained structures were shifted so the COM of the molecules was in the origin of the frame of reference and this structure was defined as the ($\phi = 0°$, $\theta = 0°$) orientation. With this definition the first residue in the sequence of each protein will have the following (ϕ, θ) angles: (21.4°, 85.2°) for HSA and (132.1°, 46.4°) for Fib.

The units of the simulations are: lengths (L) in nm, energy (ε) in $k_B T \approx 4.15 \times 10^{-21}$ J taking a temperature of $T = 300$ K, for the mass unit (M) we selected the average mass of the 20 residues (ca. 110 Da) hence in our simulations all residues have a mass of 1. The values of ε_i and σ_i can be found in Table 7.2 and as mentioned in Sect. 7.2.1 we will only consider hydrophobic NPs with $\varepsilon_s = 1$ and $\sigma_s = 0.35$ nm.

NPs with negative surface charges as well as neutral NPs were considered. For the negatively charged ones, a surface charge density of −0.02 C/m^2 was used. As explained in Sect. 7.2.1, the charged surfaces are modelled by individual point charges. The surface density of these charged beads ($\sigma_c = N_e/R^2$) was set to 4 nm^{-2} for all the simulations, which gives e.g. a $N_e = 100$ for a NP of $R = 5$ nm. Then, we assumed that each bead carries a charge of $-0.39e$, where e is the elementary charge. As we are considering physiological conditions, we use $\lambda_B = 0.73$ nm and $\lambda_D = 1$ nm. Residue charges at this condition are $+e$ for LYS and ARG, $-e$ for ASP and GLU, and $+0.5e$ for HIS. The rest of the residues are neutral.

The only free parameters of the model are $\rho\varepsilon_{es}$ in Eq. (7.2), and the parameterisation was done by systematically changing its value to match experimental data of adsorption of Lysozyme on hydrophobic surfaces reported by Chen et al. [43]. The native structure for our CG model of Lysozyme was obtained from the PDB ID: 2LYZ. With $\rho\varepsilon_{es} = 1.972 k_BT/\text{nm}^3$ we obtain a value of $-7.6k_BT$ for the adsorption energy (very close to the experimental reported value of $-7.9k_BT$).

To validate the parameterization, the adsorption energy of Myoglobin (PDB ID: 1MBN used for the CG model) was calculated using the same value of $\rho\varepsilon_{es}$ obtained from the parameterisation. In this way, a value of $-6.1k_BT$ was found for the adsorption energy of Myoglobin. This value is slightly lower that the experimental value of $-7.6k_BT$ also reported by Chen et al. [43] but reproduces the trend that Myoglobin adsorbs slightly weaker than Lysozyme to a hydrophobic surface.

7.2.4 Protein Adsorption Energies

Results for the adsorption energies calculated using Eq. (7.8) as a function of NP radius are shown in Fig. 7.3. The results show that HSA adsorbs stronger as the radius of the NP increases until the energy reaches a minimum value (Fig. 7.3a). For small NPs, the combination of the size effect (increasing R increases the van der Waals interactions) with the availability of residues to interact with the surface ensures that the proteins adsorb stronger (more negative values) as the radius is increased. Then, after a value of radius around 50 nm, the E_{ad} starts to converge to the value corresponding to a flat surface as the van der Waals interactions and the number of residues close to surfaces do not change significantly by increasing R. We performed calculations for NPs of R up to 500 nm and confirmed that the adsorption energy indeed converges to the slab values. For the Fib molecule the situation is different (Fig. 7.3b). In this case the adsorption energy decreases as a function of R at least until the biggest radius studied here (R = 100 nm) and it is lower than for the adsorption onto a flat surface. The big size of the Fib molecule (ca. 45 nm on its longest axes) makes that for at least until R = 100 nm the combined effects of curvature and number of residues that interact with the surface are still noticeable. The effect of the charge is more important for the HSA that for the Fib. HSA charge is overall negative, so the electrostatic interactions contribution is mainly repulsive increasing the values of the E_{ad}. On the other hand, the Fib molecule's charge is positive and the electrostatic interactions tend to increase the adsorption of the Fib onto a negative surface. In neither of the proteins the maximum contribution of electrostatic interactions was more than $3k_BT$.

The systematic sampling employed for the calculation of the adsorption energies can also be used to identify the most favourable orientations for adsorption and to

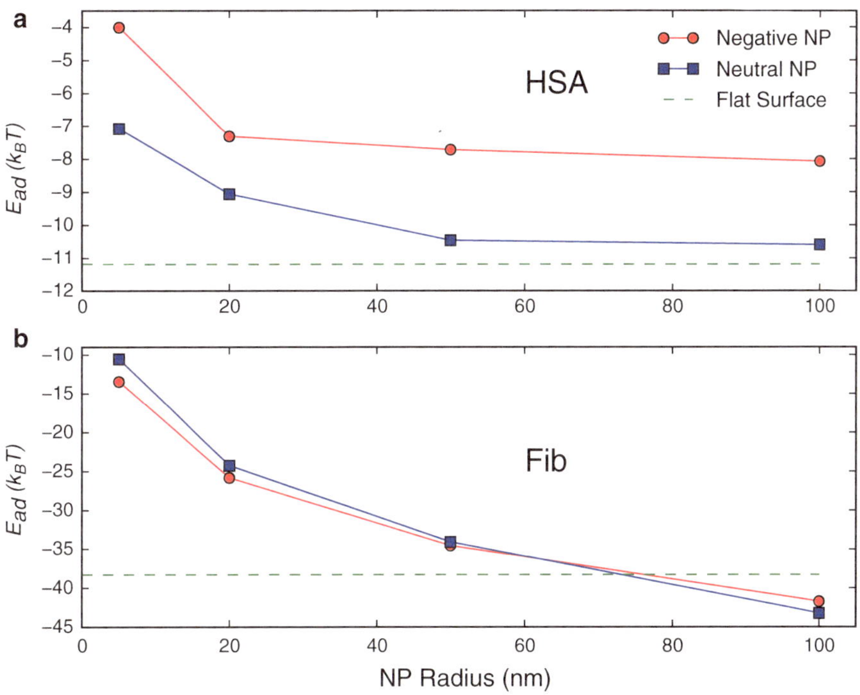

Fig. 7.3 Adsorption energies as a function of the NP radius for the proteins studied and the two types of surface charge: Negative: −0.02 C/m^2 and Neutral: no charge. Subfigure (**a**) shows the results for HSA and (**b**) for Fib. The dashed lines show the adsorption energy for the case of a flat surface

study how the charge and/or the radius of the NP influence the protein orientation. Figure 7.4 shows a surface map of the adsorption energy as a function of the angles θ and ϕ for HSA. Each panel is for a radius of 5 or 100 nm and for a neutral or a negatively charged surface. The surfaces are complex in structure showing an energy landscape with several local minima with differences less than $1k_BT$. It is also important to notice that the maps have large areas with adsorption energies of $-6k_BT$ or lower. Our results show that HSA will strongly adsorb at physiological conditions and room temperature and that orientational changes after adsorption are energetically favourable. Comparison of different panels in Fig. 7.4 shows that radius has only a small effect on the preferred orientations, while the NP surface charge density has a minor impact on the preferred orientations.

A different scenario is observed for Fib. Figure 7.5 shows colour maps of Fib adsorption energy for two radii for neutral and charged surfaces. In this case, the maps depend on the radius of the NP (compare Fig. 7.5a with Fig. 7.5b or Fig. 7.5c with Fig. 7.5d) but change very little between the charged and uncharged surface (compare Fig. 7.5a with Fig. 7.5c or Fig. 7.5b with Fig. 7.5d). As we already noticed for HSA, the charge has a small effect on the total adsorption energy so we do not expect that it

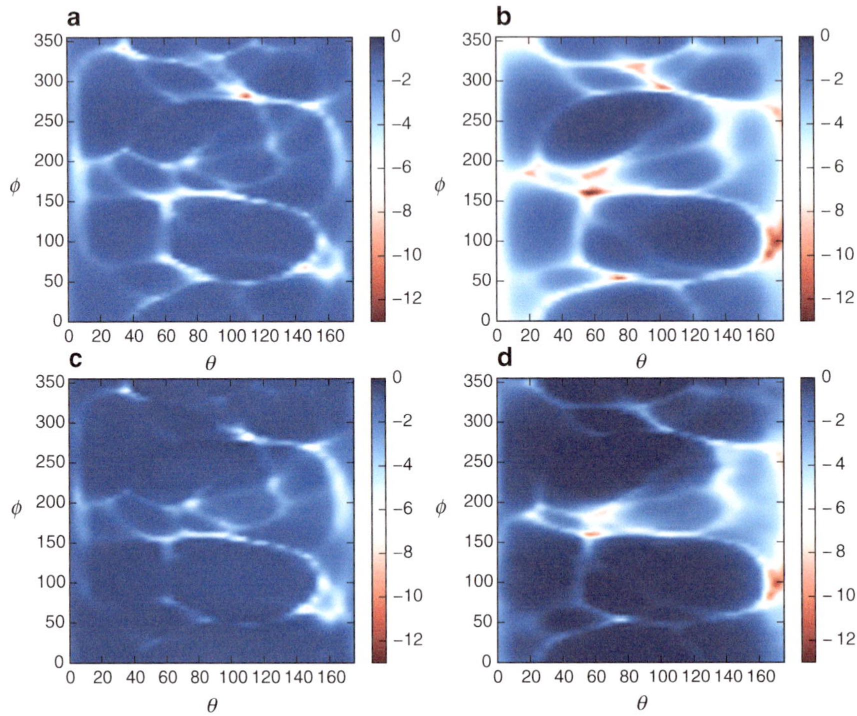

Fig. 7.4 Adsorption energy maps for HSA. (**a**) Neutral particle of $R = 5$ nm. (**b**) Neutral particle of $R = 100$ nm. (**c**) Negatively charged particle of $R = 5$ nm. (**d**) Negatively charged particle of $R = 100$ nm

would dramatically change the energy maps. The radius and the surface curvature seem to be more important as for big proteins (like Fib) a larger NP allows a more extensive contact and thus influences the preference for protein orientations (or NP binding pockets). In Fig. 7.6, we show the most favorable orientations for Fib on a neutral surface for two different NP radii. For the small NP (Fig. 7.6a), Fib has its adsorption energy minimum in a configuration where the NP interacts with a relatively small segment of the molecule. Meanwhile, for a large NP, Fib tends to bind in a completely different orientation (Fig. 7.6b). Now the most favourable orientation is the one with the longest axis of the Fib molecule along the surface.

A straightforward conclusion from the above data is that the bigger the protein, the stronger it will bind to a NP. This result agrees with the experimental observation reported by De Paoli et al. [44], which shows that the binding association constant on citrate-coated gold NPs (which can be considered as moderately negative hydrophobic NP) depends mainly on the size of the protein (they studied HSA, Fib and other blood proteins). It is interesting also to compare our results with the simulations of NP corona formation reported by Vilaseca et al. [45]. Using CG MD simulations, they found that for a flat surface at long times the most abundant protein adsorbed was Fib, then Immunoglobulin-γ (of intermediate size between HSA and

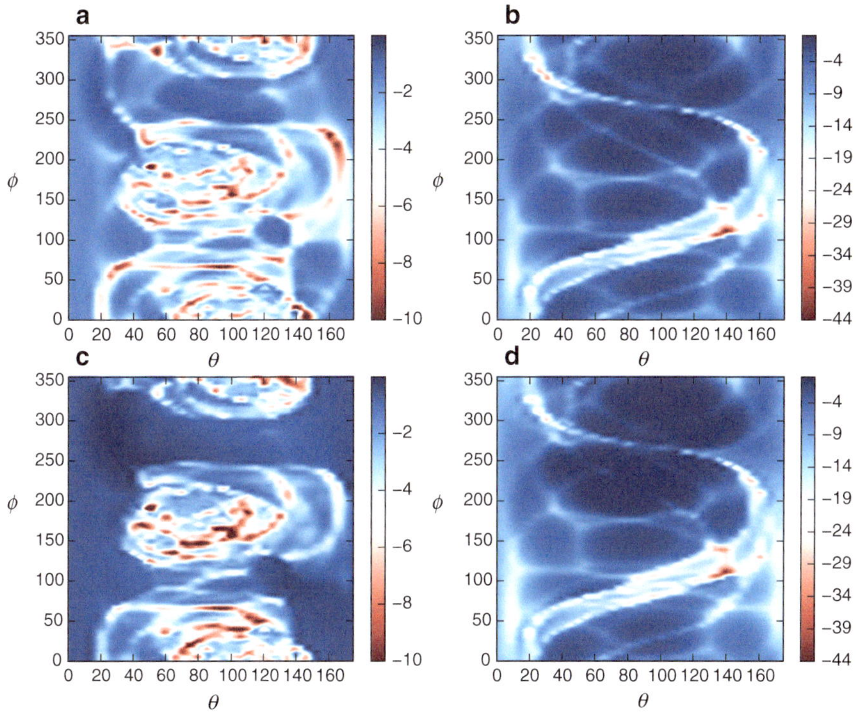

Fig. 7.5 Adsorption energy maps for Fib. (**a**) Neutral surface and $R = 5$ nm. (**b**) Neutral surface and $R = 100$ nm. (**c**) Negatively charged surface and $R = 5$ nm. (**d**) Positively charged surface and $R = 100$ nm

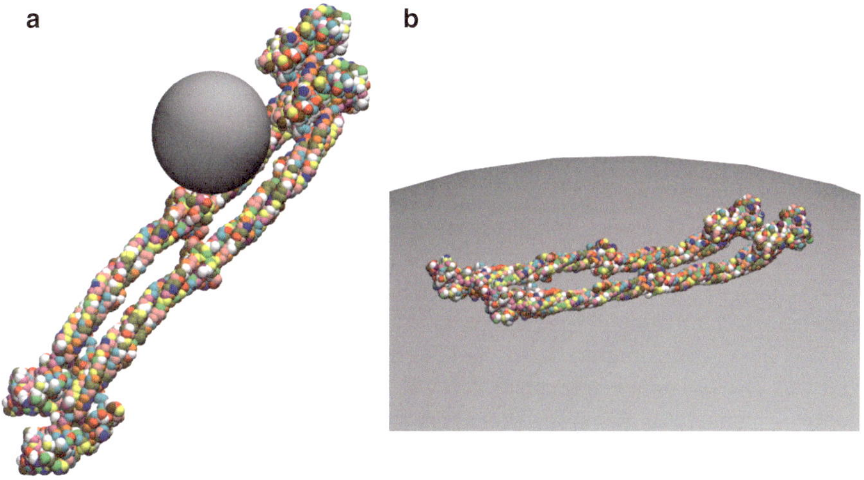

Fig. 7.6 Fib most favorable orientation for adsorption on a neutral surface: (**a**) $R = 5$ nm and (**b**) $R = 100$ nm

Fib) and at last HSA. We should note that the adsorption energies calculated in this work will be a good predictor of the equilibrium composition of the NP-protein corona while at short times other factors such as the protein sizes and their concentrations have to be considered to predict the corona composition.

We should note that the approach presented here is justified for small NPs but may be difficult to use with large NPs for two main reasons. First of all, the interaction potential resulting from integration of the van der Waals forces over the volume of the NP will become long-range in that case. Here, we explicitly assumed that the particle is hydrophobic, so that there a non-negligible attraction will be felt by the protein far away from the NP (up to tens of nanometers), well beyond the typical range of the interactions of individual molecules. In our example, the prefactor $\varepsilon_{s,i}$ scales the interactions, so that the attraction is strongest between the hydrophobic NP and the hydrophobic residues. We performed the energy calculation without any cut-off but had to limit the interaction radius in the MD simulation below to 6 nm. Second, it is not a priori clear whether the same hydrophobicity coefficient $\varepsilon_{s,i}$ can be used to describe the interaction of the bulk material of the NP with the protein as we determined for the surface beads. While the interaction at small distances is modified by water structuring at the surface, the long-range van der Waals force should not be affected by the local effects. Therefore, the coefficients for hydrophilic materials may underestimate the attraction between the NP and biomolecules. To overcome this limitation, one needs to treat the bulk of the NP differently from the surface layer. From this point of view, it would be reasonable to introduce a two-layer model of a NP, where the surface layer takes into account hydration and the attraction of biomolecules to the NP is not underestimated due to the short cut-off. To include the attraction in full, one must use cut-off distances of at least particle diameter, i.e. 5 nm for 2.5 nm NP, 10 nm for 5 nm NP, etc. The main issue to be solved in future modelling is how to increase the cut-off of the NP bulk material interaction with the biomolecule beads in common simulation codes without affecting dramatically the computational cost.

7.2.5 *Validation of the Methodology*

We now test our CG methodology with predictions of full-atomistic MD simulation. We model adsorption of small plasma protein Ubiquitin (Ubi) to a flat TiO_2 surface. The reasons for choosing Ubi for the validation was due to the its small size (only 76 residues) and known folded structure, which allows us to perform full atomistic simulations in a reasonable amount of time.

The Ubi crystal structure was obtained from the PDB (PDB ID file 1Ubi [46]) and was coarse-grained as explained above (see Fig. 7.7). To be able to directly compare against full atomistic simulations, in this case the interactions potentials between the 20 different residues and the surface were obtained by performing full atomistic simulations of the adsorption of each of the 20 aminoacids and then performing an inverse Monte Carlo calculation.

Fig. 7.7 CG model of Ubi (PDB ID: 1Ubi [46]). In our model each residue in the protein is represented by one bead

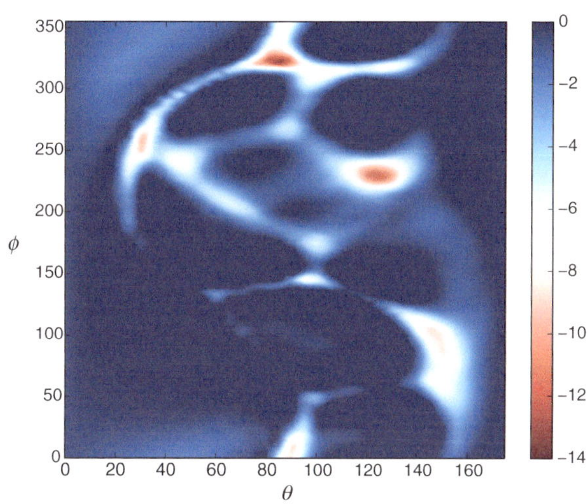

Fig. 7.8 Adsorption energy maps for Ubi adsorbing into a TiO_2 slab

With the CG methodology we obtained the total adsorption energy of $-10.7k_BT$ for Ubi and the adsorption map is shown in Fig. 7.8. The surface obtained shows two major minima and a number of local minima. It also predicts that orientational changes are favorable after adsorption as some of the minima are connected by an energy landscapes with rather small barriers (less than $3k_BT$).

To analyse the validity of our model and to understand the dynamical behavior of the adsorption process we performed a series of full atomistic simulations. We used the VESTA program [47] to construct a 5 × 20 × 32-supercell of the TiO_2 rutile unit cell. The coordinates were rotated so that the normal of the TiO_2 slab, corresponding to the (100) surface, was oriented along the z-direction. The box was elongated in the z-direction and periodicity was assumed in all directions. The final size of the simulation box was then 9.466 × 9.184 × 12 nm. Covalent bonds were added to all Ti-O pairs within a 2 Å-cutoff. We used force field parameters for the TiO_2 slab from a recent parameterisation study [48]. The same force field was use to calculate the CG potential interactions between the surface and the 20 amino acids. The same folded Ubi structure as for the CG model was used and inserted above the TiO_2 slab. The TiO_2-Ubi system was solvated by insertion of 25,817 water molecules around the protein and the slab, and the final system contained 99,802 atoms. The system was energy minimized for 1000 steps and then equilibrated at constant temperature (300 K) and pressure (1 bar) for 100 ps using Berendsen's weak scaling algorithms [49], with relaxation constant $\tau =$ 1 ps in both cases. The temperature coupling was applied independently to the TiO_2 slab and to the rest of the system. The pressure tensor must be kept anisotropic due to the solid TiO_2 slab, but the off-diagonal components of the compressibility tensor (and the reference pressure tensor) were set to zero to enforce a rectangular simulation box. The diagonal elements of the compressibility tensor were set to 5×10^{-7} bar^{-1} in the lateral directions (bulk TiO_2) and 5×10^{-5} bar^{-1} in the normal direction (bulk water). The box vectors relaxed 2–4 % during equilibration.

In a first simulation, we placed the protein in the ($\phi = 0°$, $\theta = 0°$) orientation close to the surface and followed the dynamics for 440 ns at constant volume and 300 K. The Nose-Hoover thermostat [50, 51] with the coupling constant $\tau = 5$ ps was used to ensure proper sampling of the ensemble when controlling the temperature. The simulation was run in parallel using 512 cores and frames were kept every 5 ps.

The trajectory obtained showed that the protein motion was diffusive in the bulk water for about 25 ns until making contact with the TiO_2 slab. Then the Ubi molecule attached to the surface and remained adsorbed for the rest of the simulation. To study the stability of the structure of the protein during adsorption we calculated the root-mean-square-deviation (RMSD) as a function of time and the results are shown in Fig. 7.9. The RMSD remained at a low constant value of ca. 0.2 nm^2 during the simulation, i.e. no unfolding occurred in the adsorbed state. This results clearly confirms that for the adsorption of Ubi on TiO_2 a rigid body model for the protein structure is well justified.

A detailed study of the simulation trajectory revealed that the protein motion could be characterized by four states, and that adsorption occurs through a two-step mechanism (Fig. 7.10). First, the protein diffuses freely in the bulk water. Second,

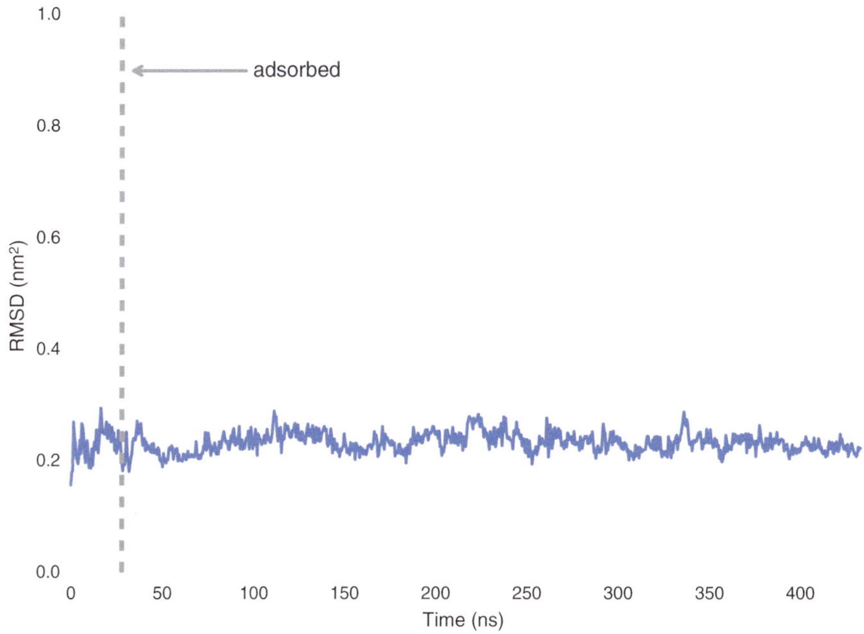

Fig. 7.9 The root-mean-square deviation (*RMSD*) of Ubi during the simulation with respect to the PDB reference structure. No unfolding occurs and the RMSD is 0.15 nm^2 throughout the simulation, which is the same as found in simulations of the folded structure in bulk water

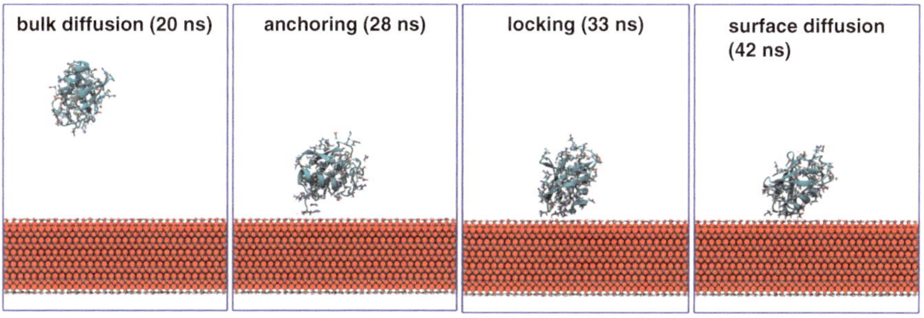

Fig. 7.10 Four distinct stages were identified during adsorption. (**a**) The protein diffuses in the bulk water. (**b**) The N-terminus of Ubi anchors to the solvation layer of the TiO_2 slab. (**c**) The protein rotates and locks to the solvation layer through GLN40 and GLN31. (**d**) The protein diffuses on the solvation layer in the locked conformation

the C-terminus of Ubi the protein contacts the TiO_2 surface and provides a lock for the protein to the first solvation layer. Third, Ubi rotates and locks into position on the surface. Fourth, the protein diffuses on the surface in the locked orientation.

The adsorption process is relatively fast once the first surface contact is initiated. The anchoring of the C-terminus to the first solvation layer occurs in about 5 ns and the locking is completed after 10 additional nanoseconds. The residues at the

C-terminus are ARG 74, GLY 75 and GLY 76. The anchoring is initiated when the charged end of ARG74 contacts the surface perpendicularly. The contact is not with the bare surface, but with the first solvation layer, which is strongly bound directly to the surface. At this point the rest of the protein diffuses in the bulk until (ca. 5 ns) GLU 40 can contact the surface, which leads to the protein being locked into an adsorbed orientation after 10 ns. The locking procedure consists of Ubi first connecting GLN40 to the surface, followed by a second connection through GLN31. The two glutamines (GLN31 and GLN40) form a bridge that stabilizes the orientation of the protein and no more changes in orientation occur for the rest of the simulation.

The residues involved in the anchor-lock mechanism are arginine and glutamine. These have been identified by potential of mean force calculations [48] of isolated side chain fragments (together with aromatic side chains) to be the strongest binders to TiO_2. In both cases, the NH_2-group of the end of the amino acid approaches the surface in a perpendicular orientation but can then rotate to maximize the interactions with the solvation layer.

The "upright" position of the protein in the adsorbed state suggests that it does not correspond to a free energy minimum. Since the orientation does not change over 400 ns, it is likely that there are high free energy barriers associated with the orientation changing into another free energy basin. To map all values of ϕ and θ, more sampling of the protein adsorption is needed. This could either be done in a repetitive fashion from different starting configurations or with enhanced sampling techniques such as metadynamics.

As for Ubiquitin we do not observe any unfolding within several hundreds of nanoseconds, we can conclude that our rigid protein model for studying adsorption is justified at least for some conditions: small NPs, non-metallic particles and small and compact proteins such that the adsorption energies are within few tens of k_BT. For other situations, one should evaluate the energy to decide whether the model is sufficient. In general, conformational changes can be an important factor for the adsorption dynamics process [44]. This assumption can be relaxed by e.g. using a Gō–Type model (see [34] for a review on CG models of proteins). Furthermore, as the methodology we presented is computationally very efficient and can provide information about the structure of NP-protein complexes, it can be used as an exploring tool to perform more sophisticated and computationally demanding calculations.

7.3 Coarse-Grained Model of a Lipid Bilayer

Any attempt to simulate with some molecular details but at length and time scale involve in the uptake of NPs through a cell membrane must rely on a CG model of the main constituents of the biological membranes. In this section we describe a methodology to systematically CG a lipid bilayer and lipid bilayer containing Cholesterol from the results of full atomistic simulations.

Table 7.3 Composition of the simulations used for the CG of lipids mixtures

System	I. DMPC-Cholesterol	II. DMPC-DOPS	III. DOPS-Cholesterol
Number of DMPC	30	30	–
Number of DPPS	–	30	30
Number of Cholesterol	30	–	30
Number of water	2000	2000	2000
Number of Na+	–	30	30

7.3.1 Molecular Simulations of Various Lipid-Cholesterol Mixtures

We started the CG procedure by performing all-atom molecular dynamics simulations for three lipid mixtures: (i) 1,2-dimyristoyl-sn-glycero-3-phosphatidylcholine with Cholesterol (DMPC + CHOL); (ii) 1,2-dioleoyl-sn-glycero-3-phosphatidylserine with Cholesterol (DOPS + CHOL); (iii) 1,2-dimyristoyl-sn-glycero-3-phosphatidylcholine with 1,2-dioleoyl-sn-glycero-3-phosphatidylserine (DMPC + DOPS). The composition of this systems are reported in the Table 7.3. In each simulation, the starting state was generated randomly and energy was minimized afterwards. Then a short 1 ns NVT simulation at density 1 g/cm^3 was carried out, which was followed by a 100 ns equilibration simulation in NPT-ensemble and a production stage of 400 ns. The Slipids force field was used [52, 53]. Other simulation parameters: time step 2 fs; Nose-Hoover isotropic thermo/barostat with temperature 303 K, pressure 1 bar, relaxation times 0.1 and 1 ps for thermostat and barostat respectively; all bonds were constrained by Links algorithm; particle-mesh Ewald with Fourier spacing 1 Å and tolerance parameter 10^{-5}. The configurations were saved in the trajectory each 10 ps. The atomistic simulations were performed using the Gromacs simulation engine (v. 4.5) and a rigid TIP3P water model.

7.3.2 Mapping of Atomistic to Coarse-Grained Trajectories: From Residue to Beads

The atomistic trajectories obtained in the simulations were mapped onto CG trajectories, and radial distribution functions between sites of the CG models have been determined. As shown in Fig. 7.11, 10 beads for representation of DMPC molecule were used at the CG level (3 beads instead of each of the two hydrocarbon tails, 4 beads instead of the head group including esters), 14 beads for DOPS molecule (5 beads instead of each hydrocarbon tail with specific distinguishing of the beads with double bond and beads uniting 3 or 4 methylene groups, and 4 beads instead of the head group), 5 beads for CHOL molecule, and Na+ ions as a single bead were used. Water was not included into CG model but its effect was included into

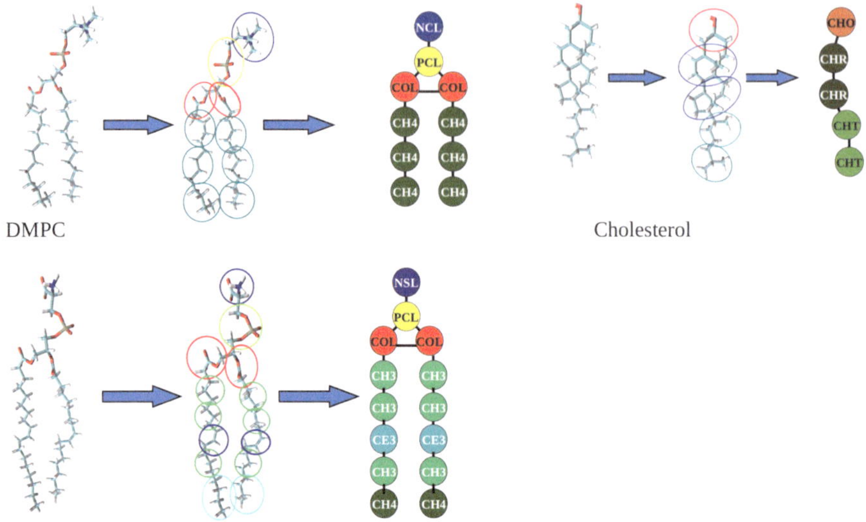

Fig. 7.11 Mapping of systems at an atomistic level to a CG level where each residue of the atomistic system is replaced by a bead for DMPC, Cholesterol and DOPS (1,2-dioleoyl-sn-glycero-3-phosphatidylserin) molecules. CG sites of the same type are given by the same color

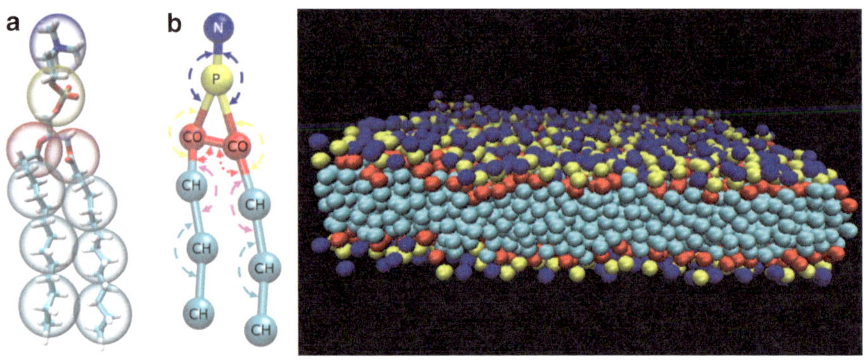

Fig. 7.12 Simulation snapshot of a single CG DMPC lipid molecule (*left*) and of self-assembled DMPC bilayer 15 × 15 nm containing 762 lipids (*right*)

solvent-mediated potentials. Figure 7.12 shows a snapshot of CG DMPC bilayer, which is spontaneously formed in a CG lipid system.

The radial distribution functions (RDF) between CG sites obtained after coarse-graining of the atomistic trajectories were used to compute effective potentials defining interactions in the CG models using the inverse Monte Carlo method. The RDF were computed for each pair of different CG sites and were used as an input to compute effective CG potentials which reproduce the RDFs. Computations of effective potentials were done for the same compositions of the systems I, II, and III as

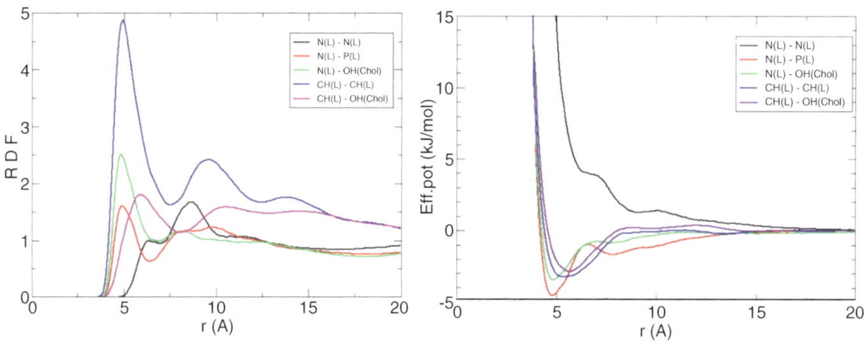

Fig. 7.13 Extraction via Inverse Monte Carlo of CG site-site pair potentials from atomistic RDF's. Five site-site RDFs (*left*) and corresponding effective potentials involving DMPC CG sites of total 28 for DMPC-CHOL mixture (*right*) are shown

the respective atomistic simulations listed in Table 7.3. The software package MagiC [54] was used. In the inversion process, the first 20 iterations have been carried out using iterative Boltzmann inversion, followed by 30–40 iterations using the inverse Monte Carlo algorithm.

More specifically, the RDF's have been determined between beads involved in "non-bonded" interactions, that is between CG sites belonging different molecules or the same molecule but separated by more than two bonds. Also, reference distribution functions for the bond lengths and bending angle distribution functions were determined for the all CG sites relevant for the three types of considered molecules. Then the calculated RDFs and bonded reference distribution functions were used to calculate parameters of the corresponding CG potentials. This is a multistage process, from a high resolution system description to a low resolution one. Monte Carlo computer simulations of the CG system DMPC + CHOL using Metropolis method (MagiC package) were carried out. The parameters were calculated using a two-step iteration technique: first, the iterative Boltzmann inversion method was performed to calculate a set of intermediate parameters; second, the inverse Monte Carlo algorithm was used to calculate the final set of parameters. The final parameters of the CG potentials for DMPC + CHOL mixture have been calculated (see Fig. 7.13).

7.3.3 Validation of the Lipid Coarse-Grained Model

The interaction potentials obtained for the CG models using the inverse Monte Carlo technique were validated by comparison with atomistic simulations. Figure 7.14 shows radial distribution functions between some selected sites of DMPC lipid and Cholesterol computed in CG and atomistic simulations of a mixture of 30 DMPC lipids, 30 Cholesterol molecules and 1800 waters. The result shows a perfect

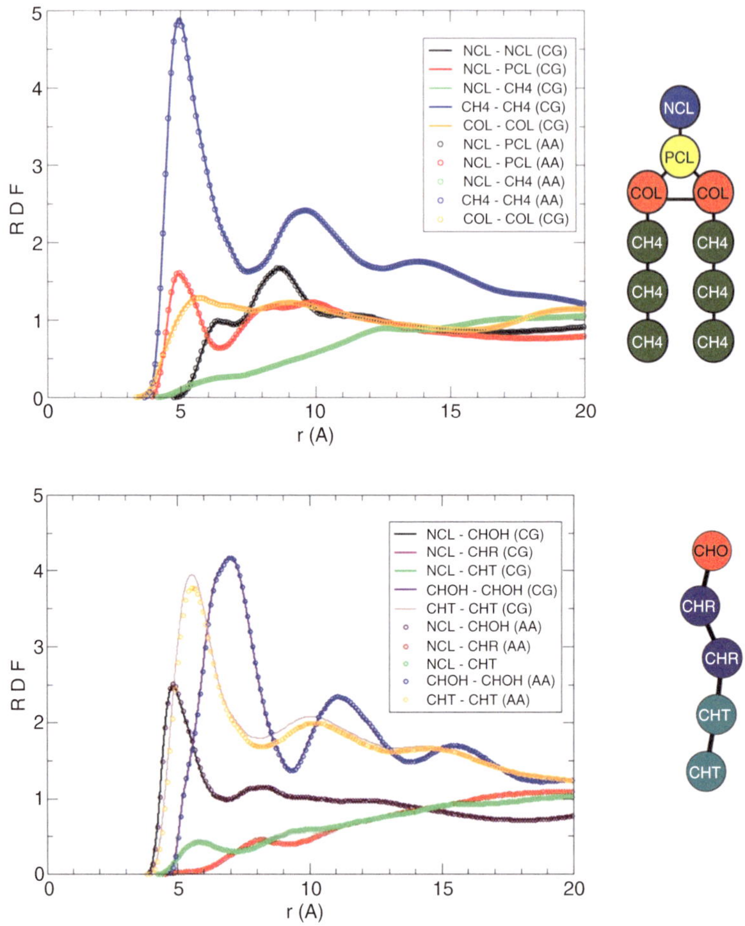

Fig. 7.14 RDFs between different sites of DMPC (*top graph*) and Cholesterol (*bottom graph*) molecules, see site definitions to the right of the graph. Atomistic (*points*) and CG (*lines*) simulations were carried out for a mixture of 30 DMPC, 30 Cholesterol, and 1800 water molecules (CG simulation were without explicit water but in a box of the same size as atomistic simulations including water)

coincidence of the RDFs, which justifies the approximations made and the quality of the CG model.

Figure 7.12 shows a snapshot of CG DMPC bilayer, which is spontaneously formed in a CG lipid system. We have carried out a number of simulations of flat lipid bilayers composed of CG lipid models representing other lipids which can be built from the CG sites presented in Fig. 7.11. These simulations were carried out at zero-tension conditions within the atomistic and CG models. Table 7.4 shows comparison of some properties not related to the RDFs obtained within atomistic and CG simulations of a piece of bilayer composed of 128 DMPC lipids. Very good agreement is observed for the average area per lipid (which is one of the most

Table 7.4 Comparison of the properties of the DMPC bilayer obtained from the full atomistic simulations and the CG model

	Area per lipid	Compressibility	Tail order parameters	
	(Å)	(10^{14} N/nm)	(1)	(2)
Atomistic	60	1.9	0.57	0.52
CG	59	2.5	0.56	0.52

Table 7.5 Average areas per lipid. Comparison of simulation results computed in CG simulations and experiments at T = 303 K

	Area per lipid (Å^2)	
Lipid	Sim	Exp
DMPC (14:0/14:0 PC)	59.0	60.5 [56]
SOPC (18:0/18:1n9 PC)	60.4	61.1 [57]
DOPC (18:1n9/18:1n9 PC)	62.0	67.4 [58]
DSPC (18:0/18:0)	43.5[a]	44.5[a,b] [55]

[a]Bilayer in gel phase
[b]Evaluated in atomistic MD simulations

important parameters for a lipid bilayer) and for the tail order parameter, and a reasonably good agreement for the bilayer compressibility. Especially important is the agreement for the order parameter, which shows that orientational fluctuations of the lipid tails are the same in atomistic and CG models.

Table 7.5 shows average areas per lipid obtained in CG simulations carried out in conditions of zero tension and experiment for a number of lipids. Except DMPC, other lipids included in this table were not used in the parameterization of the CG potentials. The models for these lipids were built from appropriate sites of DMPC and DOPS lipids shown in Fig. 7.11, and the CG interaction potentials were taken as determined in IMC computations for DMPC and DOPS lipids (some of them shown in Fig. 7.13). One can see generally good agreement with experiment, though simulations show a tendency for some underestimation of the lipid area. The bilayer composed of DSPC lipids was found in the gel phase which again is in agreement with experiment (the temperature of gel phase transition for DSPC is 55 °C). We are not aware on an experimental value of the average lipid area for the gel phase of DSPC, but it is generally accepted that average area per lipid in the gel phase is in the range 43–48 Å^2 for phosphatidylcholine lipids. Also, atomistic simulations of DSPC bilayer in a gel phase [55] reports the average lipid area of 44.5 Å^2, which is in good agreement with the result of our CG model.

7.4 NP and Bilayer Simulation

Using the methodologies described in Sects. 7.2 and 7.3, we now can construct a model to simulate the interaction of a DPMC lipid bilayer with a small hydrophobic NP and a hydrophobic NP associated with one molecule of HSA. Following is the description of the simulations and the main results.

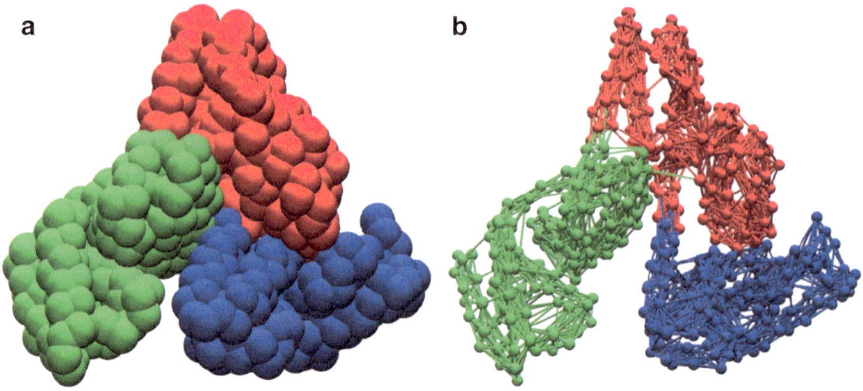

Fig. 7.15 (**a**) CG model of HSA protein used in this study. Each residue is represented by a single bead located at the position of the α-carbon. Each color represents one of the three domains of the HSA molecule. (**b**) All beads separated by less than 10 nm are connected by stiff harmonic bonds

7.4.1 Interaction Potentials and Parameters of the Simulations

The simulated systems are composed of the NP, the lipids that form the bilayer, the amino acids of the HSA, and monovalent ions that are used to resemble physiological conditions. For all interactions we assume two contributions: electrostatic and van der Waals interactions. For the electrostatic contribution, all charged beads interact through a Coulomb potential between given by:

$$U^C\left(r_{ij}\right) = \lambda_B k_B T \frac{q_i q_j}{r_{ij}} \tag{7.9}$$

where r_{ij} is the distance between the bead i and the bead j, λ_B is the Bjerrum length and q_i and q_j are the charges of the beads i and j respectively. The calculation of this long-range interaction was implemented through an Ewald summation P3M algorithm [42]. The Bjerrum length is set to 0.71 nm in all cases. Note that in the MD simulation, the background salt ions are explicitly present, so we do not need to employ the screened Coulomb potential Eq. (7.5).

For the van der Waals interactions we use the following model:

- Aminoacid – aminoacid van der Waals interactions: we do not explicitly consider interaction between any pair of aminoacids within the single protein molecule as the protein is not allowed to change conformation from the PDB crystal structure. To improve the computational efficiency instead of simulating the HSA molecule as a rigid body, we connect all residues which are separated less than 10 nm by harmonics bonds with a spring constant of $100k_BT$. Figure 7.15a shows the resulting CG model for the HSA (build according to PDB ID:1N5U), while Fig. 7.15b shows the resulting network of bonds (8059 in total).

- NP – aminoacid van der Waals interactions: we used the interaction potential defined in Eq. (7.2) and the same parameters for the aminoacids and the NP obtained by the parameterization described in Sect. 7.2.3.
- Lipid-lipid interactions: the CG models of the lipids in the bilayer and the interactions between the four different types of beads were obtained as described in Sect. 7.3.
- Lipid-NP interactions: we used the same interaction potentials as in the case of NP-aminoacid interaction Eq. (7.2) and assumed that lipid beads interact with the surface according to their hydrophobicity. We classify the lipid beads into one of two groups: head or tail. The head beads are NCL, PCL and COL (see Fig. 7.11), and they are considered to be hydrophilic with a value of $\varepsilon_i = 0.1$. The tail beads (labeled CH4 in Fig. 7.11) are hydrophobic and a value of $\varepsilon_i = 0.75$ is used for these groups. The van der Walls radius of all the lipids beads is set to σ_i = 0.6 nm.
- Lipid-aminoacid interactions: for these interactions we use the same approach as for the NP-residues and NP-lipid. The potential interaction is also based on the hydrophobicity of the beads given by the following modified 12-6 Lennard-Jones potential:

$$U_{l,i}(r) = \begin{cases} 4\varepsilon_{la}\left[\left(\frac{\sigma_{l,i}}{r}\right)^{12} - \left(\frac{\sigma_{l,i}}{r}\right)^{6}\right] + \varepsilon_{la}\left(1-\varepsilon_{l,i}\right) & r < r_c \\ 4\varepsilon_{la}\,\varepsilon_{l,i}\left[\left(\frac{\sigma_{l,i}}{r}\right)^{12} - \left(\frac{\sigma_{l,i}}{r}\right)^{6}\right] & r_c \le r \le r_{\text{cut}} \\ 0 & r > r_{\text{cut}} \end{cases} \qquad (7.10)$$

  where r is the distance from the lipid bead l to the residue i, ε_{la} is a free parameter that scales the interaction energy, $\varepsilon_{l,i}$ is the combined hydrophobicity index of lipid l and the residue i and is given by $\varepsilon_{i,s} = \sqrt{\varepsilon_i\ \varepsilon_s}, \sigma_{s,i}$ is the average van der Waals radius of residue i and the lipid $l, \sigma_{l,i} = (\sigma_l + \sigma_i)/2, r_c = 2^{1/6}\sigma_{l,i}$ and r_{cut} is the cut-off for the van der Waals interaction. As in this work we only study the applicability of the proposed methodology we do not systematically parameterize the value of ε_{la}, instead we set this scaling parameter to $0.5k_BT$ for all simulations. This values gives interactions between the lipids and the residues in the same order of magnitude as the ones reported in [59].
- Ion-ion and ion-molecule interactions: in addition to the Coulomb forces, we include excluded volume interactions by means of a WCA potential. The van der Waals radii of the ions are set to 0.2 nm.

For all simulations a NP of radius 2 nm is used with a surface charge of −0.02 C/m^2. We use NVT ensemble with the box size was 15 × 15 × 20 nm and we assume physiological conditions with monovalent salt concentration of 0.1 M (270 negative and positive ions are placed in the simulation box). For a larger NP, NPT simulation would be necessary. To keep charge neutrality, further 16 positive ions are added. The bilayer is composed of a total of 762 lipids (381 lipids in each layer). A Langevin thermostat with a friction coefficient of γ=0.05 is used and the units of mass, energy and charge are the same as described in Sect. 7.2.3. The time unit (τ) is obtained by performing a simulation of the bilayer with ions (no NP or proteins are added) and measuring the lateral diffusion constant. We obtained a value of 8×10^{-5} nm$^2/\tau$, which compared with the experimental value of 5 μm^2/s gives τ = 16 ps.

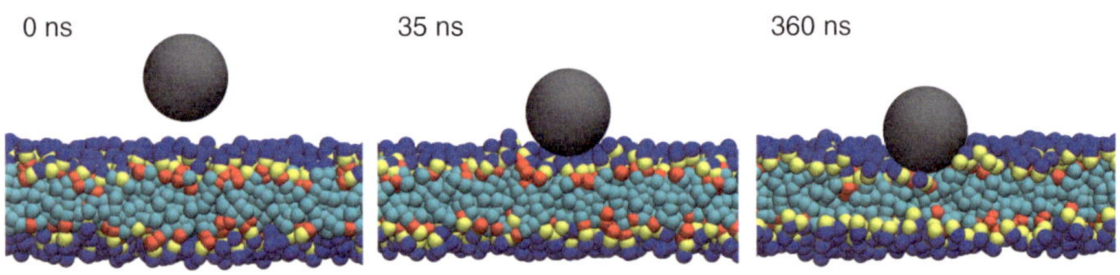

Fig. 7.16 Time sequence of simulation snapshots illustrating the interaction of a DMPC lipid bilayer with a negatively charged hydrophobic NP. The radius of the NP is 2 nm. The ions are not shown

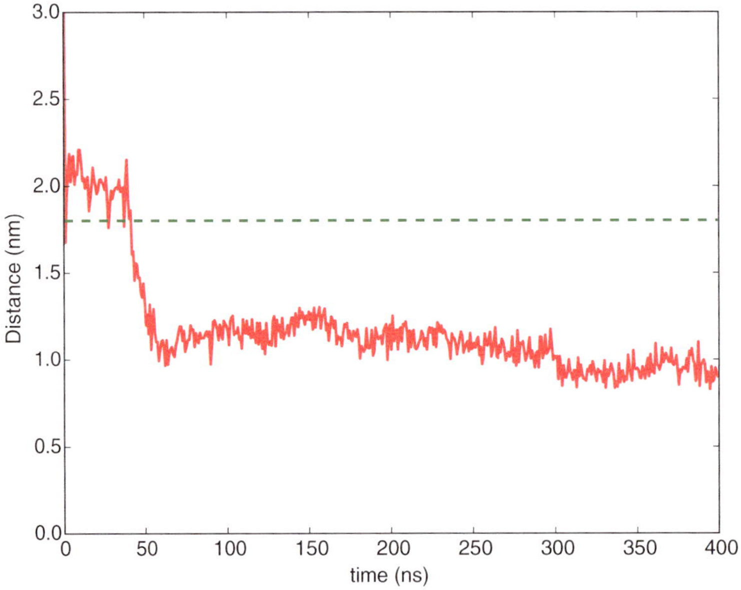

Fig. 7.17 Distance of the surface of the bare NP to the center of the bilayer as a function of time. The dashed line shows the average position of the bilayer surface

7.4.2 Simulation of a Nanoparticle in Contact with a Lipid Bilayer

To study the interaction of a bare NP with the lipid bilayer we initially position the NP close to the bilayer surface and follow the time dynamics of the system. Three snapshots are shown Fig. 7.16. From the initial state the NP adsorbs quickly to the surface of the bilayer and then penetrates to around 1 nm inside the membrane and stays strongly attached until the end of the simulation (see snapshots at 35 and 360 ns). To explore the adsorption process in more detail, the distance of the surface of the NP to the center of the bilayer was recorded and the results are shown in Fig. 7.17. The NP reaches the bilayer in a few nanoseconds and then attaches to the surface for around 50 ns (the dashed line in Fig. 7.17 marks the average position of the surface of the lipid bilayer, as defined by the position of the maximum of density of the lipid headgroups). After that, the NP starts penetrating the membrane (in a few nanoseconds). Then a slow internalization is observed until the NP reaches its final position at approximately 300 ns. The internalization of the NP is mediated by the attractions between both type of lipids and the NP. The penetration then stops (or becomes much slower) because any further displacement requires a substantial change in the membrane structure. To study long-time dynamics of the system, NP lipid wrapping and uptake one needs a bigger bilayer or/and NPT ensemble [60]. Despite of this limitation, the results obtained with our methodology agree with a recent report [61] for the absorption of a hydrophobic NP with a membrane composed of lipids and specialized receptors.

7.4.3 Simulation of a Nanoparticle-Protein Complex in Contact with a Lipid Bilayer

As we discussed above, the NP gets coated by proteins before it reaches the cell membrane, and this NP-protein complex is responsible for the final fate of the NP [62]. Considering this, we now simulate the interaction of a NP-protein complex, where protein corona is represented by a single HSA molecule. As shown in Sect. 7.2, not all orientations in which protein adsorbs onto a surface are equally probable and for our simulation of the interaction of a hydrophobic NP with a DPMC lipid membrane we first calculate the adsorption energy map of HSA onto a 2 nm of radius hydrophobic NP. The adsorption map is shown in Fig. 7.18a. We can see that, as in the cases discussed above, the energy landscape contains more than one minimum. We found the average adsorption energy of $-1.7k_BT$ and as the initial orientation for the simulation we selected the orientation $(\theta, \phi) = (145°, 110°)$, which corresponds to adsorption energy of $-2.7k_BT$ and the corresponding complex NP-HSA is shown in Fig. 7.18b.

Figure 7.19 shows a sequence of snapshots from simulation of the NP-HSA complex with the lipid bilayer. In the initial state, the protein is facing the bilayer. In the

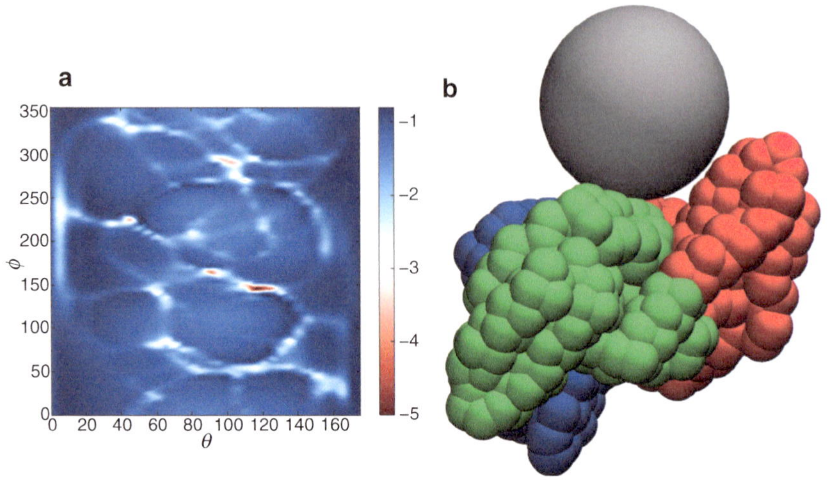

Fig. 7.18 (**a**) Initial state of the NP-HSA complex. (**b**) Surface map of the adsorption orientations of HSA onto a 2 nm negatively charged hydrophobic NP

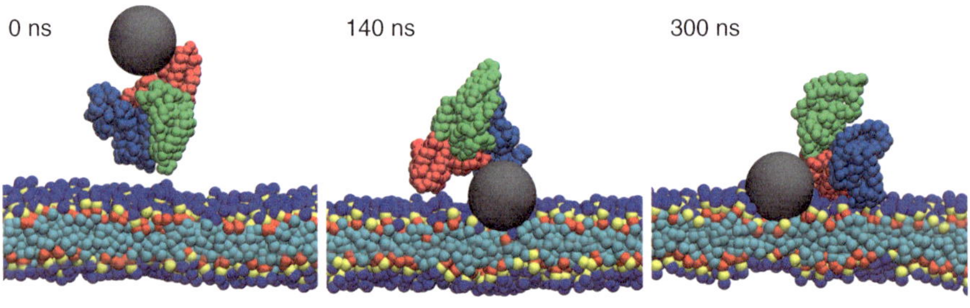

Fig. 7.19 Time sequence of the snapshots of the interaction of a DMPC lipid bilayer with a negatively charged hydrophobic NP complex with one HSA molecule. The radius of the NP is 2 nm. The ions are not shown

simulation, the HSA at first moves in front of the membrane and prevents a direct contact between the NP and the lipids. Then, the NP-HSA complex rotates so that the NP faces the bilayer and starts penetrating the membrane. Figure 7.20 shows the distance of the NP surface to the center of the membrane. The rotation is reflected in the sudden change of the position of COM of the HSA. After this quick rearrangement the NP starts the penetration while the protein stays attached to the NP for the whole simulation but moves around the surface of the NP as can be seen in the snapshot for the times 140 and 300 ns in Fig. 7.19. This movement of the HSA molecule can also be observed from the curve of the COM of the HSA curves as a function of time (Fig. 7.20).

Comparing the two simulations we see that the presence of the HSA dramatically changes the interaction of the NP with the membrane. We can envision that a NP,

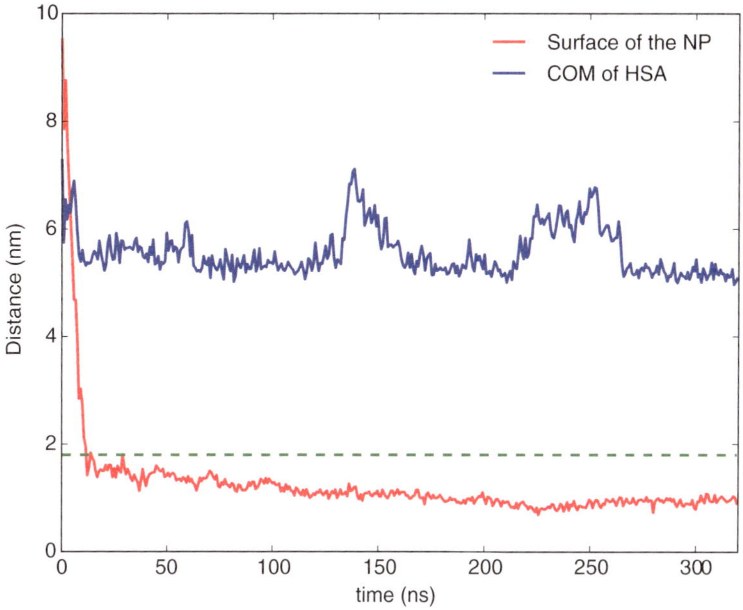

Fig. 7.20 Distance of the surface of the complex of NP with one HSA to the center of the bilayer and the COM of the HSA as a function of time. The dashed line shows the average position of the bilayer surface

fully coated NP with HSA, will not be able to penetrate into the membrane, so the coating is changing completely its biological reactivity.

7.5 Conclusions

In this work, we presented a multiscale methodology for modelling interactions at bionano interface, which is central for understanding uptake and toxicity of nanomaterials. We used systematic coarse-graining techniques to reduce the complexity of the problem by removing some degrees of freedom and focussing on the properties of interest. Since the CG models consists of about ten times less interaction centres than the atomistic model, and the solvent (water) is not modeled explicitly, simulations of the CG model take two to three orders of magnitude less CPU time compared with atomistic simulations for equal system size, or, alternatively, CG model can be used for simulations of whole proteins, small NPs and sufficiently large cell membrane fragments at the scale of tens of nanometers. We have parameterised and validated the model against experiments and all-atom MD simulations.

The technique for coarse-graining NP-protein interaction, which we presented in Sect. 7.2 can be used to calculate the binding energies for arbitrary plasma, cytosol, or membrane proteins, rank them by binding affinity to the NP and predict the content of NP protein corona. Our calculations show that the NP surface charge has a small effect on the adsorption energies in comparison to van der Waals interactions between the residues and the surface. We also find that the charge of the NP does not influence much the orientation, in which the protein prefers to adsorb. On the other hand, we have shown the size of the NP has a big effect on the adsorption energy maps, due to the amount of material involved and because the curvature of the NP determine the sections of the protein that can interact with the surface. Based on our simulations results, we can predict bigger proteins adsorb stronger on the inorganic surfaces, even for small NPs, in agreement with the Vroman effect. We have also demonstrated that a rigid protein model is justified at least for small globular proteins. In Sect. 7.3, we have parameterised a CG lipid and Cholesterol model, which reproduces the key bilayer properties of atomistic model of the same system. Finally, in Sect. 7.4, we have shown how the CG lipid and NP-protein models can be combined to model NP-cell membrane interactions and NP attachment and uptake.

References

1. Sharifi S, Behzadi S, Laurent S, Forrest ML, Stroevee P, Mahmoudi M (2012) Toxicity of nanomaterials. Chem Soc Rev 41:2323
2. Borm PJA et al (2006) The potential risks of nanomaterials. Part Fibre Toxicol 3:11
3. Johnston HJ, Hutchison GR, Christensen FM et al (2009) Identification of the mechanisms that drive the toxicity of tio2 particulates: the contribution of physicochemical characteristics. Part Fibre Toxicol 6:33
4. Nel AE, Maedler L, Velegol D et al (2009) Understanding biophysicochemical interactions at the nano-bio interface. Nat Mater 8:543
5. Verma A, Uzun O, Hu Y et al (2008) Surface-structure-regulated cell-membrane penetration by monolayer-protected nanoparticles. Nat Mater 7:588
6. Schlick T, Collepardo-Guevara R, Halvorsen LA et al (2011) Biomolecular modeling and simulation: a field coming of age. Q Rev Biol 44:191
7. Valerio LG Jr (2009) In silico toxicology for the pharmaceutical sciences. Toxicol Appl Pharmacol 241:356
8. Nigsch F, Macaluso NJ, Mitchell JB, Zmuidinavicius D (2009) Computational toxicology: an overview of the sources of data and of modelling methods. Expert Opin Drug Metab Toxicol 5:1
9. Dearden JC (2003) In silico prediction of drug toxicity. J Comput Aided Mol Des 17:119
10. Lyubartsev AP, Rabinovich AL (2011) Recent development in computer simulations of lipid bilayers. Soft Matter 7:25
11. Wong-Ekkabut J, Baoukina S, Triampo W, Tang I-M, Tieleman DP (2008) Computer simulation study of fullerene translocation through lipid membranes. Nat Nanotechnol 3:363
12. Hou WC, Moghadam BY, Westerhoff P, Posner JD (2011) Distribution of fullerene nanomaterials between water and model biological membranes. Langmuir 27:11899
13. Yang K, Ma YQ (2010) Computer simulation of the translocation of nanoparticles with different shapes across a lipid bilayer. Nat Nanotechnol 5:579
14. Monticelli L, Salonen E, Ke PC, Vattulainen I (2009) Effects of carbon nanoparticles on lipid membranes: a molecular simulation perspective. Soft Matter 5:4433

15. Izvekov S, Voth GA (2005) Multiscale coarse-graining method for biomolecular systems. J Phys Chem B 109:2469
16. Ayton GS, Noid WG, Voth GA (2007) Multiscale modeling of biomolecular systems: in serial and in parallel. Curr Opin Struct Biol 17:192
17. Lyubartsev AP, Laaksonen A (1995) Calculation of effective interaction potentials from radial distribution functions: a reverse monte carlo approach. Phys Rev E 52:3730
18. Lyubartsev AP, Mirzoev A, Chen L-J, Laaksonen A (2010) Systematic coarse-graining of molecular models by the newton inversion method. Faraday Discuss 144:43
19. Lyubartsev AP, Laaksonen A (1999) Effective potentials for ion-DNA interactions. J Chem Phys 111:11207
20. Tozzini V (2005) Coarse-grained models for proteins. Curr Opin Struct Biol 15(2):144–150
21. Bereau T, Deserno M (2009) Generic coarse-grained model for protein folding and aggregation. J Chem Phys 130(23):235106
22. Takada S (2012) Coarse-grained molecular simulations of large biomolecules. Curr Opin Struct Biol 22(2):130–137
23. Wei S, Knotts T (2013) A coarse grain model for protein-surface interactions. J Chem Phys 139(9):095102
24. Lobaskin V, Lyubartsev AP, Linse P (2001) Effective macroion-macroion potentials in asymmetric electrolytes. Phys Rev E 63:020401
25. Brunner M, Bechinger C, Strepp W, Lobaskin V, von Gruenberg HH (2002) Density-dependent pair interactions in 2D colloidal dispersions. Europhys Lett 58:926
26. Lynch I, Salvati A, Dawson KA (2009) Protein-nanoparticle interactions. What does the cell see? Nat Nanotechnol 4:546
27. Lynch I, Dawson KA, Linse S (2006) Detecting cryptic epitopes created by nanoparticles. Sci STKE 2006:14
28. Cedervall T et al (2007) Understanding the nanoparticle protein corona using methods to quantify exchange rates and affinities of proteins for nanoparticles. Proc Natl Acad Sci U S A 104:2050
29. Lindman S et al (2007) Systematic investigation of the thermodynamics of HSA adsorption to n-iso-propylacrylamide/n-tert-butylacrylamide copolymer nanoparticles. effects of particle size and hydrophobicity. Nanoletters 7:914
30. Allen LT et al (2006) Surface-induced changes in protein adsorption and implications for cellular phenotypic responses to surface interaction. Biomaterials 27:3096
31. Radke CE, Prausnitz JM (1972) Thermodynamics of multisolute adsorption from dilute liquid solutions. AIChE J 18:761
32. Lesniak A, Campbell A, Monopoli MP, Lynch I, Salvati A, Dawson KA (2010) Serum heat inactivation affects protein corona composition and nanoparticle uptake. Biomaterials 31:9511
33. Kamath P, Fernandez A, Giralt F, Rallo R (2015) Predicting cell association of surface-modified nanoparticles using protein corona structure – activity relationships (PCSAR). Curr Top Med Chem 15(18):1930–1937
34. Noid WG (2013) Perspective: Coarse-grained models for biomolecular systems. J Chem Phys 139(9):090901
35. Lopez H, Lobaskin V (2015) Coarse-grained model of adsorption of blood plasma proteins onto nanoparticles. J Chem Phys 143:243138
36. Miyazawa S, Jernigan RL (1996) Residue-residue potentials with a favorable contact pair term and an unfavorable high packing density term, for simulation and threading. J Mol Biol 256(3):623–644
37. Kim Y, Tang C, Clore G, Hummer G (2008) Replica exchange simulations of transient encounter complexes in protein-protein association. Proc Natl Acad Sci U S A 105(35): 12855–12860
38. Kim Y, Hummer G (2008) Coarse-grained models for simulations of multiprotein complexes: application to ubiquitin binding. J Mol Biol 375(5):1416–1433
39. Agashe M, Raut V, Stuart S, Latour R (2005) Molecular simulation to characterize the adsorption behavior of a Fibrinogen γ-chain fragment. Langmuir 21(3):1103–1117

40. Sun Y, Welsh W, Latour R (2005) Prediction of the orientations of adsorbed protein using an empirical energy function with implicit solvation. Langmuir 21(12):5616–5626
41. Kokh D, Corni S, Winn P, Hoefling M, Gottschalk K, Wade R (2010) Prometcs: An atomistic force field for modeling proteinmetal surface interactions in a continuum aqueous solvent. J Chem Theory Comput 6(5):1753–1768
42. Limbach H, Arnold A, Mann B, Holm C (2006) ESPResSo – an extensible simulation package for research on soft matter systems. Comput Phys Commun 174(9):704–727
43. Chen W, Huang H, Lin C, Lin F, Chan Y (2003) Effect of temperature on hydrophobic interaction between proteins and hydrophobic adsorbents: studies by isothermal titration calorimetry and the van't Hoff equation. Langmuir 19(22):9395–9403
44. Lacerda S, Park JJ, Meuse C, Pristinski D, Becker M, Karim A, Douglas J (2010) Interaction of gold nanoparticles with common human blood proteins. ACS Nano 4(1):365–379
45. Vilaseca P, Dawson K, Franzese G (2013) Understanding and modulating the competitive surface-adsorption of proteins through coarse-grained molecular dynamics simulations. Soft Matter 9:6978–6985
46. Vijay-Kumar S, Bugg C, Cook W (1987) Structure of ubiquitin refined at 1.8 resolution. J Mol Biol 194:531–544
47. Momma K, Izumi F (2011) *VESTA3* for three-dimensional visualization of crystal, volumetric and morphology data. J Appl Cryst 44:1272–1276
48. Brandt EG, Lyubartsev A (2015) Systematic optimization of a force field for classical simulations of TiO_2-water interfaces. J Phys Chem C 119:18110–18125
49. Berendsen HJC, Postma JPM, van Gunsteren WF, DiNola A, Haak JR (1984) Molecular dynamics with coupling to an external bath. J Chem Phys 81:3684–3690
50. Hoover W (1985) Canonical dynamics: equilibrium phase-space distributions. Phys Rev A 31:1695–1697
51. Nose S (1984) A unified formulation of the constant temperature molecular dynamics methods. J Chem Phys 81:511–519
52. Jämbeck JPM, Lyubartsev AP (2012) Derivation and systematic validation of a refined all-atom force field for phosphatidylcholine lipids. J Phys Chem B 116(10):3164–3179
53. Jämbeck JPM, Lyubartsev AP (2013) Another piece of the membrane puzzle: extending Slipids further. J Chem Theory Comput 9(1):774–784
54. Mirzoev A, Lyubartsev AP (2013) MagiC: software package for multiscale modeling. J Chem Theory Comput 9(3):1512–1520
55. Qin S-S, Yu ZW, Yu Y-X (2009) Structural characterization on the gel to liquid-crystal phase transition of fully hydrated DSPC and DSPE bilayers. J Phys Chem B 113:8114–8123
56. Kučerka N, Liu Y, Chu N, Petrache HI, Tristram-Nagle S, Nagle JF (2005) Structure of fully hydrated fluid phase DMPC and DLPC lipid bilayers using X-ray scattering from oriented multilamellar arrays and from unilamellar vesicles. Biophys J 88:2626–2637
57. Koenig BW, Strey HH, Gawrisch K (1997) Membrane lateral compressibility determined by NMR and X-ray diffraction: effect of acyl chain polyunsaturation. Biophys J 73(4): 1954–1966
58. Kučerka N, Nagle JF, Sachs JN, Feller SE, Pencer J, Jackson A, Katsaras J (2008) Lipid bilayer structure determined by the simultaneous analysis of neutron and X-Ray scattering data. Biophys J 95(5):2356–2367
59. Bereau T, Wang Z-J, Deserno M (2014) More than the sum of its parts: Coarse-grained peptide-lipid interactions from a simple cross-parametrization. J Chem Phys 140(11):115101
60. Lin J, Zhang H, Chen Z, Zheng Y (2010) Penetration of lipid membranes by gold nanoparticles: Insights into cellular uptake, cytotoxicity, and their relationship. ACS Nano 4(9):5421–5429
61. Hong-Ming D, Yu-Qiang M (2014) Computer simulation of the role of protein corona in cellular delivery of nanoparticles. Biomaterials 35(30):8703–8710
62. Monopoli M, Aberg C, Salvati A, Dawson KA (2012) Biomolecular coronas provide the biological identity of nanosized materials. Nat Nanotechnol 7(12):779–786

Chapter 8
Biological Surface Adsorption Index of Nanomaterials: Modelling Surface Interactions of Nanomaterials with Biomolecules

Ran Chen and Jim E. Riviere

Abstract Quantitative analysis of the interactions between nanomaterials and their surrounding environment is crucial for safety evaluation in the application of nanotechnology as well as its development and standardization. In this chapter, we demonstrate the importance of the adsorption of surrounding molecules onto the surface of nanomaterials by forming biocorona and thus impact the bio-identity and fate of those materials. We illustrate the key factors including various physical forces in determining the interaction happening at bio-nano interfaces. We further discuss the mathematical endeavors in explaining and predicting the adsorption phenomena, and propose a new statistics-based surface adsorption model, the Biological Surface Adsorption Index (BSAI), to quantitatively analyze the interaction profile of surface adsorption of a large group of small organic molecules onto nanomaterials with varying surface physicochemical properties, first employing five descriptors representing the surface energy profile of the nanomaterials, then further incorporating traditional semi-empirical adsorption models to address concentration effects of solutes. These Advancements in surface adsorption modelling showed a promising development in the application of quantitative predictive models in biological applications, nanomedicine, and environmental safety assessment of nanomaterials.

Keywords Nanoparticles • Surface interactions • BSAI • Bicorona • Physicochemistry • In situ characterization

R. Chen
Institute of Computational Comparative Medicine, Nanotechnology Innovation Center of Kansas State, Kansas State University, Manhattan, KS 66506, USA

J.E. Riviere (✉)
Institute of Computational Comparative Medicine, Department of Anatomy and Physiology, Kansas State University, Manhattan, KS 66506, USA

L. Tran et al. (eds.), *Modelling the Toxicity of Nanoparticles*, Advances in Experimental Medicine and Biology, Vol. 947, DOI 10.1007/978-3-319-47754-1_8

8.1 Introduction

The interest in the interactions between biological molecules and artificial materials has existed long before the introduction of synthetic nanomaterials into the world. It is a consensus in the medical device and biomaterials community that, when inserted into biological environment, most artificial devices or materials start to adsorb biomolecules native to that environment. Such adsorption is often manifested in changes in surface physicochemical properties, bioidentity, bioavailability, and toxicity of the device or material; as well as the change in structure and biological function of the adsorbed molecules. Research has shown, by manipulating surface functionalization, one can either reduce or enhance the adsorption of biomolecules to serve desired purposes [1–3].

With the vast production and wide application of nanomaterials in consumer products [4] and biomedical fields [5–7], the interactions of nanomaterials with biomolecules raise concerns that such interactions could alter or impair the normal functionality of the molecules. For example, the binding of proteins to nanomaterials could induce changes in the secondary structures and thus affect the stability and folding of the proteins; alternatively, the adsorption or binding to membrane receptors could induce immune responses in the cell. On the other hand, the adsorption of biological macromolecules also modifies the surface functionalization of the nanomaterials by forming a layer or layers of biomolecules known as a biocorona, or protein corona (molecules are proteins), upon introduction into a physiological environment.

The formation and unique composition of the biocorona literally defines the biological identity of the nanomaterials and determines how the host system sees them, and in turn affects their bioavailability, biodistribution, therapeutic efficacy, and toxicity [8–12]. In terms of scientific fundamentals, the one key process to understand the complexity of such intertwined matrix of biological systems and artificial materials is adsorption, and the factors which represent the physicochemical properties of both nanomaterials and biomolecules can be quantitated as profiles of signature characteristics to determine their respective identities. In this chapter, we review the fundamentals of surface adsorption, discuss the application of statistical modelling in extracting aforementioned signature characteristics, the formulation of the Biological Surface Adsorption Index (BSAI) model, and how traditional semi-empirical adsorption models can be adopted for its correction at varied adsorbate concentrations.

This chapter is organized as follows: introduce the role of biomolecular adsorption in biomedical applications of various materials of a wide range of sizes, followed by presentation of adsorption models that have been used to characterize and predict these phenomena. Specifically in the area of nanomaterial application, we show the landscape of current study on biocorona, as well as its profound impact on the understanding of the application and implication of artificial nano-sized devices or materials in biological systems. We further discuss the physical foundation of these phenomena, driven by intermolecular forces and thermodynamics, and introduce statistical modelling methods used to describe the surface interaction in terms of physicochemistry. Secondly, we introduce our protocol for chemical analysis of surface adsorption of ~30 small organic molecules onto ~20 types of nanoparticles with various surface functionalizations, and then we develop our BSAI models using statistical tools based on our experimental data collected from chemical analysis. We also describe our most recent efforts to improve this model by incorporation of higher order adsorbate concentration terms in the original BSAI model equation, as well as the inclusion of one of the traditional semi-empirical adsorption models – Langmuir Adsorption – to address the possible "crowding effect" present at high solute concentrations. Thirdly, we examine the predictive power of both the original and modified BSAI models, including high order term corrections and Langmuir incorporated versions. Moreover, a molecular dynamics (MD) study is presented to show its computational capabilities of adsorption prediction by comparing to experimental data. Finally, we present examples of the application of Principal Component Analysis (PCA) on these surface adsorption models for the characterization and clustering of the nanomaterials based on their surface physicochemical properties and adsorption profiles, we also provide an outlook of applying such profiling techniques in the field of nanomaterial risk assessment and prediction of toxicological effects.

8.2 Biomolecular Adsorption

The topic of biomolecular adsorption onto solid surfaces, as a widely observed and complex phenomenon, has attracted enormous research interest in the medical and pharmaceutical sciences, analytical chemistry, bioengineering, biophysics, and even environmental sciences. Bioengineers often find themselves in a situation where they have to deal with thrombosis caused by the adsorption of protein and other biological molecules on the surface of medical implants in contact with bodily fluids which is abundant in protein [13–16]; analytical devices like sensors and micro protein chip may face failure due to the accumulated adsorption of biological molecules on their surfaces [17–19]. On the other hand, adsorption of specific type of biological molecules may have desirable effect, including increased biocompatibility, suppressed immune response, or binding with specific receptors to selectively target cell types with those receptors expressed on their surfaces [20, 21].

In the era of nanotechnology, biomolecular adsorption remains a common puzzle, only with increased complexity from the diverse physicochemical properties of nanoparticle surfaces, much more enhanced mobility due to their extremely small sizes and thus significantly increased rates of biomolecular interactions. Upon nanomaterial entrance into biological or ecological systems, they first encounter numerous and varied biological or organic macromolecules, adopting a more dynamic and convoluted mode of movement in those systems compared to bulk surfaces. Due to their extremely small sizes, it is much easier for them to be in contact with various subsystems or organs within an organism. They readily interact with blood proteins if injected as drug carriers, gastric mucus if ingested, and pulmonary surfactant lipids if inhaled, transformations which alter their subsequent uptake by a wide range of tissue or cell types which may be determined by the type of molecules coated on the particles surfaces. On the other hand, such interaction could also change the structure or impair the function of adsorbed biological molecules; the proteins adsorbed may not fold normally because their secondary is altered and become very unstable.

8.2.1 Intermolecular Forces Determines Biological Adsorption

In biological systems, the interactions of biomolecules with synthetic nanomaterials are governed by the same principles which determine those between colloidal particles. These intermolecular forces are loosely categorized as follows.

1. Electrostatic interactions result in attraction or repulsion depending on the charge of the particle/molecule, or re-orientation for partially charged/zwitterionic particles. Another important factor in considering electrostatic force is with ions, which in high ionic strength biological fluid tend to shield electrostatic interactions, decreasing the active distance to nanometers.
2. Interactions involving sharing of electron clouds, such as hydrogen bonding (H-bonding), which is universal in biological aqueous environment and π interactions often found between carbon-based nanomaterials and aromatic residue-rich proteins. Since H-bonding is essential for protein-protein interaction, including polymerization of protein complexes by their monomers or dimers, H-bonding between biomolecules and synthetic particles could disrupt such natural occurring biochemical process. [22] The π-stacking is an attractive interaction between aromatic carbon rings resulting in two most energetically preferred configurations: parallel and T-shaped. It is often observed that the aromatic rings of amino acids orient parallel to the aromatic rings of carbon sheets or fullerene cages. Such interaction could disrupt the native secondary and tertiary structures of proteins [23–26].
3. Electrodynamic interaction, or dispersion force, sometimes loosely referred to as van der Waals (vdW) force. vdW is a short-range interaction which decreases rapidly as the participating atoms move away from each other. The vdW force between two atoms is usually very weak, however a large contact area sometimes can be accomplished through the deformation of the proteins; thus the vdW force can be drastically enhanced. On the other hand, the vdW force also tends to maximize the contacting area to suppress the attractive potential energy between the atoms, so the complementary shapes of the NP and proteins can sometimes dictate the affinity of their binding.
4. Entropic forces include depletion force and hydrophobic force. In a biological system, water molecules play an essential role through hydrogen-bonding: the bonded water network exclude anything that is nonpolar and does not form

hydrogen bonds together to maximize their entropy, or create an additional energy barrier for particles to interact with each other by creating a hydration layer on the surfaces of hydrophilic particles. In aqueous solution, biomolecules form such structures such that their hydrophobic moieties are hidden from water, and their hydrophilic components are present at the surfaces. Most NPs entering a biological system possess partially hydrophobic surfaces, including the NPs with hydrophilic surface coating that are usually incomplete. The hydrophobic interactions between proteins and NPs tend to either integrate the NP to the hydrophobic core of the protein, or unfold the protein to expose its hydrophobic residues. Research has demonstrated that hydrophobic nanoparticles such as carbon nanotubes (CNT) can insert themselves into the hydrophobic core of proteins forming stable structures, sterically inhibiting normal binding with other proteins required for their functions [27].

Characterizing and quantitating as a surface force profile the abilities of nanomaterials to engage in various interactions could shed light on the design, engineering, and synthesizing of nanostructures to either enhance or diminish their ability to adsorb certain types of molecular species to modify their bioidentity or biodistribution for their intended purposes, and reduce their toxicity. However in experimental practices, it is extremely difficult to quantitatively discriminate these forces from adsorption measurements of the combined effects. MD computer simulations are sometimes used to access or predict the interaction between nanoparticles and proteins; however the construction of more complex particles and larger molecules, and the calculation of large group of particles can be extremely expensive relative to computer time. Also, simulated models of nanoparticles are often a generic and idealized version of a particular particle type; they typically do not reflect the batch/lot or manufacturer specific differences among the same type of materials. Thus, MD simulation might not be ideal for quality control, risk assessment or surface characterization on a large scale.

8.2.2 Adsorption Models

Adsorption models have been developed by treating adsorption and desorption at equilibrium as dynamic processes, which depend on the energy of adsorption determined by the interaction potential with the surface. Naturally, at equilibrium of an adsorption process, the rates of adsorption and desorption are equal, assuming the interaction potential is the same at every possible interaction site on the surface for the same type of molecule (uniform surface assumption), we have Langmuir adsorption:

$$\theta = \frac{KC_e}{1 + KC_e}, \tag{8.1}$$

where θ is the ratio of number of occupied interaction sites to number of total sites, C_e is the equilibrium concentration of the adsorbate. K is Langmuir constant, which is temperature and adsorption energy dependent: $K \propto e^{\left(\frac{-\Delta E}{RT}\right)}$, where $-\Delta E$ is the adsorption energy, R is gas constant, and T is temperature in Kelvin. The application of the Langmuir model is largely limited by its assumptions to a certain range of concentration and temperature. Numerous methods have been proposed to make corrections to the model to generalize the assumptions especially where it fails to fit the experimental data.

For heterogeneous surfaces, which have varying interaction potentials, it is possible generalize the adsorption model by superimposing a set of Langmuir equations, each corresponding to different interaction energy:

$$\theta = \sum_i a_i \frac{KC_e}{1+KC_e}, \ \sum_i a_i = 1, \tag{8.2}$$

where a_i is the fraction of sites associated with a specific interaction potential. This type of generalization can expand the applicability of the model over a wider range of experimental conditions with more drastically varying temperatures and concentrations.

Under many circumstances, the trend of experimental adsorption data gives non-zero slope beyond the point where adsorption sites are predicted to be reaching saturation. This is due to a secondary or multiple layer adsorptions forming on top of the first layer, resulting from the longer range of intermolecular forces between the adsorbate molecules and the adsorbent surface, or the interactions between the adsorbed and free adsorbate molecules. In a multilayer adsorption system, each layer can be treated as a new effective adsorption surface with a different adsorption potential, thus the distinction between different layers comprised a generalized monolayer adsorption model. With the assumption of (1), the layers after the first have the same adsorption potential, and those layers of adsorption are treated as liquefaction of the adsorbate, (2), the number of layers goes to infinity when the concentration approaches saturation, by assuming the rates of adsorption and desorption are equal between adjacent layers, Brunauer, Emmett, and Teller proposed BET adsorption model:

$$\theta = \frac{B\frac{C_e}{C_s}}{\left(1-\frac{C_e}{C_s}\right)\left(1-\frac{C_e}{C_s}+B\frac{C_e}{C_s}\right)}, \tag{8.3}$$

where C_s is the solubility of the adsorbate, and B is BET constant. The concentration term $\frac{C_e}{C_s}$ is defined to be chemical activity, which plays an important role in one of our modified BSAI surface adsorption models developed in a later section and serves as a parameterized level of chemical saturation in the aqueous phase.

A more relevant adsorption energy based model was proposed by Dubinin, Polanyi, and Radushkevich, originally for porous structures. In Polanyi theory [28–31], the ratio of occupied adsorption site over total possible site is related to adsorption potential energy by: $ln\theta = -k(\Delta E)^2$, where k is a adsorbate specific parameter. The adsorption potential energy defined as $\Delta E = -RTln\left(\frac{C_e}{C_s}\right)$, then $ln\theta = -k\left[RTln\left(\frac{C_e}{C_s}\right)\right]^2$. Polanyi model theoretically relates the partition of solute molecules between the solution and the adsorbent to the adsorption potential energy of the adsorbent surfaces.

8.2.3 Quantitative Structure-Activity Relationship (QSAR) Approach

As mentioned in Sect. 8.2.2, in the Polanyi model, parameter k is adsorbate specific, which can presumably be related to different physicochemical properties of the adsorbate molecules due to their different chemical structure. Thus, by selecting a few types of adsorbate molecules with varying physicochemical properties and compare their adsorption characteristics, one can deduce the relative contributions of those properties to the adsorptions. In Quantitative Structure-Activity Relationship (QSAR), the fundamental assumption is that the chemical activity of molecules is a function of their chemical structure. As a result, physicochemical structural properties can be parameterized, and introduced into a statistical model thereby generating an index, or molecular descriptor, for each parameter corresponding to one of the chemical properties. Such indices then can be used to predict the adsorption of a particular type of adsorbate.

QSAR has been applied successfully in drug design, pharmacokinetics, toxicological prediction and medicinal chemistry by quantifying the atomic, structural and topological features of a target molecule. A simple Linear Free Energy Relationship (LFER) model can stablish a linear connection between a parameterized chemical or physical interaction and a few descriptors describing relevant physicochemical properties of participating parties. Molecular connectivity indices and the solvatochromic parameters are the most widely used sets molecular descriptors for the prediction of chemical/biological activities. Molecular connectivity indices are molecular structure-based, sometimes also called topological indices; they are usually calculated using the molecular graph of a chemical compound. In contrast, solvatochromic parameters are comprised of indices that represent the relatively scaled abilities of the chemical compound to interact with other molecules through intermolecular forces, such as Coulomb force, dispersion force, hydrogen-bonding, polarity, etc. As previously described, those forces are exactly the ones deemed important in the interpretation of biological molecular adsorption onto nanomaterials. One study tested molecular connectivity-based QSAR models and compared them to solvatochromic parameters-based LFER models [32], the results showed that the QSAR models were slightly more predictive, but the parameters used in the models were less physically interpretable as they failed to explicitly correlate physicochemical properties of nanomaterials with their biological or environmental behavior.

8.3 Biological Surface Adsorption Index (BSAI)

The interaction between nanomaterials and biological systems focuses on the interface between a solid nanoparticle surface and an aqueous media containing various ionic species and biological macromolecules [33]. In this complex system, the nanoparticle surface can be dramatically altered by solvation, the adsorption of those molecules and ions, forming a unique solid-liquid interface, significantly different from what was observed from bulk materials and larger devices due to their much increased surface to volume ratio, and thus drastically different adsorption energy profiles. The biological surface adsorption index (BSAI) is a novel approach to characterize surface adsorption energy of nanomaterials that is the primary force behind nanoparticle aggregation, protein corona formation, and other complex interactions of nanomaterials within biological systems [34–36]. This method has helped in understanding of fundamental interactions between nanomaterials and environmental pollutant molecules such as natural organic matter, or functional groups and amino acid residues on biomacromolecules which may cause the formation of biocorona [11, 12, 37–39]; the latter considered as the determining factor of bio-identity, bio-availability, and toxicity of nanoparticles in biological systems. Five quantitative nanodescriptors were selected to represent the surface adsorption forces (hydrophobicity, hydrogen bond, polarity/polarizability, and lone-pair electrons) of the nanomaterial interaction with biological components. BSAI nanodescriptors are intrinsic properties of nanomaterials useful not only for QSAR statistical modelling for the prediction of adsorption, but also for the characterization of nanoparticle surfaces through their physicochemical intermolecular forces to generate a fingerprint mapping the categorization and profiling of nanomaterials.

8.3.1 BSAI Model

Similar to QSAR, which can generate linear regression coefficients utilizing molecular descriptors to predict surface adsorption; in the BSAI model the descriptors are used to generate nanodescriptors to characterize the physiochemical properties of surfaces of nanomaterials through the interaction with and adsorption of various probe organic compounds.

The BSAI approach was formulated based on the elements of fundamental molecular forces that contribute to biomacromolecule adsorption, and other biological processes: Coulomb force (charged particles), London dispersion (hydrophobic interactions), hydrogen-bond acidity and basicity, dipolarity/polarizability and lone-pair electrons [40]. Direct experimental determination is possible only for Coulomb forces through zeta potential measurements. The magnitude of the other four molecular interaction parameters can be calculated using the proposed BSAI approach. In theory, the adsorption process of organic molecules onto the surfaces of nanomaterials is determined by each type of molecular interactions. These interactions can be mathematically obtained from a set of probe compounds with diverse physicochemical properties covering a reasonably large chemical space. The adsorption coefficients (k) of the probe compounds can be measured by experimental quantification of the probe compounds adsorbed on the surfaces after equilibrium of interaction is reached. The log$k$ values are scaled to a set of known molecular descriptors (Abraham's descriptors) of the probe compounds using multivariate linear regression (MLR) to generate a group of nanodescriptors which can be interpreted as the contributions from each type of molecular interactions. In principle, for the purpose of predictive modelling, different sets of molecular descriptors (either empirical or theoretical) developed in QSAR studies can be used for the BSAI approach [41]. But to retain clear physical relevance and interpretability, we employed Abraham solute descriptors as molecular descriptors. Abraham solute descriptors were chosen also because of their successful use in biological-related context. The model is expressed as:

$$\log k_i = c + rR_i + p\pi_i + a\alpha_i + b\beta_i + vV_i, i = 1,2,3,\ldots,n \tag{8.4}$$

where k_i is the adsorption coefficient, n is the number of compounds used as probes, and c is the regression constant. Five variables $R_i, \pi_i, \alpha_i, \beta_i, V_i$ are the molecular descriptors of the ith probe compound, where R_i is the excess molar refraction representing molecular force of lone-pair electrons, π_i is the polarity/polarizability parameter, α_i and β_i are the hydrogen-bond acidity and basicity respectively, and V_i is the McGowan characteristic volume describing hydrophobic/lipophilicity interactions. The BSAI nanodescriptors are the regression coefficients r, p, a, b, v reflecting the differential compound-nanomaterials interactions.

In our approach, the nanodescriptors for a given nanoparticle are obtained using multiple linear regression analysis of the [logk, r, p, a, b, v] matrix. The Abraham solute descriptors [$R_i, \pi_i, \alpha_i, \beta_i, V_i$] were generated using Absolv module provided by either the Absolv program in the ADME Suite software or the ADME online service (Advanced Chemistry Development Inc., Toronto, Canada). The regression analysis was performed by SAS or JMP Pro (SAS Institute Inc., Cary, NC).

The conceptualization of the BSAI approach was inspired by the successful prediction of biological activities of small molecules using octanol–water partition coefficient (log $K_{o/w}$) [42, 43]. The value of log $K_{o/w}$ measures the contribution of hydrophobic interactions in biological processes [44]. However, it is difficult to measure the log $K_{o/w}$ values for nanomaterials, because it is difficult to form stable suspensions for most nanomaterials in both polar and non-polar liquid. The key thrust of the BSAI approach is to measure a set of log $K_{o/w}$ equivalent parameters for nanomaterials that can be used for quantitative model development. Correlation of the log $K_{o/w}$ values and the adsorption coefficients (log k) of the probe chemicals on

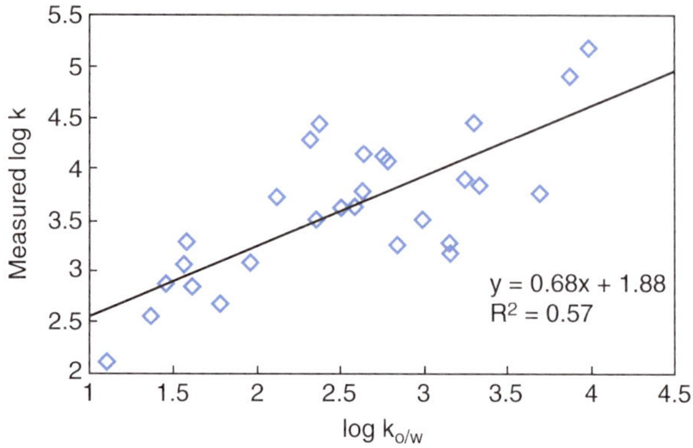

Fig. 8.1 Correlation of the adsorption coefficients (log *k*) of the probe compounds on MWCNTs with their log$K_{o/w}$ values, suggesting that lipophilicity is significant in the adsorption process (R^2 = 0.57) but is not the only factor (Adapted by permission from Macmillan Publishers Ltd.: [Nature Nanotechnology] (Ref. [34]), copyright (2010))

multiwalled carbon nanotubes (MWCNTs) is shown in Fig. 8.1. This indicates that hydrophobicity has significant contribution toward the adsorption process, with a correlation slope of 0.68 and R^2 of 0.57, but the scattered data points indicate that lipophilicity is not the only significant factor, with other molecular interactions possibly playing an important role in surface adsorption processes. This observation is consistent with the literature [45]. Thus the BSAI index can be viewed simply as a multivariate partition coefficient quantitating a nanomaterial's surface properties in terms of biologically-relevant molecular interaction forces.

8.3.2 *Chemical Analysis*

8.3.2.1 Adsorption of Nanomaterials with Chemical Probes

The adsorption experiments are conducted by incubating nanomaterial with a standard solution containing the probe compounds of known concentrations. The concentration of probe compounds should be kept as low as possible in order to reduce the concentration effects and interactions among the probe compounds, however this is restricted by the limits of detection and quantification of the instruments. In this case, only 'parts per billion' levels of the probe compounds (individual concentrations) were required in this study due to analytical sensitivity of the SPME/GC–MS method. Specifically, 2 mg of nanoparticles was added to 200 μL deionized water in a 2 mL glass vial. The vial was then vortexed to for uniform nanoparticle suspensions, then 1 mL of working solution (W1, W5, W10, W25 or W50) containing probe compounds of different concentrations were added to the vial. The vial was then sealed immediately with a Teflon-lined septa cap to prevent evaporation. For nanomaterials that are received in aqueous suspensions, a volume of the aqueous suspensions containing 2 mg of solid nanomaterial was mixed with certain volume of working solution to get the same final concentrations as solid nanomaterials above. The mixtures were then put under vigorous shaking for 5 h until equilibrium condition was reached. The equilibrium was verified by monitoring the kinetics of the adsorption. The particles were then removed from solution by either centrifugation or filtration.

8.3.2.2 Analysis of Chemical Probes by SPME and GC/MS

Solid phase microextraction (SPME) in combination with gas chromatography / mass spectrometry (GC/MS, Agilent GC-QQQ 7000B) was employed to determine the concentration changes of probe components before and after adsorption with nanomaterials. In SPME, a poly-dimethylsiloxane/divinylbenzene (PDMS/DVB) membrane coated fiber was used for the extraction of probe compounds from the liquid phase. The extraction time was 20 min. In GC/MS, separation was performed on a 30 m × 0.25 mm (i.d.) × 0.25 μm (df) HP-5MS capillary column (Agilent, Palo Alto, CA). The column oven was programmed as follows: the initial temperature was 40 °C and held for 1 min, ramped at 20 °C/min to 60 °C and 2 °C/min to 100 °C, held at 100 °C for 2 min, then ramped at 20 °C/min to 200 °C and 40 °C/min to 270 °C, and finally held at 270 °C for 3 min. The injection port was maintained at 280 °C for using PDMS/DVB fibers. The injection model was Pulsed/Splitless and desorption time was 5 min. Agilent GC-QQQ/MassHunter Workstation was used for data acquisition. The equilibrium concentrations (C_e) were directly determined using Qualitative Analysis software (Agilent). The surface concentration of adsorbed probe compounds was determined as $C_{ad} = \frac{V_0\left(C_0 - C_e\right)}{m}$, where V_0 is the total volume in the vial, C_0 is the concentration of a probe compound prior to adsorption, and m is the mass of nanoparticles present in the suspension. The adsorption constant of a given compound onto a particular type of particle is the ratio of surface concentration (C_{ad}) versus the equilibrium concentration (C_e)

$$k = \frac{C_{ad}}{C_e} = \frac{V_0\left(C_0 - C_e\right)}{mC_e} \tag{8.5}$$

8.3.3 *Model Results and Validation*

Figure 8.2 shows the nanodescriptors of MWCNT representing the relative contributions of the four types of molecular interactions are depicted . The term for hydrophobic force (v = 4.18) is a strong contributor. This is consistent with the result obtained when the log k values were correlated with log $K_{o/w}$ values of the probe compounds alone. Hydrogen-bond basicity (b = −2.78) is the second most significant factor, but has a negative value, which suggests that the MWCNT surface is

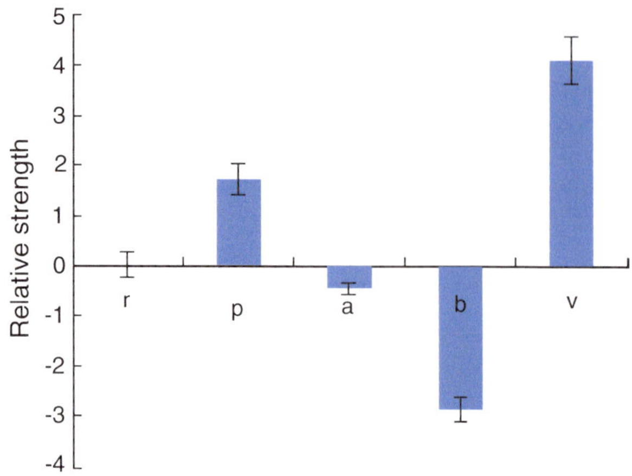

Fig. 8.2 The five nanodescriptors [*r*, *p*, *a*, *b*, *v*] measured by the BSAI approach, representing the five major molecular interactions in nanoparticle adsorption processes: lone-pair electrons, polarity/polarizability, hydrogen-bond donor, hydrogen-bond acceptor and London dispersion, respectively. The nanodescriptors of MWCNTs are depicted with standard errors of the regression analysis. Positive values indicate that the nanoparticle surfaces have stronger interaction potentials with the chemicals or biomolecules, and negative values indicate the molecular interactions are stronger in the aqueous phase (Adapted from by permission from Macmillan Publishers Ltd.: [Nature Nanotechnology] (Ref. [34]), copyright (2010))

less likely to donate protons (to the probe chemicals) than water at the liquid–solid interface. Hydrogen-bond acidity ($a = -0.37$) has a slightly negative value, indicating that the proton acceptor strength of the MWCNT surface is slightly weaker than water. The third strong factor is the dipolarity/polarizability (p = 1.75), which can be explained by the abundant π-electron clouds on the MWCNT surface [9]. The lone-pair electrons ($r = 0.043$) appear to have a minimal effect, which could be due to the fact that the π-electron cloud may shield the lone-pair electrons in MWCNTs disabling their ability to contribute toward the adsorption processes. The BSAI approach provides both rational interpretations for the molecular interactions and five quantitative physicochemical parameters, which can be used for the characterization of the relative strengths of the molecular interactions of the nanomaterials in the adsorption processes. It is worthy to note that hydroxyl and carboxyl derivatization of MWCNTs significantly increased its suspension stability in aqueous media, while their surface adsorption property was not significantly altered. This could be due to the fact that the main polar derivatization sites occurred at the ends of MWCNT while the adsorption property of tube surfaces was not significantly altered. Therefore, carbon nanotubes may be envisioned as drug carriers *via* tube surface adsorption of the drugs [46]. This also explains one of the differences between carbon nanotubes and asbestos, which has hydrophilic surfaces that could not form a strong adsorbed protein layer [47–50].

Plots of the predicted log k values of the five nanodescriptors for MWCNT versus the measured log k values are shown in Fig. 8.3. A linear correlation was

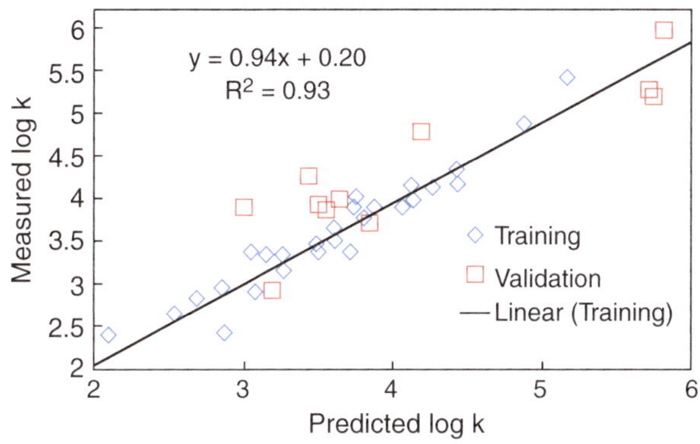

Fig. 8.3 Predicted versus measured log *k* values of the training probe and validation compounds. The predicted log *k* values correlate well ($R^2 = 0.93$) with the measured values of the training probe compounds. The data points of the validation compounds lying close to the regression line reveal the robustness of the predictive model (Adapted from by permission from Macmillan Publishers Ltd.: [Nature Nanotechnology] (Ref. [34]), copyright (2010))

obtained, with a correlation coefficient (R^2) of 0.93. This demonstrated that the five nanodescriptors $[r, p, a, b, v]$ of the BSAI approach provide a much better prediction of the adsorption affinity of MWCNTs than lipophilicity (log $K_{o/w}$) alone (Fig. 8.1). The robustness of the model was studied by internal cross-validation using the leave-one-out (LOO) and leave-many-out (LMO25%) techniques [51, 52], resulting in validation coefficients $Q^2_{LOO}$of 0.888 and $Q^2_{LMO25\%}$ of 0.883 for MWCNT. All of the cross-validation coefficients were greater than 0.7, revealing the robustness of the predictive model [51].

External validation was conducted using 12 different compounds on MWCNT (measured log *k* values and solute descriptors are given in Table 8.1). The external validation coefficient (Q^2_{ext}) of the model was 0.78, suggesting satisfactory predictive capability for the external validation compounds [51, 53].

A Williams plot (Fig. 8.4) was used to verify the applicability of the model, showing the leverages of the probes (diagonal elements of the Hat matrix) versus the Euclidean distances of the compounds to the models measured by standardized and cross-validated residuals [51, 53]. If the standardized and cross-validated residual of a compound is greater than three standard deviation units ($\pm 3\sigma$), the compound will be treated as outliers. If the leverage, or hat-value of the compound is greater than the warning leverage ($h > h^*$), it suggests that the compound is very influential in the model. The warning leverage is defined as $h^* = 3(N + 1)/n$, where N is the number of independent variables in the predictive model ($N = 5$ in Eq. (8.2)) and n is the number of observations, or number of probe compounds ($n = 28$ in this study). All the training probe and validation compounds are within the chemical domain defined by the training probe compounds ($\pm 3\sigma$ and $h^* = 0.64$), suggesting no outliers, and that the predictive capability of the model is reliable.

The BSAI model was also applied to various types of metal or metal oxides nanomaterials; we illustrate with AlOOH I (aluminum hydroxide oxide) as an

Table 8.1 Measured and predicted logk values for the validation compounds on MWCNT

	Validation Compounds	Measured log k	R	π	α	β	V	Predicted log k
1	2-chlorophenol	3.18	0.84	0.84	0.33	0.3	0.897	2.97
2	o-dichlorobenzene	3.56	0.83	0.85	0	0.1	0.961	3.93
3	m-dichlorobenzene	3.65	0.83	0.84	0	0.05	0.961	4.05
4	p-dichlorobenzene	3.51	0.86	0.88	0	0.1	0.961	3.99
5	hexachloroethane	3.42	0.86	0.74	0	0.15	1.124	4.29
6	nidrobenzene	3.64	0.83	1.26	0	0.21	0.89	4.05
7	isophorone	3.01	0.54	0.76	0	0.45	1.24	3.96
8	2,4-dichlorophenol	3.84	0.98	0.93	0.5	0.23	1.019	3.78
9	1,2,4-trichlorobenzene	4.19	0.96	0.95	0	0.03	1.083	4.82
10	2-chloronaphthalene	5.73	1.42	1.1	0	0.17	1.207	5.23
11	2,4-dinitrotoluene	5.82	1.12	1.77	0	0.31	1.205	5.99
12	azobenzene	5.72	1.38	0.86	0	0.41	1.48	5.28

Adapted with permission from (Ref. [35]), copyright (2011) American Chemical Society
The predicted logk values were obtained by plugging the five solute descriptors $[R, \pi, \alpha, \beta, V]$ of a given validation compounds into the predictive model for MWCNT. The solute descriptors were generated by the Absolv program in ADME Suite software, Version 4.95 (Advanced Chemistry Development Inc., Toronto, Canada)

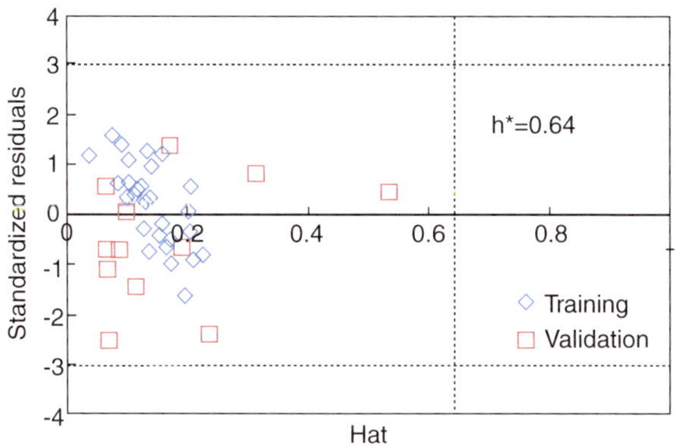

Fig. 8.4 Williams plot for verifying the applicability of Eq. (8.2). Twelve new compounds were used to validate the model. Training (blue diamonds) and validation (pink triangles) compounds are within the chemical domain (dotted lines) defined by the training probe compounds ($\pm 3\sigma$ and $h^* = 0.64$), suggesting no outliers and the predictivity of the model is reliable (Adapted from by permission from Macmillan Publishers Ltd.: [Nature Nanotechnology] (Ref. [34]), copyright (2010))

example. The 5 nanodescriptors of AlOOH I nanoparticles are indicated in Fig. 8.5. Those descriptors were calculated using MLR on adsorption coefficients along with the molecular descriptors of a wide range of probe chemicals at the lowest concentration (W1) used in this study. The relative magnitude of the descriptors (*a, b, v* values significantly different from 0) suggested that the binding of the molecules

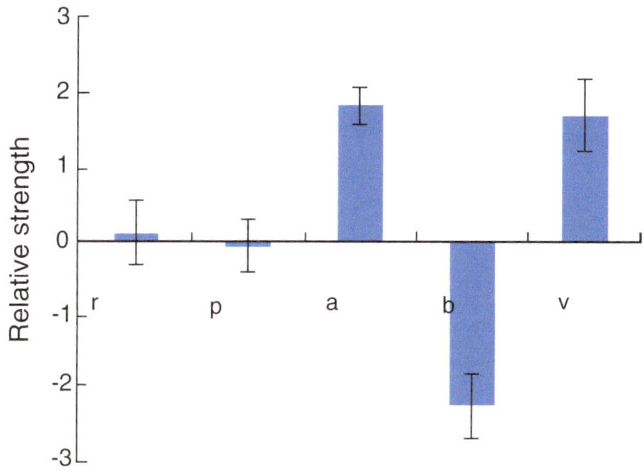

Fig. 8.5 Nanodescriptors [r,p,a,b,v] measured by the BSAI modelling representing the five major molecular interaction forces in the nanoparticle adsorption processes: lone-pair electrons, polarity/polarizability, hydrogen-bond donor, hydrogen-bond acceptor, and London dispersion, respectively. The nanodescriptors of AlOOH I were calculated from adsorption data averaged over four repeats. Positive value indicates the corresponding interaction tends to pull chemical molecules from the aqueous phase, while negative value indicate it is preferred the molecule stays in the aqueous environment (Adapted with permission from (Ref. [36]), copyright (2014) American Chemical Society)

onto nano surfaces received most significant contribution from hydrogen bonding and hydrophobic interactions. Williams plot (Fig. 8.6) was used to test the applicability of the model, indicating randomly distributed studentized residuals below and above 0 in the range from -2.5σ to 2.5σ, Hat values all smaller than $h^* = 0.72$. Further tests were run by randomly separating the data into two groups: a training group on which the model was built, and an external testing group using which the predictive capabilities were confirmed. Cross validation was conducted using leave-one-out (LOO) method on model built based on the training data set. Although the model showed acceptable goodness-of-fit ($R^2 = 0.86$), Fig. 8.7 shows that neither internal cross validation ($Q^2_{cv} = 0.66$) nor external testing were ideal.

The feasibility of the BSAI approach and some applications of nanodescriptors have been described using both carbon-based and metal oxide nanomaterials as examples. Predictions such as those made for MWCNTs can also be made for other nanomaterials if their five nanodescriptors are measured using the BSAI approach. The BSAI nanodescriptors for a few additional nanomaterials including silver (AgP: powder and Ag50: colloid), TiO_2, ZnO, CuO, NiO, Fe_2O_3, SiO_2, C_{60} (powder), nC_{60} (colloid), MWCNT and hydroxylated MWCNT (CNTOH) were measured using the same experimental protocols and probe compounds used for MWCNT listed in Table 8.2. However, for some of the nanomaterials, the adsorption of some of the probe compounds was too weak to be measured accurately; therefore, these compounds were excluded from regression analyses. This lack of adsorption of specific probes is reflective of these nanomaterial surface interactions. The number of

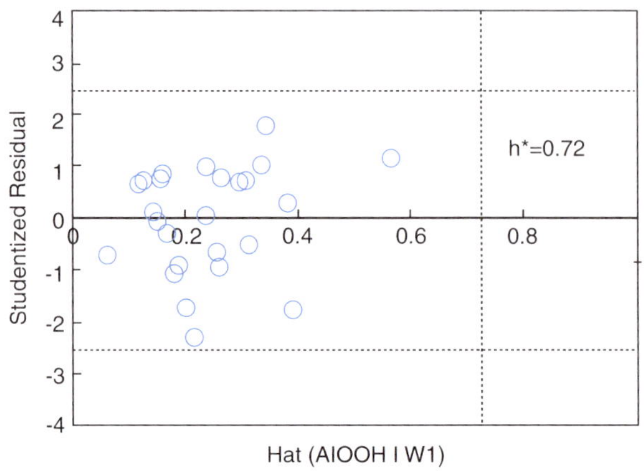

Fig. 8.6 Williams Plot (h* = 0.72), studentized residuals are randomly distributed below and above x-axis, and within the range of 2.5σ (σ is the standard deviation unit). Hat values are all smaller than h* (h* = 3(N + 1)/n = 3(5 + 1)/25) (Adapted with permission from (Ref. [36]), copyright (2014) American Chemical Society)

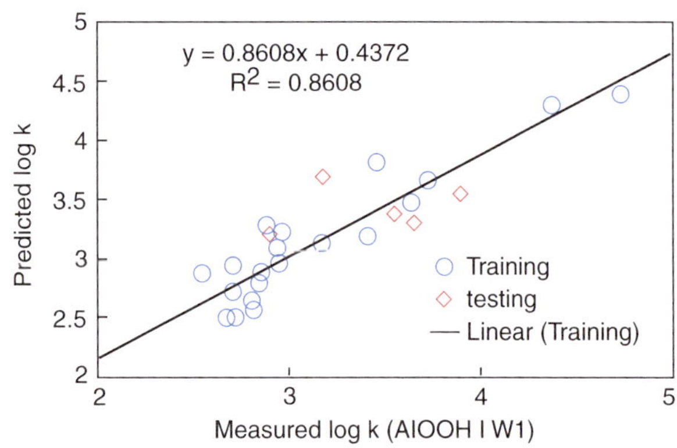

Fig. 8.7 Five observations were randomly chosen as testing set (*blue circles*), the rest were used to build the model (*red diamonds*) (Adapted with permission from (Ref. [36]), copyright (2014) American Chemical Society)

compounds (*n*) used in the regression analysis and the R^2 value for each of the nanomaterials are also given (Table 8.2). The nanodescriptors for a given nanomaterial can be used to construct the predictive model for the nanomaterial similar to the predictive model for MWNCT using data from entry 7, Table 8.2. A few oxide and metal nanomaterials were also included for diversity, while their nanodescriptor values reflect larger errors as seen in the low regression R^2 values due to the weak adsorption of the probe compounds on these nanomaterials. A different set of probe compounds optimized for these metal oxides would be needed to be developed for optimal characterization.

Figure 8.8 shows the five nanodescriptors obtained across each of the nanomaterials. The nanodescriptors of tested nanomaterials provide fingerprints for comparisons on their physical chemical properties. An irregular pattern of the nanodescriptors was observed in the case of NiO nanoparticles because of its specific chemisorption of

Table 8.2 BSAI nanodescriptors of nanomaterials

Entry	Nanomaterial	n	R^2	r	p	a	b	v	PC-1	PC-2
1	s-MWCNT	28	0.92	0.09	2.34	−0.44	−3.06	5.26	1.14	−0.09
2	C_{60} (powder)	28	0.91	0.5	−0.75	−1.23	−2.89	2.85	−0.89	−1.93
3	MWCNT-OH	28	0.9	0.7	2.37	−0.28	−2.45	4.63	0.74	−0.74
4	l-MWCNT	28	0.92	−0.15	2.88	−0.47	−2.79	6.63	1.57	0.26
5	MWCNT8nm-COOH	28	0.93	−0.10	1.85	0.18	−3.23	6.64	1.39	0.86
6	MWCNT15nm-COOH	28	0.92	0.64	2.68	−0.17	−3.03	5.81	1.51	−0.54
7	MWCNT40nm-COOH	28	0.95	0.41	2.3	−0.51	−3.59	6.59	1.82	−0.70
8	carbon hollow sphere	28	0.91	−0.47	3	−0.21	−4.07	7.16	2.42	1
9	r-graphene oxide	28	0.92	0.68	0.38	−0.79	−1.13	3.05	−1.27	−1.49
10	Ag-silica	22	0.69	−0.99	0.52	0.35	−1.19	0.85	−2.00	2.42
11	Ag-carbon	20	0.74	0.14	0.26	0.21	−1.81	2.02	−1.30	0.52
12	C_{70}TGA	28	0.95	−0.16	−0.59	−1.63	−2.79	1.15	−1.45	−1.39
13	$nC_{60}(OH)_{20}$	28	0.84	−0.45	0.64	0.32	−3.86	2.49	0.02	1.49
14	$nC_{60}(OH)_{32}$	28	0.98	0.08	−0.03	−0.66	−3.34	4.57	0.08	−0.60
15	SiO_2	26	0.78	0.13	−0.33	−0.27	−1.20	0.78	−2.23	−0.07
16	TiO_2	26	0.84	0.1	−0.13	0.57	−1.94	1.39	−1.56	0.99

Adapted with permission from (Ref. [35]), copyright (2011) American Chemical Society

The number of probe compounds (n) used for the multiple linear regression analysis to obtain five nanodescriptors $[r, p, a, b, v]$ for each of the nanomaterials with a regression coefficient (R^2). PC-1 and PC-2 are the two components obtained in principal component analysis of the five nanodescriptors for two-dimensional presentation of the molecular interaction forces across the 16 nanomaterials

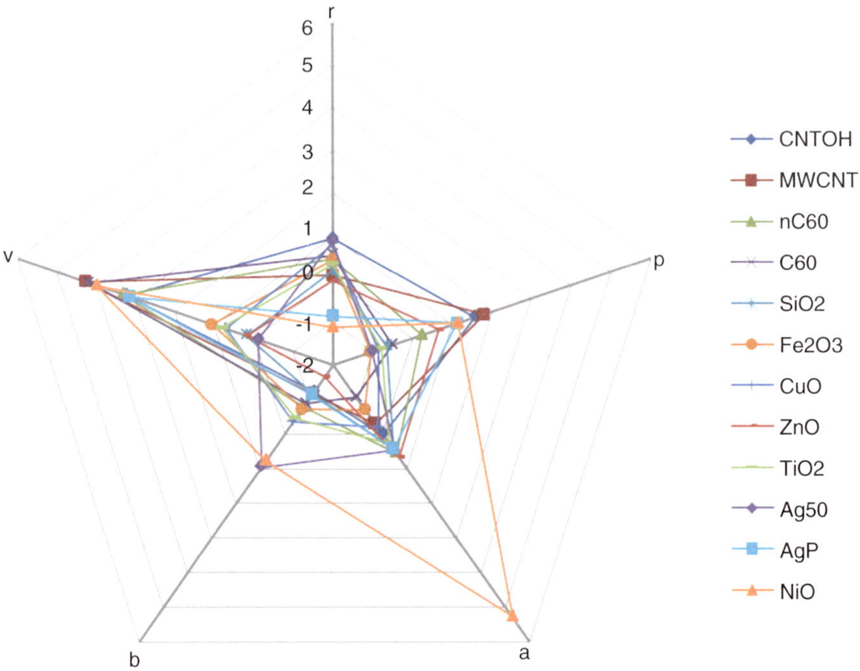

Fig. 8.8 The nanodescriptors $[r, p, a, b, v]$ are regression coefficients representing the relative molecular interaction strengths of the nanomaterials. The nanodescriptors of NiO nanoparticles showed an irregular pattern because of its unique chemisorption of phenol-derivative probe compounds (Adapted from by permission from Macmillan Publishers Ltd.: [Nature Nanotechnology] (Ref. [34]), copyright (2010))

phenol derivatives, suggesting that compounds having specific chemical interactions with a given nanomaterial should not be used as probe compounds for that nanomaterial. In turn, the BSAI approach can be used to identify specific interactions of chemicals or biomolecules that would act as outliers in the predictive model.

8.3.4 Profiling of Nanoparticle Surface Physiochemical Properties by Principal Component Analysis (PCA)

Surface physicochemical properties of nanomaterials in terms of how likely they engage in different surface interactions can be quantitatively compared using those five nanodescriptors, however such comparison can be counterintuitive. Principal component analysis (PCA) provides an efficient way of reducing the dimension from five two by projecting most of the variance to the first two principal components. PCA was performed by orthogonally transforming the five-dimensional data set into two principal components, with the first principal component accounting for

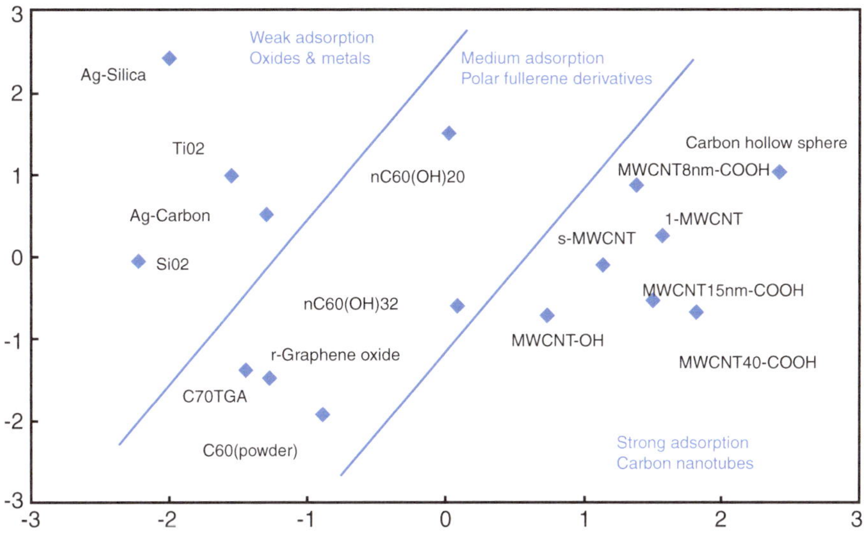

Fig. 8.9 Nanoparticle scattering plot by two principal components. The two components were obtained by principal component analysis of the five nanodescriptors of 16 nanomaterials listed in Table 8.2. The 16 different nanomaterials were clustered roughly to three zones by their surface adsorption properties (Adapted with permission from (Ref. [35]), copyright (2011) American Chemical Society)

as much of the variability in the data as possible and the second component containing the highest variance possible under the constraint that it be orthogonal to the first component [54]. The two components (PC-1 and PC-2) generated by PCA from the nanodescriptors of the 16 nanomaterials are listed in Table 8.2. Figure 8.9 shows a two-dimensional plot of the two principal components (PC-1 vs. PC-2). The 16 nanomaterials can be roughly divided into three groups: strong adsorption nanomaterials including carbon hollow spheres, carbon nanotubes and their derivatives, medium adsorption nanomaterials including C_{60} (powder), polar derivative of fullerenes and graphene oxides, and weak adsorption nanomaterials including oxide and metal nanoparticles. This two-component reduction of the five descriptor index provides a simpler comparison of nanomaterial surface properties, which may be useful for the categorization of biological effects based on surface properties.

These nanodescriptors can be directly correlated with the biological activity in weight-based dose quantities, since they are obtained based on weight concentrations of the nanomaterials. This overcomes the difficulties in measuring surface area of nanomaterials due to their possibly assumed aggregated or agglomerated states under biological conditions. In fact, the surface area measured using the conventional BET method [55] by nitrogen gas adsorption in dry environment may have less physical relevance in biological conditions due to various interactions with their aqueous environment including agglomeration and aggregation caused by hydrophobic interactions and hydrogen bonding [56, 57]. For example, pristine fullerene in powder form is located in the polar derivative zone despite its hydrophobic nature, because single fullerene particles form larger aggregates with specific surface area significantly smaller than molecular C_{60}; therefore, their affinity for hydrophobic small organic or biological molecules is significantly reduced. In fact, the BSAI nanodescriptors reflect the real aggregation state of the measurement in the solution, not the idealized size or particle sizes in purified forms before dose preparation. This has a significant advantage in predicting effects under *in situ* biological conditions.

8.4 Predictive Concentration-Dependent BSAI Model

The process of adsorption of small organic or larger biological molecules onto nanoparticle surfaces is usually complex and concentration-dependent. In particular, the concentrations of the probe compounds (partially as a result of multiple layer adsorptions) could affect the values of the adsorption coefficients. In the chemical analysis described in previous sections, the probe compounds were used at extremely low concentrations in the adsorption experiments to reduce these concentration effects. We adopted a realistically applicable method to avoid the experimental or instrumental error inherent to very low probe concentrations: we calculated the nanodescriptors for a given nanomaterial from binding coefficients experimentally obtained at serial concentrations, and then extrapolated them to infinitely low concentration to obtain a set of theoretical nanodescriptors in an ideal solution [58]. The dependences of the nanodescriptors collected from MWCNT on solute concentrations are presented in Fig. 8.10. The extrapolated values at the ideal condition (infinitely low concentration) of the five nanodescriptors $[r, p, a, b, v]$ for MWCNTs are 0.21, 1.17, −0.69, −3.24 and 4.31, respectively. These mathematically extrapolated values would be generally applicable to derive predictive models without the concentration effects [34]. However, a more reasonable approach would be directly incorporating a concentration-related term into the model.

8.4.1 Concentration Effects and Concentration-Corrected Model

The BSAI model was initially formulated at the condition of low probe concentrations to achieve nanoparticle surface characterization by avoiding influences of probe chemicals concentrations because of possible nonlinear adsorption isotherms at higher concentrations, as well as cross-chemical interactions. Experimental results show that the BSAI nanodescriptors have a nonlinear dependence on probe chemical concentrations, and such concentration effects arise from the increased probability of interactions among molecules of both the same species and different species as the concentrations increase. Thus previous experiments were limited at extremely low probe chemical concentrations to eliminate the potential

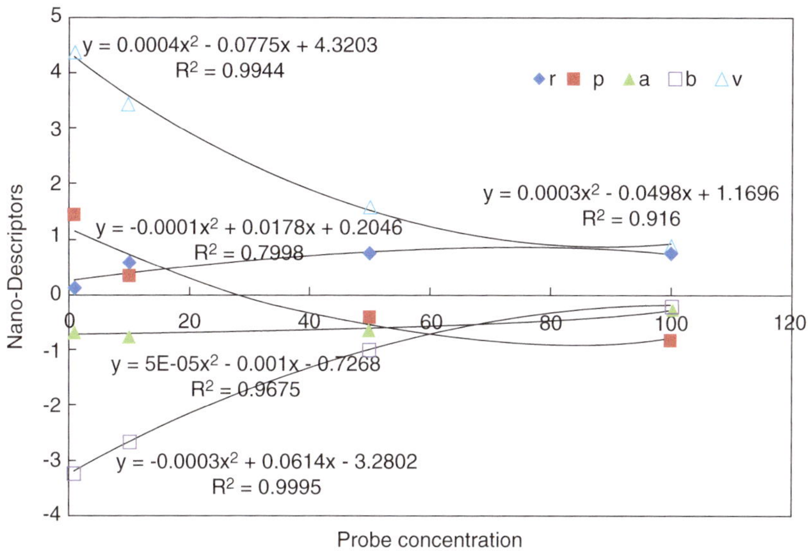

Fig. 8.10 Extrapolation of the nano-descriptors to ideal water solutions. The profiles of the nano-descriptors of MWCNT versus concentrations are given, which were measured with serial concentrations (1×, 10×, 50× and 100×) of probe compounds. Polynomial trend lines and fitting equations were given in the figure, from which the theoretical nano-descriptors in ideal solution were obtained by extrapolating the probe concentration to infinity low (Nx→0). Adapted from by permission from Macmillan Publishers Ltd.: [Nature Nanotechnology] (Ref. [34]), copyright (2010).

concentration effects for better understanding of the surface physicochemical properties of the particles. However, such concentration effects should not be overlooked for the purpose of prediction on chemical binding on the nanomaterials, since from a realistic perspective, concentration of contaminants can vary across unpredictable ranges with respect to environmental applications of nanomaterials. Previous studies were conducted using a LFER-based modelling strategy on the binding of select aromatic organic molecules with granular activated carbon, and validation of linear relationships between the descriptors and a concentration-related term was successful only at extremely low concentrations. [59] However, a different study showed that quadratic polynomial relation between the descriptors and concentration terms was effective to quantitatively incorporate the probe concentration effects before the subsequent LFER modelling. This approach was especially effective when such model was applied to the prediction of adsorption of various chemical species of interests [60]. But this method requires a separate regression analysis, which means significantly more concentration-dependent adsorption data to ensure the quality of the fitting of the isotherm are needed.

A polynomial dependence of nanodescriptors on concentrations is incorporated in the model: $\log k_i = c + rR_i + p\pi_i + a\alpha_i + b\beta_i + vV_i, i = 1, 2, 3, \ldots, n$, where

$$c = lT^2 + mT + n$$

$$r = l_r T^2 + m_r T + n_r$$

$$p = l_p T^2 + m_p T + n_p$$

$$a = l_a T^2 + m_a T + n_a$$

$$b = l_b T^2 + m_b T + n_b$$

$$v = l_v T^2 + m_v T + n_v \quad (8.6)$$

$T = \log \frac{C_e}{C_s}$, where C_e is the equilibrium concentration of a probe in the solution after adsorption, and C_s is the solubility of the probe, chemical activity = $\frac{C_e}{C_s}$ is used as a parameterized level of chemical saturation in the aqueous phase.

The new regression coefficient becomes: $[l, m, n, l_r, \ldots, m_v, n_v]$ (polynomial indices). In the process of regression, these coefficients were directly incorporated in the model, so there is no need for a separate quadratic polynomial model after the initial LFER-based modelling. This newly constructed polynomial model is ideally suited for predictions of the binding of various organic chemical species at different concentration levels.

For each nanoparticle, a set of nanodescriptors were calculated using MLR analysis on the matrix of $[\log k, r, p, a, b, v]$. The Abraham solute molecular descriptors $[R_i, \pi_i, \alpha_i, \beta_i, V_i]$ were obtained using Absolv module provided by ADME online service (Advanced Chemistry Development Inc., Toronto, Canada). MLR analyses were then performed using JMP Pro (SAS Institute Inc., Cary, NC) to obtain polynomial indices. Internal cross-validation using leave-one-out (LOO) technique were conducted to test the robustness of the models, typically the model is considered robust when $PRESS\ RMSE < 1$ and $Q^2_{CV} > 0.7$.

8.4.2 Results and Validation of Concentration-Corrected Model

Modelling results with direct incorporation of the concentration-related term $T = \log \frac{C_e}{C_s}$ is presented in this section to demonstrate the role of chemical concentration in the binding of these chemicals. MLR analyses on the complete set of data were performed before the reconstruction of the BSAI index $[r, p, a, b, v]$ using Eq. (8.6). AlOOH nanoparticles, along with various carbon-based and metal/metal oxide materials, were presented here as an example. Figure 8.11 shows the dependences of these descriptors on T. As expected, different trend of concentration-dependences was observed for the descriptors of different nanomaterials. Ranges of those descriptors indicated that, most significant positive contribution toward the binding is from the hydrophobic force (v, 0.82–5.06), showing a trend of decrease with the increase

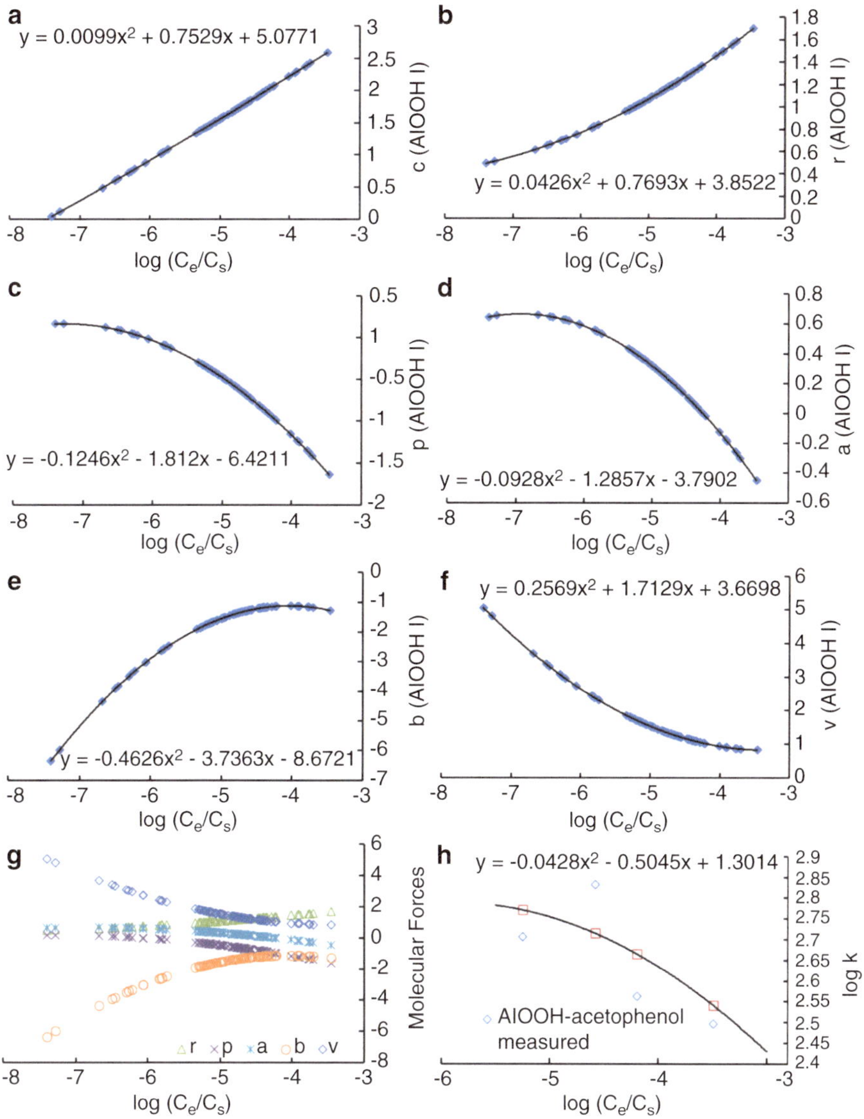

Fig. 8.11 Nanodescriptors of AlOOH I obtained from the modified model were shown to be polynomial functions of T = $\log(C_e/C_s)$ (**a**), regression constant c; (**b**), nanodescriptor r of the excess molar refraction R; (**c**), nanodescriptor p of the effective solute dipolarity and polarizability π; (**d**), nanodescriptor a of the effective solute hydrogen-bond acidity α; (**e**), nanodescriptor b of the effective solute hydrogen-bond basicity β; (**f**), nanodescriptor v of the McGowan characteristic volume V; (**g**), comparison of the dependences on concentration among the descriptors; (**h**), concentration dependent of predicted log k follows quadratic polynomial function (AlOOH I -acetophenol) (Adapted with permission from (Ref. [36]), copyright (2014) American Chemical Society)

of equilibrium concentration C_e. This trend could be explained by the competition for nanoparticle surfaces among different organic molecules present in the mixture. A larger v term is reasonable for particles with both hydrophobic and hydrophilic sites on their surfaces, because for these types of particles, interactions between hydrophobic regions or moieties on the particle surfaces and the organic molecules likely make significant contribution to the binding. While the molecules being driven to bind with less hydrophobic and more hydrophilic sites on the particle surfaces could be the reason for decreases in v term as the concentrations increase. However, further increase in concentration could still cause increase of v, because the spatial rearrangements of the molecules on the surfaces, or multiple layer adsorption induced by intermolecular forces can still increase the density of molecules bound to hydrophobic sites. Hydrogen-bond basicity (b, −6.35 to −1.14) starts with a negative contribution with large magnitude at low concentration, which means, compared to water molecules, the surface is less likely to accept protons from the probe chemicals and form hydrogen bonds. Then it increases with the increase of concentration, probably because surface defects such as oxidation, which could potentially accept protons, can host more molecules as the concentration increases and the proton-accepting ability of the oxygen atom in water molecules are saturated.

Cross validation using LOO method were carried out ($Q^2_{cv} = 0.83$ for AlOOH I, the rest of regression coefficients and validation results of some example nanoparticles are listed in Table 8.3), External validation were employed by randomly choosing 20 % of the entire observations as the testing set and the rest as the training set to build the model (data presented in Table 8.4).

Obvious nonlinearity was found in most of the nanodescriptors – concentration relationships. But some have negligible quadratic terms (e.g. constant c in the case of AlOOH I), which means the concentration dependence of these descriptors can reasonably well approximated using linear functions within the concentration ranges. Such concentration dependence was actually demonstrated by another study over a different and smaller range of chemical activities (T ranging from −1 to −3), and the material discussed was granular activated carbon [59]. However, due to magnified surface chemical effect, such approach was probably not well suitable for the nano scale adsorbent judging by on the results presented here. Nonetheless in particular cases, the model can be significantly simplified yet still accurate enough by eliminating the quadratic term in polynomial model within the specific range of concentration.

Further examination was conducted to test the interpolation of this polynomial model to another concentration range that is not included in model training process, as well as extrapolation to different chemicals. Specifically, the data were divided into three groups: training, validation, and testing. The validation sets consisted of results from the second highest concentration used in our experiments (W10). The testing sets consisted of results from environmental pollutants, nitrobenzene and chlorophenol, at four concentration levels. The models were then built using training set. Stepwise regression based on minimal AICc criteria ($AICc = AIC + \frac{2k(k+1)}{n-k-1}$, where n is the sample size, k is the number of parameters in the model) was employed for each particle separately for

Table 8.3 Regression from polynomial models and cross-validated using LOO method

	AlOOH I	BaSO4 NM220	TiO2 NM105	ZnO NM110	SiO2. Amino	SiO2. Naked	ZrO2. PEG	AG50. EO	AG50. PVP	AG200. PVP	sMWCNT	Fullr C60	MWNT -OH	MWNT 50um	MWNT-COOH 10–20 nm	MWNT-COOH 8 nm	MWNT-COOH30–50 nm
n	5.08	−1.90	−11.53	−4.87	7.13	−8.14	−8.46	−11.23	−7.51	−2.57	−16.07	−1.48	5.35	−18.93	0.06	−10.62	−2.51
l	0.01	−0.25	−0.87	−0.48	0.25	−0.65	−0.63	−0.79	−0.57	−0.29	−0.82	−0.26	0.37	−0.92	0.17	−0.67	−0.40
m	0.75	−2.07	−6.93	−3.62	2.29	−5.33	−5.18	−6.65	−4.84	−2.37	−7.34	−1.54	3.26	−8.19	1.08	−5.83	−2.25
l_r	0.04	0.35	−0.16	−0.08	−0.05	−0.17	−0.04	−0.15	−0.11	−0.26	0.22	0.37	−0.29	−0.49	−0.54	0.44	−0.29
m_r	0.77	2.61	−1.01	−0.56	−0.34	−1.09	−0.16	−1.30	−0.46	−2.10	2.07	2.40	−2.14	−4.34	−4.62	4.44	−2.42
n_r	3.85	5.74	−0.14	0.15	0.25	−0.64	1.40	−1.29	0.95	−2.60	5.92	4.13	−2.12	−8.39	−8.58	11.87	−3.80
l_p	−0.12	−0.02	0.41	−0.10	−0.18	0.22	0.14	0.58	0.06	−0.20	0.26	−0.34	0.15	0.59	0.51	0.04	0.71
m_p	−1.81	−0.37	2.70	−0.99	−1.54	1.45	0.90	4.23	−0.13	−2.24	2.51	−2.43	1.53	5.44	4.98	0.07	5.92
n_p	−6.42	−2.15	3.07	−3.33	−3.92	0.99	0.05	6.03	−2.68	−6.90	6.12	−5.29	4.34	12.94	12.26	−1.09	12.17
l_a	−0.09	0.11	0.16	0.45	0.36	0.56	0.17	0.06	0.21	0.00	−0.56	−0.42	−0.83	−0.48	−1.19	−0.48	−1.44
n_a	−1.29	0.96	0.74	4.03	3.48	5.16	0.85	−0.38	1.75	−0.41	−6.44	−5.05	−9.10	−6.02	−12.45	−6.36	−15.61
m_a	−3.79	2.13	−0.07	9.05	8.29	11.40	0.28	−3.34	3.50	−1.88	−19.39	−15.35	−25.30	−19.17	−33.27	−21.52	−42.95
l_b	−0.46	0.01	0.09	0.08	0.33	0.26	−0.19	−0.20	−0.07	−0.58	−0.05	0.06	−0.33	−0.50	−0.35	0.24	0.21
m_b	−3.74	−0.15	0.48	0.40	2.88	1.99	−1.83	−2.35	−0.92	−5.42	0.16	0.19	−1.73	−4.27	−2.09	3.47	2.82
n_b	−8.67	−1.57	−0.83	−1.75	4.54	2.02	−6.41	−7.87	−4.21	−13.81	0.26	−2.33	−2.43	−9.95	−4.16	10.25	6.91
l_v	0.26	0.02	0.50	0.65	−0.09	0.46	0.55	0.32	0.60	0.93	0.56	0.33	−0.03	0.89	0.07	0.20	0.10
m_v	1.71	0.41	4.14	5.07	−1.08	3.79	4.57	3.37	5.13	8.09	4.29	2.37	−1.50	7.37	−0.19	0.68	−0.74
n_v	3.67	2.27	9.70	11.18	−1.68	8.71	10.87	9.34	12.25	18.43	10.49	6.62	−2.58	18.34	1.83	1.19	−1.62
R^2	0.90	0.82	0.87	0.87	0.82	0.83	0.89	0.89	0.86	0.87	0.88	0.95	0.92	0.93	0.94	0.90	0.93
PRESS RMSE	0.24	0.23	0.27	0.28	0.22	0.29	0.34	0.32	0.29	0.28	0.41	0.23	0.36	0.39	0.39	0.37	0.42
Q^2_{CV}	0.83	0.72	0.75	0.78	0.72	0.70	0.78	0.80	0.75	0.74	0.78	0.91	0.86	0.85	0.86	0.76	0.80

Adapted with permission from (Ref. [36]), copyright (2014) American Chemical Society

Table 8.4 Regression from polynomial models, with 20 % of the observations randomly chosen as test set

	AlOO H I	BaSO4 NM220	TiO2 NM105	ZnO NM110	SiO2. Amino	SiO2. Naked	ZrO2. Amino	ZrO2. PEG	AG50. EO	AG50. PVP	AG200. PVP	sMW CNT	Fullr C60	MWNT -OH	MWNT 50um	MWNT-COOH 10–20 nm	MWNT-COO H8 nm	MWNT-COOH 30–50 nm
n	5.84	−0.98	−12.65	−3.31	7.17	−8.03	−2.56	−0.28	−12.34	−7.04	−5.71	−21.03	−3.68	5.09	−21.06	−6.60	−16.11	2.07
l	0.08	−0.12	−0.93	−0.46	0.25	−0.71	−0.40	−0.23	−0.88	−0.56	−0.48	−1.14	−0.45	0.33	−1.03	−0.43	−0.98	−0.28
m	1.22	−1.37	−7.46	−3.22	2.28	−5.61	−2.83	−1.39	−7.30	−4.67	−3.97	−9.95	−2.89	3.03	−9.18	−3.12	−8.44	−0.70
l_r	0.12	0.57	−0.02	0.07	−0.05	−0.10	−0.03	−0.04	−0.16	−0.06	−0.30	0.09	0.34	−0.33	−0.55	−0.91	0.78	−0.24
m_r	1.48	4.19	0.24	0.52	−0.38	−0.69	−0.39	−0.15	−1.28	−0.02	−2.17	1.86	2.30	−2.47	−5.06	−7.96	7.74	−2.15
n_r	5.18	8.20	2.64	1.93	0.17	−0.17	−0.65	1.27	−1.05	2.06	−2.39	7.26	4.06	−2.62	−10.42	−15.81	19.85	−3.58
l_p	−0.08	−0.05	0.27	−0.05	−0.30	0.25	0.08	0.09	0.74	0.01	−0.16	0.18	−0.45	0.14	0.67	1.06	−0.11	0.06
m_p	−1.35	−0.41	1.41	−0.60	−2.39	1.75	0.67	0.44	5.49	−0.60	−2.08	1.47	−3.31	1.49	6.28	9.31	−1.54	−0.09
n_p	−5.36	−1.65	−0.07	−2.58	−5.53	1.53	0.65	−0.99	8.38	−3.79	−6.90	3.12	−7.05	4.33	15.33	20.34	−5.35	−1.66
l_a	−0.34	0.10	0.22	0.31	0.31	0.46	0.26	0.12	0.11	0.14	0.02	−0.48	−0.15	−0.95	−0.45	−1.66	−0.58	−0.82
n_a	−3.86	0.92	1.32	2.63	2.66	4.21	1.93	0.50	0.15	1.06	−0.23	−5.64	−2.52	−10.17	−5.76	−17.96	−7.57	−9.85
m_a	−10.22	1.98	1.28	5.54	5.41	9.37	3.21	−0.20	−2.01	1.92	−1.59	−17.34	−9.37	−27.82	−18.63	−49.01	−25.07	−29.59
l_b	−0.62	0.01	0.25	0.13	0.29	0.23	0.20	0.06	−0.15	0.01	−0.68	−0.14	0.17	−0.56	−0.21	0.09	0.11	0.45
m_b	−5.04	−0.49	1.99	0.75	2.58	1.76	1.56	0.50	−2.10	−0.02	−5.93	−0.24	1.31	−3.32	−1.62	1.69	2.32	4.54
n_b	−11.27	−3.01	2.72	−1.11	4.12	1.75	1.52	−1.33	−7.73	−2.02	−14.05	0.27	0.37	−4.95	−4.13	4.71	7.99	9.62
l_v	0.14	−0.28	0.50	0.47	0.02	0.48	0.33	0.13	0.21	0.60	1.11	1.11	0.60	0.15	0.93	0.38	0.41	0.19
m_v	0.72	−1.51	4.26	3.46	−0.29	3.85	2.28	0.64	2.51	5.07	9.51	8.29	4.30	−0.24	7.74	2.16	2.38	−0.02
n_v	1.80	−0.72	10.05	7.53	−0.29	8.36	4.91	2.53	7.62	11.85	20.97	17.28	9.79	−0.72	19.18	5.72	4.61	−0.30
R^2	0.92	0.79	0.87	0.90	0.80	0.86	0.84	0.90	0.90	0.86	0.84	0.89	0.94	0.90	0.93	0.95	0.90	0.93
R^2_{test}	0.81	0.80	0.80	0.73	0.82	0.63	0.76	0.80	0.84	0.69	0.92	0.75	0.95	0.93	0.84	0.88	0.76	0.77

the reduction of model complexity, and then manual adjustments were taken to obtain maximum unification possible unification across different nanoparticles. The validation sets were then used to test interpolation capability of the model, i.e. the predictive capability on the adsorption of the compounds at a different concentration; testing sets were used to test the extrapolation capability, i.e. predictive power on the adsorption of environmental contaminant chemical compounds. Select results are shown in Table 8.5: model complexity was successfully reduced from 17 parameters to 10 for metal or oxide particles (five left columns) and 6 for carbon-based particles (five right columns). For most of the particles $R^2>0.8$, and $R^2_{vlidation}$ and $R^2_{testing}$ are all larger than 0.7. Figure 8.12 clearly shows successful predictive abilities of the reduced models in cases of AlOOH I ($R^2 = 0.86$, $R^2_{validation} = 0.84$, $R^2_{testing} = 0.79$) and MWNT-OH ($R^2 = 0.86$, $R^2_{validation} = 0.75$, $R^2_{testing} = 0.79$).

8.4.3 *Profiling of Nanoparticle Surface Physiochemical Properties Using Concentration-Corrected Model*

From the regression analysis described above, sets of 17-dimensional indices were obtained, describing the surfaces physicochemical properties of those particles used in experiments. For the purpose of characterization and categorization of the physicochemical properties the dimension needs to be reduced. PCA was conducted in order to mathematically convert the 17-dimensional vectors comprised of polynomial indices into a set of new vectors that are orthogonal to each other with the first two accounting for maximum variance possible. Then comparisons based on those two first vectors, or components were made among the nanomaterials. The results clearly indicated the capability of separation or categorization between metal/oxide and carbon-based materials for both the original BSAI model and the polynomial model (Fig. 8.13).

8.4.4 *Infinite Dilution Adsorption Descriptors – Incorporating Traditional Adsorption Models*

One drawback of the polynomial model is its large number of descriptors obtained from the MLR analyses, which could limit its application for the interpretation of different physicochemical interactions engaged in the binding process. In addition, according to the clustering analyses results, clear separation could be made only for particles with drastically different surfaces. Such restrictions could reduce the model's value in providing guidelines in nanomaterials designing and safety assessment. A different method, involving approximation at the low concentration using Langmuir physiosorption model, is used to address this concentration effect. The underlying theory of such method is that the concentration effect is induced by the

Table 8.5 Reduced models for example nanoparticles

	AlOOH I	BaSO4 NM220	TiO2 NM105	ZnO NM110	SiO2. Amino	SiO2. Naked	ZrO2. Amino	ZrO2. PEG	ZrO2. TODacid	AG50. EO	AG50. citrat	AG50. PVP	AG200. PVP	sMW CNT	MWNT -OH	MWNT-COOH 10–20 nm	MWNT-COOH8 nm	MWNT-COOH 30–50 nm
n	4.37	0.82	0.41	1.10	−0.09	0.42	1.41	0.14	2.62	2.04	4.27	1.69	2.15	−0.30	−1.41	−1.02	1.88	0.41
l	0	0	0	0	0	0	0	0	0	0	0	0	0	0	0	0	0	0
m	0.54	−0.33	−0.29	−0.17	−0.46	−0.48	−0.17	−0.41	0.10	0.01	0.37	−0.13	0.03	0	0	0	0	0
l_r	0	0	0	0	0	0	0	0	0	0	0	0	0	0.07	0.10	0.06	0.02	0.05
m_r	0.63	0.33	0.10	0.23	−0.08	0.14	0.01	0.36	0.00	0.15	−0.20	0.42	0.34	0	0	0	0	0
n_r	4.00	2.28	1.55	1.99	0.33	1.43	0.34	2.73	0.49	1.74	0.25	2.71	2.86	0	0	0	0	0
l_p	0	0	0	0	0	0	0	0	0	0	0	0	0	0	0	0	0	0
m_p	−0.99	−0.28	−0.20	−0.33	0.01	−0.10	0.08	−0.26	−0.10	−0.79	−0.63	−0.65	−0.90	0	0	0	0	0
n_p	−5.04	−2.04	−1.54	−2.27	−0.48	−1.46	−0.40	−2.27	−1.29	−4.07	−3.63	−3.60	−4.80	0	0	0	0	0
l_a	0.01	0.01	0.00	0.01	0.01	0.00	0.01	0.01	0.03	0.00	0.01	0.00	0.00	0	0	0	0	0
n_a	0	0	0	0	0	0	0	0	0	0	0	0	0	0	0	0	0	0
m_a	0	0	0	0	0	0	0	0	0	0	0	0	0	−1.34	−0.99	−0.95	−1.17	−1.47
l_b	0	0	0	0	0	0	0	0	0	0	0	0	0	0	0	0	0	0
m_b	0	0	0	0	0	0	0	0	0	0	0	0	0	0	0	0	0	0
n_b	−1.42	−0.73	−1.39	−1.81	−1.57	−1.56	−1.64	−2.30	−1.82	−2.33	−1.52	−1.48	−1.30	−1.37	−0.74	−1.59	−1.43	−1.82
l_v	0	0	0	0	0	0	0	0	0	0	0	0	0	−0.01	−0.04	0.03	0.03	0.02
m_v	−0.21	0.17	0.19	0.19	0.40	0.32	−0.03	0.22	−0.03	0.35	0.29	0.18	0.35	0	0	0	0	0
n_v	0.20	1.68	2.22	2.36	3.11	2.47	1.39	2.65	1.22	2.85	2.21	2.29	2.69	3.19	3.66	3.34	1.91	2.73
R^2	0.88	0.77	0.84	0.86	0.78	0.82	0.77	0.86	0.76	0.86	0.77	0.82	0.82	0.80	0.86	0.92	0.84	0.90
$R^2_{validation}$	0.84	0.71	0.71	0.57	0.33	0.63	0.72	0.86	0.63	0.78	0.68	0.81	0.82	0.85	0.75	0.83	0.78	0.84
R^2_{test}	0.79	0.74	0.83	0.74	0.73	0.66	0.59	0.72	0.85	0.72	0.79	0.92	0.87	0.86	0.79	0.71	0.79	0.76

Metal/oxides (16 columns on left) have a different model from carbon-based particles (5 columns on right). Adapted with permission from (Ref. [36]), copyright (2014) American Chemical Society.

Adsorption data were split into three groups, the first group was used to build the model, and the other two were used as validation and test. The model was first built by minimizing AICc, then manually adjusted to reach unification within the groups of particles. Validation was performed using data from N10 W concentration, and testing was performed using data from nitrobenzene and chlorophenol at 4 different concentrations.

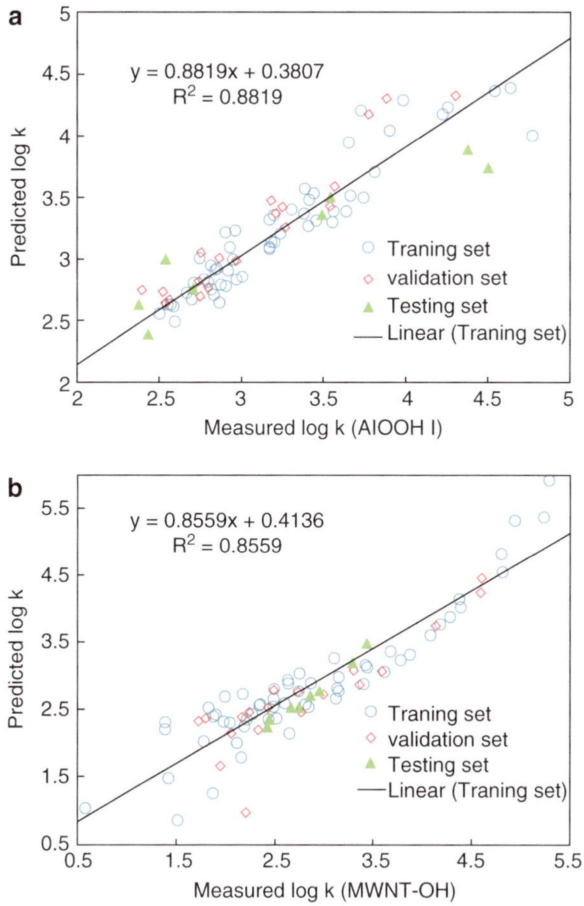

Fig. 8.12 Predicted *versus* measured log *k* values for (**a**) AlOOH I ($R^2 = 0.86$, $R_{\text{validation}}^2 = 0.84$, $R_{\text{testing}}^2 = 0.79$) and (**b**) MWNT–OH ($R^2 = 0.86$, $R_{\text{validation}}^2 = 0.75$, $R_{\text{testing}}^2 = 0.79$) (Adapted with permission from (Ref. [36]), copyright (2014) American Chemical Society)

intertwined interactions of water with chemical molecules, as well as among chemical molecules of the same or different species [34, 60], the probability of which increases as the concentration increases.

Both nanoparticle surface physicochemical properties and the probe chemical concentrations are determining factors of surface physisorption. One must be minimized in order to identify the other. Ideally, adsorption coefficient can be defined at infinitesimally low concentrations, referred to as *infinite dilution adsorption index* (k_∞):

$$k_\infty = \lim_{C_0 \to 0} \frac{C_{ad}}{C_e} \tag{8.7}$$

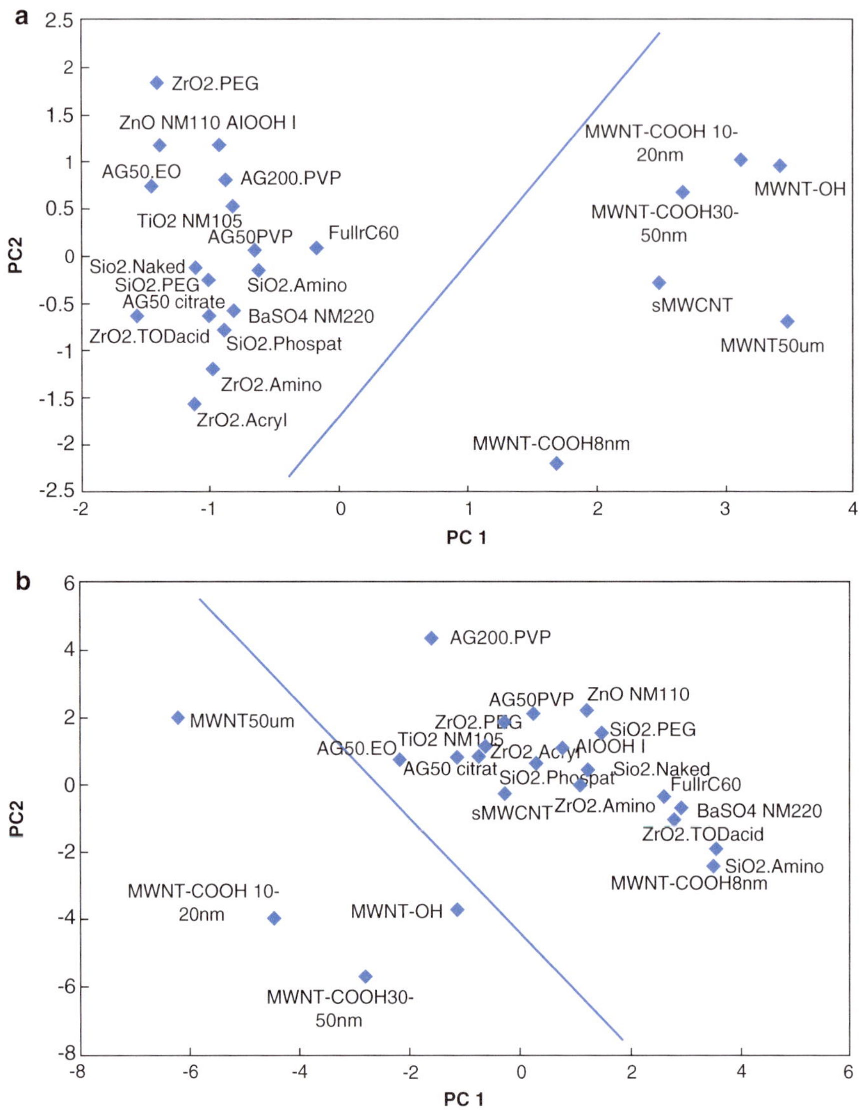

Fig. 8.13 Nanoparticle clustering plots by two principal components. The two components were obtained by principal component analysis of the five nanodescriptors and 17 polynomial indices of both metal/oxide and carbon-based nanoparticles from (**a**) original BSAI model and (**b**) polynomial model (Adapted with permission from (Ref. [36]), copyright (2014) American Chemical Society)

In the original BSAI model, very small concentrations were used to reduce the concentration effect. However, the limit of detection and quantification are usually restricted by measurement errors and instrumental conditions, ambient signal noise, or cross contamination. Such limitation can be avoided to certain extent by alternative analyses based on Langmuir physisorption theory, which relates to adsorbed concentration of a chemical to its equilibrium concentration (modified from Eq. (8.1) introduced in Sect. 8.2.2.):

$$C_{ad} = \frac{KQC_e}{1+KC_e}, \tag{8.8}$$

In this formula, Q is the adsorption capacity, or maximum possible adsorbed amount as C_e increases, and K is Langmuir equilibrium constant. At infinitesimally low concentration, $\frac{C_{ad}}{C_e} \rightarrow QK$, suggesting that $k_\infty = QK$. Therefore, an estimated value of the infinite dilution adsorption coefficient can be calculated by fitting the Langmuir model to experimental results. More importantly, such estimation seems to be insensitive to experimental errors possibly because the model fitting process using large datasets from multiple concentrations averages out the errors.

Adsorption data from four concentration groups were used for the Langmuir regression in the linearized form: $\frac{C_e}{C_{ad}} = \frac{1}{Q}C_e + \frac{1}{QK}$. Regression analyses between $\frac{C_e}{C_{ad}}$ and C_e were used to obtain an approximated adsorption coefficient k_0 at ideal condition (infinitely low concentration). Then similar BSAI modelling was conducted using logk_∞ in the same manner as the previous model, and concentration-independent nanodescriptors were obtained to achieve the elimination of chemical concentration effect. Better characterization and categorization based on surface physicochemical properties of nanomaterials can be obtained using the model built on such approximation. The whole process of BSAI approach, including experimental methods and statistical analyses, for the original model, as well as the two improved ones, is illustrated by the flow chart in Fig. 8.14.

8.4.5 Infinite Dilution Model Validation and Surface Profiling

The extrapolation of adsorption data to an infinitely low concentration can approximate an ideal aqueous environment. As a result, it is now possible to characterize and categorize the surface properties without the external influence (Table 8.6). This is an important improvement, because surface physicochemical properties of the nanoparticles are independent on probe concentration. The results of categorization of the metal/oxide nanoparticles are shown in Fig. 8.15. These results were obtained using PCA based on nanodescriptors generated from the original BSAI model (Fig. 8.15a, at the lowest experimental chemical concentration) and the Langmuir low-concentration approximation (Fig. 8.15b). The ability of that the low-concentration approximation to generate similar clustering but much better separation is revealed by comparing the two. Although excellent predictive abilities were discovered for the polynomial model discussed in previous sections, incases of both inter-concentration and inter-chemical

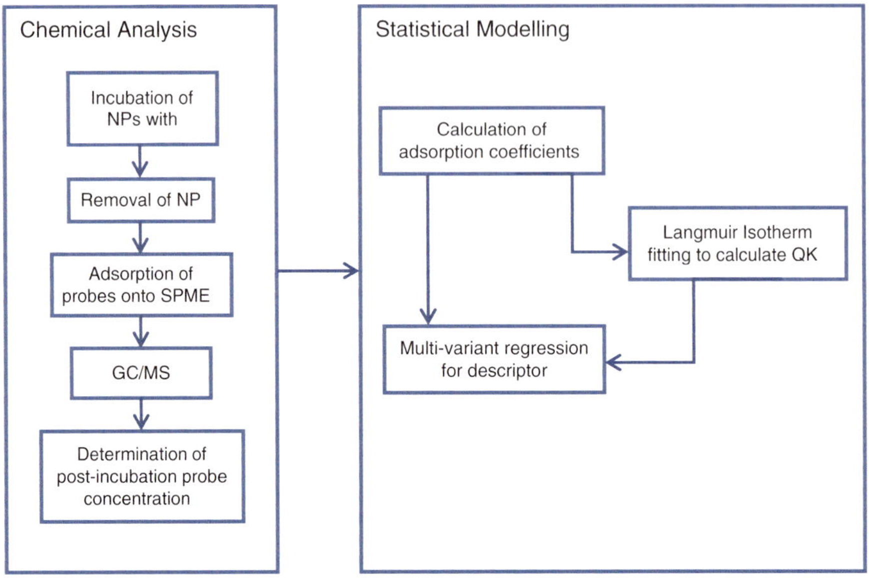

Fig. 8.14 Flow chart of complete BSAI approach (Adapted with permission from (Ref. [36]), copyright (2014) American Chemical Society)

Table 8.6 Infinite dilution adsorption descriptors obtained from Langmuir isotherm approximation

	c	r	p	a	b	v	R^2	Q_{CV}^2
AlOOH I	1.33	0.50	−0.05	1.51	−2.15	1.70	0.81	0.64
BaSO4 NM220	1.76	0.51	−0.09	0.86	−1.33	1.11	0.74	0.45
TiO2 NM105	1.52	0.51	−0.07	1.19	−1.68	1.41	0.78	0.59
ZnO NM110	1.48	0.37	−0.36	1.47	−1.88	1.88	0.74	0.51
SiO2.Amino	1.60	0.31	−0.29	1.40	−1.88	1.77	0.72	0.49
SiO2.Naked	2.07	0.61	−0.32	1.16	−1.79	1.15	0.78	0.58
SiO2PEG	2.01	0.53	−0.63	1.57	−2.13	1.44	0.80	0.60
SiO2.Phosphat	1.72	0.41	−0.38	1.47	−1.77	1.68	0.83	0.69
ZrO2.Amino	1.85	0.27	−0.63	1.63	−1.93	1.77	0.82	0.70
ZrO2.PEG	1.96	0.84	−1.03	1.52	−1.91	1.64	0.69	0.46
ZrO2.TODacid	1.88	0.31	−0.55	1.75	−2.09	1.48	0.81	0.70
AG50.citrat	2.08	0.74	−0.53	1.13	−1.75	1.34	0.73	0.48
AG200.PVP	2.40	0.39	−0.33	1.27	−1.22	0.88	0.65	0.24
sMWCNT	0.81	0.91	1.26	−0.83	−1.20	1.53	0.71	0.38
FullrC60	0.42	0.44	−1.07	−0.16	−2.11	2.52	0.88	0.75
MWNT-OH	−1.77	2.19	1.53	−0.30	−1.38	2.30	0.88	0.73
MWNT50 um	−0.20	1.07	1.32	−0.83	−0.88	2.39	0.74	0.27
MWNT-COOH 10–20 nm	−2.11	1.24	1.52	0.01	−1.73	4.02	0.92	0.83
MWNT-COOH30–50 nm	0.00	1.57	0.43	−0.44	−0.60	2.71	0.86	0.70

Adapted with permission from (Ref. [36]), copyright (2014) American Chemical Society

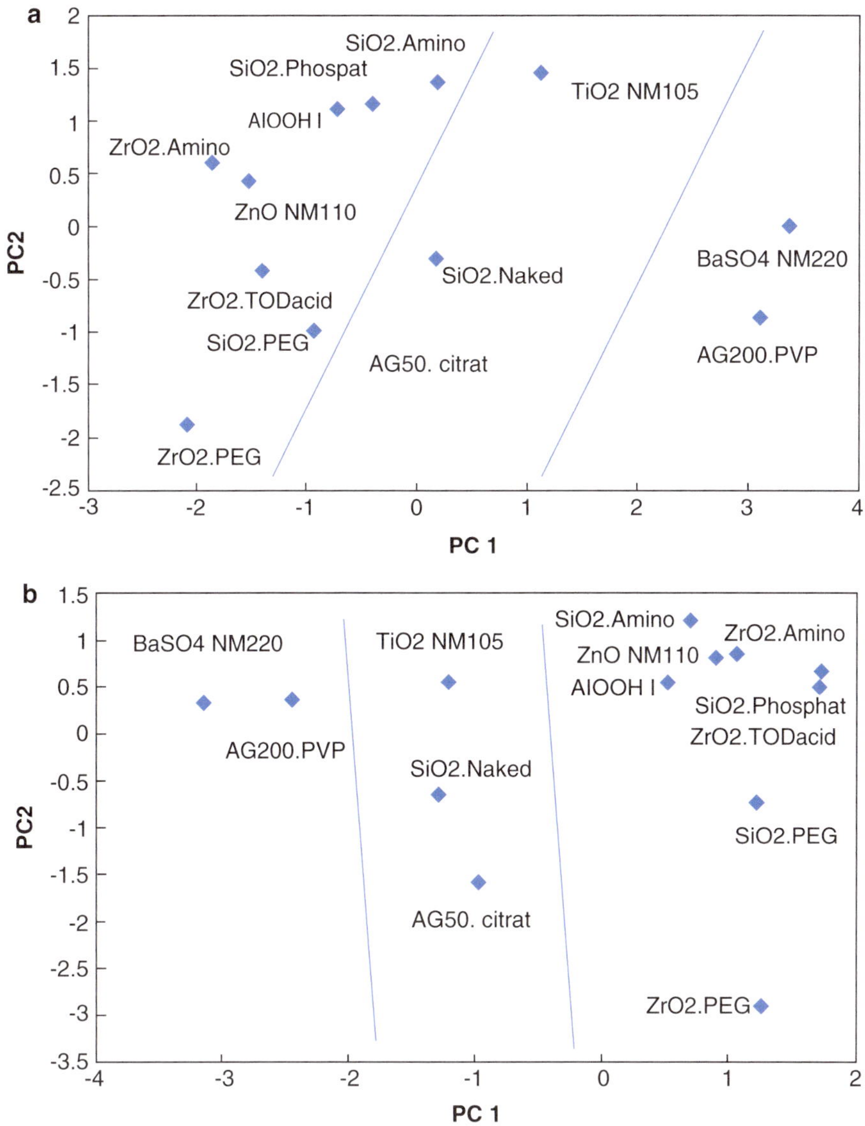

Fig. 8.15 Nanoparticle clustering plot by two principal components. The two components were obtained by principal component analysis of the five nanodescriptors of metal and oxide nanoparticles from (**a**) the original BSAI model and (**b**) the low-concentration approximation by the Langmuir model (Adapted with permission from (Ref. [36]), copyright (2014) American Chemical Society)

species predictions, the categorization ability was acceptable only in the case of separating carbon-based from metal/oxide particles, achieving no clear distinction within the group of metal/oxides. The possible reason for such performance discrepancy was because of the larger number of parameters in the polynomial model, and confounded errors from dimension reduction employed in PCA.

Table 8.7 List of adsorbate molecules considered in this work

Abbrev.	Compound Name	Substituent	Halogen
Ah	acetophenone	ketone	
bPh	biphenyl	aromatic	
BrPl	3-bromophenol	hydroxyl	X
ClAh	4-chloroacetophenone	ketone	X
ClAn	4-chloroanisole	ether	X
ClPl	3-chlorophenol	hydroxyl	X
ClT	4-chlorotoluene	alkyl	X
dMPl	3,5-dimethylphenol	alkyl, hydroxyl	
EtBa	ethylbenzoate	ester–alkyl	
EtPh	ethylbenzene	alkyl	
EtPl	4-ethylphenol	alkyl, hydroxyl	
FPl	4-fluorophenol	hydroxyl	X
mCr	m-cresol	alkyl, hydroxyl	
MeBa	methylbenzoate	ester–alkyl	
MeBl	(3-methylphenyl)methanol	alkyl, hydroxyl	
MeNh	1-methylnaphthalene	aromatic	
MMBa	methyl-2-methyl benzoate	alkyl, ester–alkyl	
Nh	naphthalene	aromatic	
NoPh	nitrobenzene	nitro	
NoT	4-nitrotoluene	alkyl, nitro	
PhAc	phenylacetate	ester–alkyl	
PhAm	4-chlorophenylamine	amine	X
PhBr	bromobenzene	–	X
PhCl	chlorobenzene	–	X
PhCN	benzonitrile	nitrile	
PhEl	2-phenylethanol	alkyl–hydroxyl	
PhI	iodobenzene	–	X
PhMl	phenylmethanol	alkyl–hydroxyl	
Pl	phenol	hydroxyl	
PrPh	propylbenzene	alkyl	
pXy	p-xylene	alkyl	
T	toluene	alkyl	

For clarity in our plots, the compounds are indicated by the short, but sometimes nonstandard abbreviations given here. The other columns indicate how the molecule differs from a benzene prototype and whether the compound is halogenated. Adapted with permission from (Ref. [61]), copyright (2015) American Chemical Society

8.4.6 *Comparative Molecular Dynamics (MD) Simulation Predictions on Adsorption*

Molecular dynamics techniques have the potential to facilitate the design and optimization of nanomaterials surfaces for applications such as drug delivery and contaminant removal, given that reliable atomistic models of nanomaterial surfaces are available as well as accurate descriptions of their interactions with relevant solutes. This section provides an example on the calculation of adsorption of small organic molecules onto CNTs. The equilibrium constants determined using MD simulation coupled with free-energy calculation techniques are directly compared to those obtained from experimental measurements. Based on the results, these calculations are highly predictive of the relative adsorption affinities of the compounds, with excellent correlation between calculated and measured values of the logarithm of the adsorption equilibrium constant.

The model of CNTs were set to best match those used in BSAI experiments [61]. Fig. 8.16a shows an example of the models used in the MD simulations. According to calculation results, the presence of four graphene sheets versus a single graphene sheet has only a mild effect on calculated adsorption constants, implying that such models are reasonable representations of any relatively flat graphenic surface, including free-standing graphene and graphene nanoplatelets. Figure 8.16b shows the potentials of mean force for a few representative compounds. At large distances, $z > 1.4$ nm, the interaction between the surface and adsorbate becomes negligible, yielding a plateau at a fixed value, which by convention is anchored to zero. On the other hand, steric interaction leads to a rapidly rising free energy as the adsorbate attempts to penetrate the surface ($z < 0.33$ nm). Given the relatively hydrophobic natures of both the small aromatics and carbon nanotubes [62], one would expect adsorption to be thermodynamically favored. Minima are found in the potentials of mean force with magnitudes of several times the thermal energy for all adsorbates, occurring at distances from $z = 0.35$–0.37 nm. The deepest free-energy well of −8.7 kcal/mol is seen for 1-methylnaphthalene, while the shallowest is for toluene. As seen in Fig. 8.16a, in all cases the lowest free-energy configuration corresponds to the adsorbates lying flat on the graphene, with the aromatic moieties parallel to surface. The fraction of adsorbed molecules is not determined solely by the depth of the free energy well, but also by its shape. For example, although 4-nitrotoluene possesses a slightly deeper minimum in Fig. 8.16b than biphenyl, the larger size of biphenyl leads to a broader well, which gives it a larger equilibrium constant. The logarithm of adsorption constants is plotted against the corresponding experimentally determined value for each compound in Fig. 8.16c. Substantial linear correlation is unmistakable and can be quantified by a Pearson correlation coefficient of $r = 0.90$. Furthermore, in absolute terms, values of log k range from 2.0 to 5.7, while those of calculated values occupy a similar range of 1.7–4.2. Thus, there is excellent consistency between experiment and simulation in the relative values for different adsorbates, and substantial agreement in an absolute sense as well.

The experimental data for comparison was obtained to construct BSAI models [34–36] that relates adsorbate physicochemical properties to their affinity for nanomaterial surfaces. In some cases, the surface adsorption descriptors could be constructed more conveniently *in silico*. While MD simulations can directly yield adsorption free energies, such calculations become rapidly more expensive and time-consuming as the systems become larger and more complicated, such as the presence of multiple chemical species, and defects or impurities in the nanomaterials. Therefore, we recognize the potential of combining information gleaned from

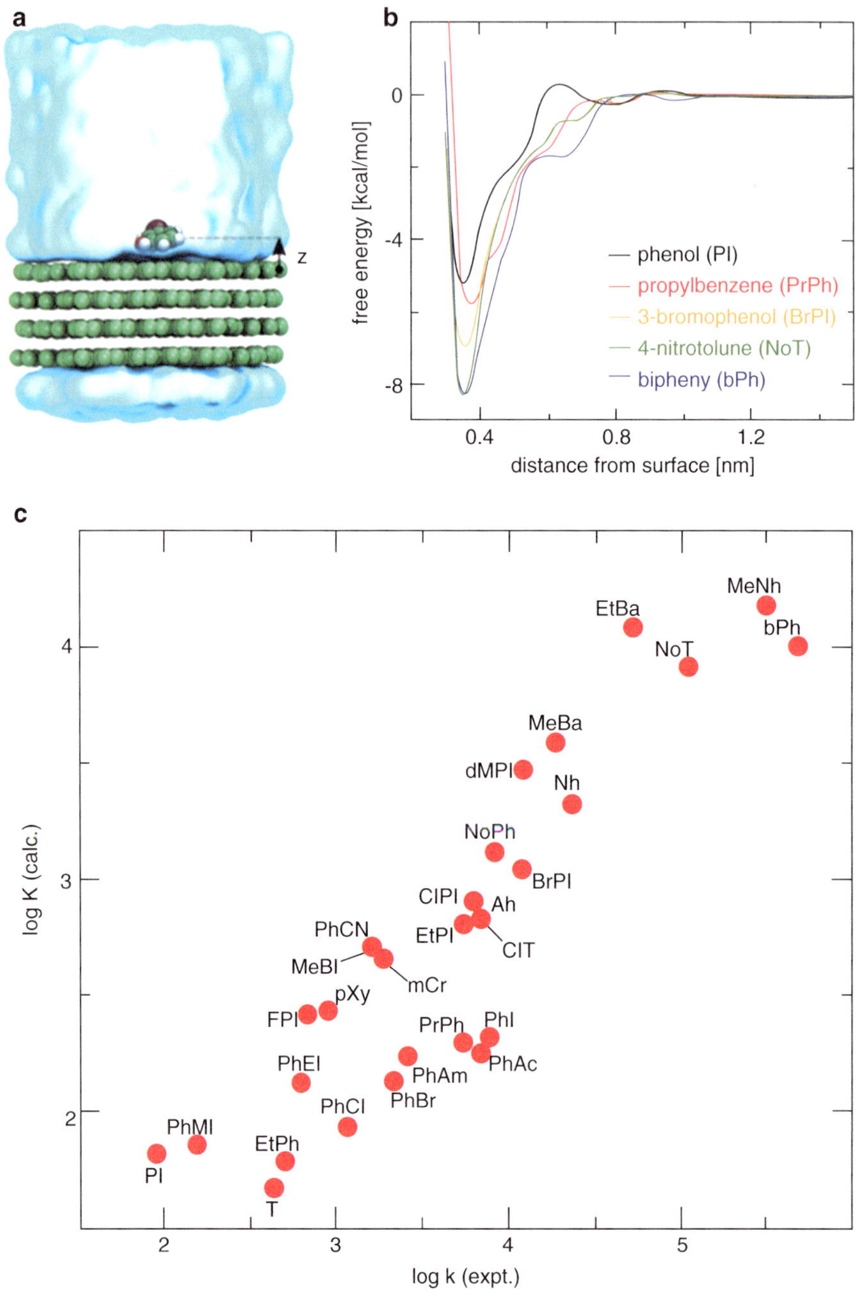
a
z
b
free energy [kcal/mol]
0
-4
-8
0.4
0.8
1.2
distance from surface [nm]
phenol (Pl)
propylbenzene (PrPh)
3-bromophenol (BrPl)
4-nitrotolune (NoT)
bipheny (bPh)
c
log K (calc.)
log k (expt.)
2
3
4
5
MeNh
EtBa
bPh
NoT
MeBa
dMPl
Nh
NoPh
BrPl
ClPl
Ah
EtPl
ClT
PhCN
MeBl
mCr
pXy
FPl
PrPh
PhI
PhAc
PhEl
PhAm
PhBr
PhCl
PhMl
EtPh
Pl
T

simulation with a structure activity approach exemplified by the BSAI surface adsorption index. The optimal strategy may be a multi-scale approach, using explicit-solvent molecular dynamics and coarser descriptions derived from it, such as the grand-canonical Monte Carlo/Brownian dynamics approach, to build predictive structure-activity models for complex heterogeneous systems. A combination of both approaches also allows for a direct testing of MD assumptions of fine structure (e.g. distribution of surface substituents) using the experimental data.

While the MD simulation method has validated existing atomistic models of graphenic and modified graphenic surfaces, there has been little systematic evaluation of models used for other nanomaterials. Existing force fields for materials such as metals and ceramics may require further evaluation to demonstrate whether they perform consistently for a wide variety of organic adsorbates. Moreover, many exciting new materials lack any force fields with even minimal validation. Thus, for MD approach to be practical, it is necessary to construct a comprehensive library of force fields suitable for interactions between nanosurfaces and organic molecules. On the experimental side, more consistent synthesis and better characterization can also help in constructing more accurate models. Thus, comparative research on experimentally and computational characterizations can shed light on the fine structure of nanomaterials surfaces on atomistic/molecular level, and the nature of their interactions in terms of physical forces and energetics.

8.5 The Outlook for BSAI Applications

Factors that determines the nature of the surface interactions, including the heterogeneity of the particle surface, the surface functional groups and the surrounding aqueous environment are the major reasons why the adsorption process of these chemicals or biomacromolecules on to particle surfaces is so complex. Elevated concentration typically causes such complexity to increases significantly due to the competitive interactions between molecules of either the same species or different species coexisted in the solution at higher concentration. When the Langmuir low

Fig. 8.16 Comparison of experiment and simulation for compound adsorption on naked carbon nanotubes. (**a**) Model for molecular dynamics simulation of adsorption of small molecules onto the surface of multi-wall carbon nanotubes. The atoms of the graphene sheets and an exemplary adsorbate (3-bromophenol) are shown as spheres, with hydrogen, carbon, oxygen, and bromine atoms shown in *white*, *green*, *red*, and *crimson*. Here, for clarity, the water is indicated by a translucent cyan surface; however, in the simulations, water molecules were represented explicitly. (**b**) Calculated free energy as a function of distance between the first graphene sheet to the center of mass of the adsorbate (the coordinate z) for exemplary adsorbates. (**c**) Comparison of the logarithm of the adsorption equilibrium constant measured in experiment and the same quantity calculated in simulation for all 29 adsorbates. The abbreviations of the compounds used here are listed in Table 8.7 (Adapted with permission from (Ref. [61]), copyright (2015) American Chemical Society) concentration approximated BSAI model was built, the requirement of low concentrations of probe chemicals were still valid to insure the applicability of the Langmuir isotherms [63]. To study the binding at much higher concentrations, models that consider solute-solute interaction or multilayer adsorption should replace the Langmuir approach [63–65]. Another potential improvement on the accuracy of the isotherm fitting can be achieved by reduce the experimental concentration intervals. The applicability of polynomial models, although with parameterized concentration, should still be experimentally restricted within the concentration range where the dependence of the descriptors on concentration can be accurately described as quadratic functions.

Other environmental factors like pH and temperature can also be incorporated. Temperature fluctuation is important in both biological systems and the environment since the energetics of those interactions can strongly influenced. In principle BSAI model can be expanded to include both factors. Measures should be taken to avoid large fluctuation in temperature, which could cause evaporation/condensation of the volatile small organic chemicals. Changes in pH should also be realistic considering the nanomaterials in biological systems can be transported from one compartment to another with different proton concentration. Significant impact on the models should be expected due to this difference, since most of the physicochemical interactions considered in the models are based on electron displacement, and can be drastically influenced by the pH of local environment. When we obtained experimental data, we used small organic molecules because their physicochemical properties can be quantitatively and semi-empirically parameterized by rather simple and small sets of descriptors. Also, the modelling using these physically interpretable or well defined descriptors would be better for the purpose of surface characterization and categorization of NPs. However, there still exist challenges of quantitatively parameterize the physicochemical properties of a complex structure such as a large protein. In order to obtain a statistically reliable model for such larger macromolecules or molecular complexes, a much larger experimental data set, as well as a much more complex set of descriptors, would be required.

8.5.1 Biological/Environmental Activity Prediction

Apparently the polynomial model could be applied for the prediction of the molecular physisorption, in cases of either environmental contaminants or biomacromolecules. Our experimental setup requires that the probe chemicals should be kept at minimal concentrations to insure the applicability of the original BSAI model and Langmuir model. The application of such model for the purpose of prediction requires that the model training concentrations should to be on the same level as the concentrations to be predicted. Because the validity of the model is actually limited by a certain range of concentration. For similar reasons, the chemicals to be predicted should fall into the same chemical space of the probes which are used for model building. Thus, in order to produce a more applicable model, the probes

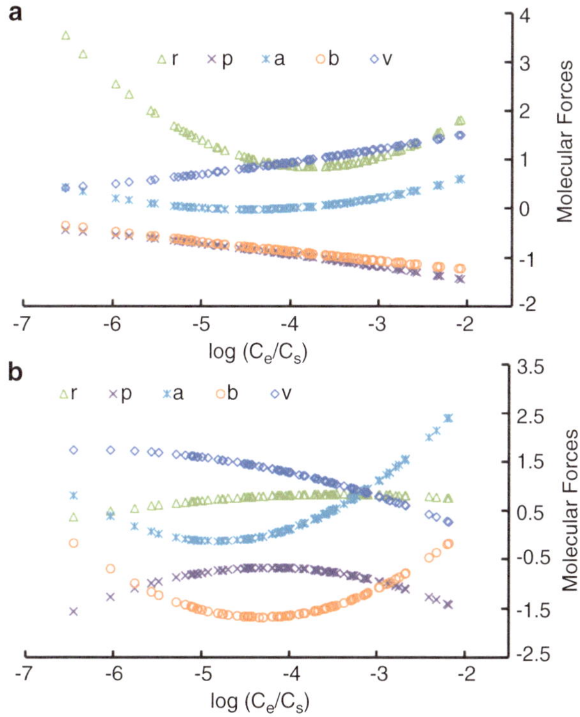

Fig. 8.17 Concentration dependency of BSAI nanodescriptors of (**a**), SiO_2.Amino and (**b**), SiO_2.Naked (Adapted with permission from (Ref. [36]), copyright (2014) American Chemical Society)

should have comparable but diverse physicochemical properties to cover a large chemical space.

The categorization using the models can render significantly different adsorption profile for particles with the same core materials but different surface. For example, in cases of SiO_2.Amino and SiO_2.Naked (Fig. 8.17), the quadratic terms had opposite signs, reversing the trend of concentration dependences. This can be explained by the fact that both the number of probe molecules in the solution and the number of active surface adsorption sites (supposedly determined by surface coating) are the determining factors in terms of concentration dependence. Comparing SiO2.Amino and SiO_2.Naked, the amino coating caused large differences in p (interactions from molecule dipolarity and polarizability), and v (hydrophobic forces). These differences could be due to the increased hydrophilic sites caused by the surface coating compared to pristine nanoparticles. Thus, by illustrating the trend of concentration dependence in BSAI descriptors, polynomial indices can make additional distinctions about the surface coating, and may be able to better predict differences in biomolecule adsorption and potential biological activity, caused by different surface coating. Another advantage of this experimental-based approach is that actual manufactured nanoparticles can be evaluated under biological relevant conditions.

In general activities of nanomaterials in biological or environmental systems, or the response of the host systems can be predicted by creating a link between the activity and the physicochemical parameters using similar QSAR-like approaches. Clustering of 50 nanomaterials was achieved using QSAR modelling by creating a characterization profile using a "*i j k*" parameter (certain biological effect is measured at dose *i* to cell type *j* and the effect is measured using assay *k*); based on the results, a screening mechanism can be established to target desired biological effects for certain cell types with certain assay [66]. Another study successfully created a model to predict smooth muscle apoptosis caused by nanoparticles using three parameters representing particle core material, surface coating, and surface charge respectively as an example of QSAR bridging biological effects and physicochemical properties [67–69]. A similar nanomaterial environmental impact (NEI) modelling framework was proposed explore and evaluate the environmental toxicity of nanomaterials using a zebrafish model system, the model can generate a summarized EZ score representing the toxic effect of nanomaterials from precious measurements [70]. In a different direction, non-linear models were employed to predict the complex phase behavior of amphiphilic nanoparticles for drug delivery; it is capable of predicting how different drugs, drug loadings, and temperatures affect nanoparticle mesophase behavior [71].

Efforts have been made in applying nanotechnology for environmental pollution detection and remediation to pesticide contamination. Many nanoparticles were proposed as sensors for the detection of organophosphorus pesticides like paraoxon, malathion, chlorpyrifos, and dichlorvos [72–74], where adsorption of the target molecule onto the nanoparticle surfaces is the key signaling process. Similarly, methods aimed at the removal of pesticides from the environment have been proposed using nanomaterials. Alumina (Al_2O_3) and magnesium oxide (MgO) nanoparticles embedded in active carbon fibers were capable of destructive adsorption of diazinon [75], zinc oxide (ZnO) or silver (Ag) – chitosan composites removed permethrin and other pesticides from water [76, 77]. Recent studies revealed that metal or metal oxide nanoparticles, such as Ag [78–80], ZrO_2 [81], or titanium dioxide (TiO_2) [82] were capable of pesticide degradation via either surface or photocatalysis. Nanoparticle based filtration and surface catalyzed degradation technologies [83] mostly rely on surface adsorptions of the contaminants onto particle surfaces, while photocatalytic methods benefit from high surface affinity between the two, since they usually utilize enhanced localized photon energy to magnify the catalytic capabilities. Prediction on adsorption of environmental contaminants has also been achieved using BSAI by analyzing the interactions of select group of nanomaterials with a variety of pesticides. Statistical modelling was conducted on the experimentally obtained adsorption data based on polynomial BSAI models [84]. These quantitative computational approaches support the application of BSAI modelling in the area of environmental contaminant detection and remediation.

Researchers have been developing applications of nanotechnology for detection and remediation of environmental contamination by organic pollutants like pesticides. The detection of organophosphorus pesticides like paraoxon, malathion, chlorpyrifos, and dichlorvos were explored using nanoparticles-based sensing methods [72–74]. In these methods, the key signaling process depends on the nanoparticle surface adsorption of the target molecule. Furthermore, researchers have also explored nanoparticle-based methods for the removal of pesticides from the environment. For example, destructive adsorption of diazinon was achieved using active carbon fibers with alumina (Al_2O_3) and magnesium oxide (MgO) nanoparticles [75], removal of permethrin and other pesticides from water were achieved using zinc oxide (ZnO) or silver (Ag) – chitosan composites [76, 77]. pesticide degradation capability via either surface or photocatalysis was also revealed for metal or metal oxide nanoparticles, including Ag [78–80], ZrO_2 [81], or titanium dioxide (TiO_2) [82]. Surface adsorptions of the contaminants is the key process for most nanoparticle-based filtration and surface catalyzed degradation technologies [83], even for methods based on photocatalysis, since it benefit from high surface affinity toward the nano surfaces, which helps taking advantage of localized photon energy magnifying the catalytic capabilities. By utilizing BSAI models to analyze the interactions of select group of nanomaterials with a variety of pesticides, prediction on the adsorption of these environmental contaminants can be made. Out latest efforts include statistical modelling on the experimentally obtained adsorption data based on polynomial BSAI models [84], validating the application of BSAI modelling in the area of environmental contaminant detection and remediation.

8.5.2 Surface Characterization, Categorization and Safety Assessment

The BSAI approach can also be used for the determination of the toxicity thresholds of nanomaterials. The strong affinity of proteins toward the nanomaterial surfaces could result in irreversible adsorption, which forms a biocompatible coating reducing the toxicity of the nanomaterials in biological systems. For example, carbon nanomaterials have very hydrophobic surface; irreversible adsorption with proteins is postulated by the BSAI characterization data, which could render such carbon nanomaterials nontoxic in biological systems. Clarification or validation from some toxicity studies of carbon nanomaterials is still needed to identify the true cause for observed toxicities reported in the literature; for example, the toxicity from leached metal ions would attribute to the defects or contaminants, not the carbon nanomaterial itself [85–87]. On the other hand, binding of proteins with certain functions or physiological roles, such as complement factors, to the surface of carbon nanomaterials could also lead to adverse immunological effects [88–90]. The adsorption of proteins could lead to protein structural changes in higher degrees such as conformation, unfolding, or new epitopes [91], which could cause new biological consequences. Clearly, there is a need for a rational approach to predict such effects.

Physicochemical characterization of nanosurfaces in terms of adsorption capabilities is also of great importance. Adsorption isotherms can be fitted according to physical adsorption models described in Sect. 8.2.2, and generate metrics reflecting surface interaction affinities. Categorization on the surface adsorption profiles were conducted based on the fitness of their isotherms to the three physical physisorption models using PCA on the coefficients of determination obtained from each of the model fittings. Our recently published [84] research shows that better separation were achieved for all ZrO_2, SiO_2, and silver groups than that obtained using polynomial BSAI indices. Furthermore, PCA was again applied to model fitting parameters obtained from Freundlich models. The separating based on these fitting parameters alone seems to be comparable to previous results using the coefficients of determination of all three models. The possible reason is that the linear range of the isotherms can be better fitted by the Freundlich models. The Freundlich models were also able to extract more accurate information for the comparison of isotherms at the low concentration range to those at relatively higher and saturating concentrations. ZrO_2 particles seem to be significantly better separated than in the case of polynomial models, possibly because their adsorption isotherms presented very distinctive patterns for a few pesticides, such as relatively lower affinity for tetrachlorom-xylene indicated by the trend toward saturation compared to the linear trend for some other pesticides.

Since the low concentration approximation of BSAI descriptors is able to largely exclude factors other than particle forces, including inter-molecular interactions among different solutes and water, the clustering on those descriptors is expected be able to more accurately categorize nanomaterials surfaces. One potential application would be predicting biological or environmental behaviors of unknown nanomaterials by categorization using low concentration approximated BSAI descriptors together with nanomaterials with known biological or environmental behavior. Particles clustered within the same group should attract similar types of organic or biological molecules. Since the biological impact of nanomaterials are largely determined by their affinity for biomacromolecules including proteins and subcellular structures like lipid bilayers, materials with the same bio-identity, biodistribution or toxicity may likely be found within the same cluster [92]. A similar model was proposed for the characterization of biological activities of nanomaterials: small organic molecules were replaced with larger biomacromolecules, the adsorption of proteins and the content of formed protein corona were used as fingerprints to create surface characterization profiles of nanomaterials, the adsorption of proteins can be predicted similarly, and the materials can be categorized according to their profiles. However due to the complexity of protein and other macromolecules, at this stage the model is still rather primitive to achieve more accurate prediction [93].

Multiple potential applications could be conceived in the field of nanomaterial environmental safety assessment. An assessment tool based on BSAI approach would be particularly powerful in quantitatively characterizing and categorizing the biological identity of various engineered nanomaterial. A nanomaterial's BSAI indices would allow initial predictions of the material's interactions to contaminants in aqueous environments such as aquifers, surface ponds and lakes, the ability to predict nanomaterial binding to environmental contaminants is demonstrated in this chapter as an example. Such approach could also find application in environmental remediation where one could define an optimal BSAI with high affinity for a specific contaminant in a defined environment and then using statistical clustering, identify an appropriate nanomaterial. In order to continue the development and application of these indices, the correlation between BSAI-characterized nanomaterials and known biological and environmental endpoints is crucial.

8.6 Summary

In summary, we reviewed the physical nature and impact of biomolecular adsorption in general, introduced the physicochemical molecular interactions, explored mathematical models that have been developed to describe the adsorption. Furthermore, we presented a new approach of statistical modelling (BSAI) that utilized large sets of experimental nanomaterial chemical probe adsorption data and modified modelling approaches to generate more advanced BSAI models, including a polynomial model, and a low concentration approximated model. These improved models possess either improved predicting or better categorizing abilities. Specifically, better prediction results were achieved by the polynomial models over varied concentration ranges in cases of both inter-chemical species and inter-concentration predictions. Better separation, even among the group of nanomaterials with similar chemical compositions (e. g. metals & oxides), was achieved by low concentration approximated models. These modified models can be utilized either as standardized characterization method, or to build a comprehensive database of descriptors from tested nanomaterials combined with data on their biological behavior. Such databased could make significant contribution to the prediction on bio-identity due to biomolecule adsorption and their possible cellular uptake pathways for new and untested nanomaterials.

Identification of new nanomaterials with desirable biological effects can be achieved through clustering analysis on their surface descriptors in combination with known nanomaterials documented the database. In another direction of future development, pharmacokinetic models can be incorporated to BSAI modelling to better describe or predict the fate of nanomaterials at the whole organism level, e.g. concentration in different organs, clearance rate, etc. We have shown that these improvements toward the original BSAI approach can better assist *in silico* quantitative safety assessment for nanomaterials to provide guidelines for the engineering of novel nanomaterials with potential applications in diagnostics and therapeutics by targeting specific surface descriptors.

References

1. Unsworth LD, Sheardown H, Brash JL (2005) Protein resistance of surfaces prepared by sorption of end-thiolated poly(ethylene Glycol) to gold: effect of surface chain density. Langmuir 21:1036–1041
2. Tilton RD, Robertson CR, Gast AP (1991) Manipulation of hydrophobic interactions in protein adsorption. Langmuir 7:2710–2718

3. Gubala V, Siegrist J, Monaghan R, O'Reilly B, Gandhiraman RP, Daniels S, Williams DE, Ducrée J (2013) Simple approach to study biomolecule adsorption in polymeric microfluidic channels. Anal Chim Acta 760:75–82
4. An Inventory of Nanotechnology-based Consumer Products Introduced on the Market. http://www.nanotechproject.org/cpi/. Accessed 26 Nov 2014
5. Serra A, Filippo E, Re M, Palmisano M, Vittori-Antisari M, Buccolieri A, Manno D (2009) Non-functionalized silver nanoparticles for a localized surface plasmon resonance-based glucose sensor. Nanotechnology 20:165501
6. Yigit MV, Zhu L, Ifediba MA, Zhang Y, Carr K, Moore A, Medarova Z (2011) Noninvasive MRI-SERS imaging in living mice using an innately bimodal nanomaterial. ACS Nano 5:1056–1066
7. Willets KA, Van Duyne RP (2007) Localized surface plasmon resonance spectroscopy and sensing. Annu Rev Phys Chem 58:267–297
8. Sund J, Alenius H, Vippola M, Savolainen K, Puustinen A (2011) Proteomic characterization of engineered nanomaterial-protein interactions in relation to surface reactivity. ACS Nano 5:4300–4309
9. Lynch I, Dawson KA (2008) Protein-nanoparticle interactions. Nano Today 3:40–47
10. Lynch I, Cedervall T, Lundqvist M, Cabaleiro-Lago C, Linse S, Dawson KA (2007) The nanoparticle-protein complex as a biological entity; a complex fluids and surface science challenge for the 21st century. Adv Colloid Interface Sci 134–135:167–174
11. Chen R, Choudhary P, Schurr RN, Bhattacharya P, Brown JM, Ke PC (2012) Interaction of lipid vesicle with silver nanoparticle-serum albumin protein corona. Appl Phys Lett 100:013703
12. Chen R, Radic S, Choudhary P, Ledwell KG, Huang G, Brown JM, Ke PC (2012) Formation and cell translocation of carbon nanotube-fibrinogen protein corona. Appl Phys Lett 101:133702
13. Vroman L, Mattson JS, Smith CA (1974) Surface charge, protein adsorption, and thrombosis. Science 184:585–586
14. Li Y, Neoh KG, Kang E-T (2004) Plasma protein adsorption and thrombus formation on surface functionalized polypyrrole with and without electrical stimulation. J Colloid Interface Sci 275:488–495
15. Missirlis PYF, Lemm DW (1991) Protein adsorption and thrombus formation. In: Missirlis PYF, Lemm DW (eds) Modern aspects of protein adsorption on biomaterials. Springer Netherlands, Dordrecht, pp. 219–248
16. Pitt WG, Park K, Cooper SL (1986) Sequential protein adsorption and thrombus deposition on polymeric biomaterials. J Colloid Interface Sci 111:343–362
17. Lim C, Slack S, Ufer S, Lindner E (2004) Protein adsorption to planar electrochemical sensors and sensor materials. Pure Appl Chem 76:754–764
18. Frederix F, Bonroy K, Reekmans G, Laureyn W, Campitelli A, Abramov MA, Dehaen W, Maes G (2004) Reduced nonspecific adsorption on covalently immobilized protein surfaces using poly(ethylene Oxide) containing blocking agents. J Biochem Biophys Methods 58:67–74
19. Russo CJ, Passmore LA (2014) Controlling protein adsorption on graphene for cryo-EM using low-energy hydrogen plasmas. Nat Methods 11:649–652
20. Andrade JD, Hlady V (1986) Protein adsorption and materials biocompatibility: a tutorial review and suggested hypotheses. In: Biopolymers/non-exclusion HPLC; advances in polymer science. Springer, Berlin/Heidelberg, pp. 1–63
21. Le X, Poinern G, Jai r E, Ali N, Berry CM, Fawcett D (2013) Engineering a biocompatible scaffold with either micrometre or nanometre scale surface topography for promoting protein adsorption and cellular response. Int J Biomater 2013:e782549
22. Ratnikova TA, Govindan PN, Salonen E, Ke PC (2011) In vitro polymerization of microtubules with a fullerene derivative. ACS Nano 5:6306–6314
23. Zuo G, Zhou X, Huang Q, Fang H, Zhou R (2011) Adsorption of villin headpiece onto graphene, carbon nanotube, and C60: effect of contacting surface curvatures on binding affinity. J Phys Chem C 115:23323–23328

24. Zuo G, Kang S-G, Xiu P, Zhao Y, Zhou R (2013) Interactions between proteins and carbon-based nanoparticles: exploring the origin of nanotoxicity at the molecular level. Small 9:1546–1556
25. Zuo G, Fang H, Zhou R (2011) Nanotoxicity : Exploring the Interactions Between Carbon Nanotubes and Proteins. In: Carbon Nanotubes-Growth and Applications. Naraghi M (ed). InTech: Rijeka, Croatia
26. Tournus F, Latil S, Heggie M, Charlier J-C (2005) π-stacking interaction between carbon nanotubes and organic molecules. Phys Rev B 72:075431
27. Zuo G, Huang Q, Wei G, Zhou R, Fang H (2010) Plugging into proteins: poisoning protein function by a hydrophobic nanoparticle. ACS Nano 4:7508–7514
28. Prakash J, Nirmalakhandan N, Speece RE (1994) Prediction of activated carbon adsorption isotherms for organic vapors. Environ Sci Technol 28:1403–1409
29. Nirmalakhandan NN, Speece RE (1993) Prediction of activated carbon adsorption capacities for organic vapors using quantitative structure-activity relationship methods. Environ Sci Technol 27:1512–1516
30. Urano K, Omori S, Yamamoto E (1982) Prediction method for adsorption capacities of commercial activated carbons in removal of organic vapors. Environ Sci Technol 16:10–14
31. Reucroft PJ, Simpson WH, Jonas LA (1971) Sorption properties of activated carbon. J Phys Chem 75:3526–3531
32. Apul OG, Wang Q, Shao T, Rieck JR, Karanfil T (2013) Predictive model development for adsorption of aromatic contaminants by multi-walled carbon nanotubes. Environ Sci Technol 47:2295–2303
33. Nel AE, Mädler L, Velegol D, Xia T, Hoek EMV, Somasundaran P, Klaessig F, Castranova V, Thompson M (2009) Understanding biophysicochemical interactions at the nano–bio Interface. Nat Mater 8:543–557
34. Xia X-R, Monteiro-Riviere NA, Riviere JE (2010) An index for characterization of nanomaterials in biological systems. Nat Nanotechnol 5:671–675
35. Xia XR, Monteiro-Riviere NA, Mathur S, Song X, Xiao L, Oldenberg SJ, Fadeel B, Riviere JE (2011) Mapping the surface adsorption forces of nanomaterials in biological systems. ACS Nano 5:9074–9081
36. Chen R, Zhang Y, Darabi Sahneh F, Scoglio CM, Wohlleben W, Haase A, Monteiro-Riviere NA, Riviere JE (2014) Nanoparticle surface characterization and clustering through concentration-dependent surface adsorption modeling. ACS Nano 8:9446–9456
37. Ke PC, Lamm MH (2011) A biophysical perspective of understanding nanoparticles at large. Phys Chem Chem Phys 13:7273–7283
38. Podila R, Chen R, Ke PC, Brown JM, Rao AM (2012) Effects of surface functional groups on the formation of nanoparticle-protein corona. Appl Phys Lett 101:263701
39. Wen Y, Geitner NK, Chen R, Ding F, Chen P, Andorfer RE, Govindan PN, Ke PC (2013) Binding of cytoskeletal proteins with silver nanoparticles. RSC Adv 3:22002–22007
40. Stone A (2013) The theory of intermolecular forces. Oxford University Press
41. Karelson M (2000) Molecular descriptors in QSAR/QSPR. Wiley-Interscience, New York
42. Hansch C, Fujita T (1964) ρ-σ-π analysis. A method for the correlation of biological activity and chemical structure. J Am Chem Soc 86:1616–1626
43. Hansch C (1993) Quantitative structure-activity relationships and the unnamed science. Acc Chem Res 26:147–153
44. Giaginis C, Tsantili-Kakoulidou A (2008) Alternative measures of lipophilicity: from octanol–water partitioning to IAM retention. J Pharm Sci 97:2984–3004
45. Pan B, Xing B (2008) Adsorption mechanisms of organic chemicals on carbon nanotubes. Environ Sci Technol 42:9005–9013
46. Kostarelos K, Bianco A, Prato M (2009) Promises, facts and challenges for carbon nanotubes in imaging and therapeutics. Nat Nanotechnol 4:627–633
47. Shvedova AA, Kisin ER, Porter D, Schulte P, Kagan VE, Fadeel B, Castranova V (2009) Mechanisms of pulmonary toxicity and medical applications of carbon nanotubes: two faces of janus? Pharmacol Ther 121:192–204

48. Bianco A, Kostarelos K, Prato M (2005) Applications of carbon nanotubes in drug delivery. Curr Opin Chem Biol 9:674–679
49. Beg S, Rizwan M, Sheikh AM, Hasnain MS, Anwer K, Kohli K (2011) Advancement in carbon nanotubes: basics, biomedical applications and toxicity. J Pharm Pharmacol 63:141–163
50. Zhang Y, Bai Y, Yan B (2010) Functionalized carbon nanotubes for potential medicinal applications. Drug Discov Today 15:428–435
51. Gramatica P (2007) Principles of QSAR models validation: internal and external. QSAR Comb Sci 26:694–701
52. Dearden JC, Cronin MTD, Kaiser KLE (2009) How not to develop a quantitative structure–activity or structure–property relationship (QSAR/QSPR). SAR QSAR Environ Res 20:241–266
53. Gramatica P, Giani E, Papa E (2007) Statistical external validation and consensus modeling: a QSPR case study for K_{oc} prediction. J Mol Graph Model 25:755–766
54. Jolliffe IT (2002) Principal component analysis. Springer Science & Business Media, New York
55. Kaufhold S, Dohrmann R, Klinkenberg M, Siegesmund S, Ufer K (2010) N_2-BET specific surface area of bentonites. J Colloid Interface Sci 349:275–282
56. Pettibone JM, Cwiertny DM, Scherer M, Grassian VH (2008) Adsorption of organic acids on TiO_2 nanoparticles: effects of pH, nanoparticle size, and nanoparticle aggregation. Langmuir 24:6659–6667
57. Gilbert B, Ono RK, Ching KA, Kim CS (2009) The effects of nanoparticle aggregation processes on aggregate structure and metal uptake. J Colloid Interface Sci 339:285–295
58. Do DD (1998) Adsorption analysis: equilibria and kinetics. Imperial College Press, London
59. Shih Y-H, Gschwend PM (2009) Evaluating activated carbon-water sorption coefficients of organic compounds using a linear solvation energy relationship approach and sorbate chemical activities. Environ Sci Technol 43:851–857
60. Zhao Q, Yang K, Li W, Xing B (2014) Concentration-dependent polyparameter linear free energy relationships to predict organic compound sorption on carbon nanotubes. Sci Rep 4:3888–3888
61. Comer J, Chen R, Poblete H, Vergara-Jaque A, Riviere JE (2015) Predicting adsorption affinities of small molecules on carbon nanotubes using molecular dynamics simulation. ACS Nano 9:11761–11774
62. Leenaerts O, Partoens B, Peeters FM (2009) Water on graphene: hydrophobicity and dipole moment using density functional theory. Phys Rev B 79:235440
63. Ruthven DM (1984) Principles of adsorption and adsorption processes. John Wiley & Sons, New York
64. Wu W, Chen W, Lin D, Yang K (2012) Influence of surface oxidation of multiwalled carbon nanotubes on the adsorption affinity and capacity of polar and nonpolar organic compounds in aqueous phase. Environ Sci Technol 46:5446–5454
65. Yang K, Xing B (2010) Adsorption of organic compounds by carbon nanomaterials in aqueous phase: polanyi theory and its application. Chem Rev 110:5989–6008
66. Shaw SY, Westly EC, Pittet MJ, Subramanian A, Schreiber SL, Weissleder R (2008) Perturbational profiling of nanomaterial biologic activity. Proc Natl Acad Sci 105:7387–7392
67. Epa VC, Burden FR, Tassa C, Weissleder R, Shaw S, Winkler DA (2012) Modeling biological activities of nanoparticles. Nano Lett 12:5808–5812
68. Winkler DA, Burden FR, Yan B, Weissleder R, Tassa C, Shaw S, Epa VC (2014) Modelling and predicting the biological effects of nanomaterials. SAR QSAR Environ Res 25:161–172
69. Le T, Epa VC, Burden FR, Winkler DA (2012) Quantitative structure–property relationship modeling of diverse materials properties. Chem Rev 112:2889–2919
70. Liu X, Tang K, Harper S, Harper B, Steevens J, Xu R (2013) Predictive Modeling of Nanomaterial Exposure Effects in Biological Systems. Int J Nanomedicine 8:31–43
71. Le TC, Mulet X, Burden FR, Winkler DA (2013) Predicting the complex phase behavior of self-assembling drug delivery nanoparticles. Mol Pharm 10:1368–1377

72. Liu G, Lin Y (2005) Electrochemical sensor for organophosphate pesticides and nerve agents using zirconia nanoparticles as selective sorbents. Anal Chem 77:5894–5901
73. Zhang W, Asiri AM, Liu D, Du D, Lin Y (2014) Nanomaterial-based biosensors for environmental and biological monitoring of organophosphorus pesticides and nerve agents. Trends Anal Chem 54:1–10
74. Zhou J-H, Deng C-Y, Si S-H, Wang S-E (2010) Zirconia electrodeposited on a self-assembled monolayer on a gold electrode for sensitive determination of parathion. Microchim Acta 172:207–215
75. Behnam R, Morshed M, Tavanai H, Ghiaci M (2013) Destructive adsorption of diazinon pesticide by activated carbon nanofibers containing Al_2O_3 and MgO nanoparticles. Bull Environ Contam Toxicol 91:475–480
76. Moradi Dehaghi S, Rahmanifar B, Moradi AM, Azar PA (2014) Removal of permethrin pesticide from water by chitosan–zinc oxide nanoparticles composite as an adsorbent. J Saudi Chem Soc 18:348–355
77. Saifuddin N, Nian CY, Zhan LW, Ning KX (2011) Chitosan-silver nanoparticles composite as point-of-use drinking water filtration system for household to remove pesticides in water. Asian J Biochem 6:142–159
78. Bootharaju MS, Pradeep T (2012) Understanding the degradation pathway of the pesticide. Chlorpyrifos by noble metal nanoparticles. Langmuir 28:2671–2679
79. Sreekumaran NAN, Pradeep T A method for decontaminating water containing pesticides. EP1715947 B1, August 22, 2012
80. Manimegalai G, Shanthakumar S, Sharma C (2014) Silver nanoparticles: synthesis and application in mineralization of pesticides using membrane support. Int Nano Lett 4:1–5
81. Basahel SN, Ali TT, Mokhtar M, Narasimharao K (2015) Influence of crystal structure of nanosized ZrO_2 on photocatalytic degradation of methyl orange. Nanoscale Res Lett 10:73
82. Barakat NAM, Nassar MM, Farrag TE, Mahmoud MS (2013) Effective photodegradation of methomyl pesticide in concentrated solutions by novel enhancement of the photocatalytic activity of TiO_2 using $CdSO_4$ nanoparticles. Environ Sci Pollut Res 21:1425–1435
83. Lowry GV (2007) Nanomaterials for groundwater remediation. In: Environmental nanotechnology: applications and impacts of nanomaterials. McGraw-Hill Professional, New York
84. Chen R, Zhang Y, Monteiro-Riviere NA, Riviere JE (2016) Quantification of nanoparticle pesticide adsorption: computational approaches based on experimental data. Nanotoxicology 10:1118–1128
85. Lam C, James JT, McCluskey R, Arepalli S, Hunter RL (2006) A review of carbon nanotube toxicity and assessment of potential occupational and environmental health risks. Crit Rev Toxicol 36:189–217
86. Pulskamp K, Diabaté S, Krug HF (2007) Carbon nanotubes show no sign of acute toxicity but induce intracellular reactive oxygen species in dependence on contaminants. Toxicol Lett 168:58–74
87. Rana S, Yeh Y-C, Rotello VM (2010) Engineering the nanoparticle–protein interface: applications and possibilities. Curr Opin Chem Biol 14:828–834
88. Salvador-Morales C, Flahaut E, Sim E, Sloan J, Green ML, Sim RB (2006) Complement activation and protein adsorption by carbon nanotubes. Mol Immunol 43:193–201
89. Rybak-Smith MJ, Sim RB (2011) Complement activation by carbon nanotubes. Adv Drug Deliv Rev 63:1031–1041
90. Andersen AJ, Robinson JT, Dai H, Hunter AC, Andresen TL, Moghimi SM (2013) Single-walled carbon nanotube surface control of complement recognition and activation. ACS Nano 7:1108–1119
91. Lynch I, Dawson KA, Linse S (2006) Detecting cryptic epitopes created by nanoparticles. Sci Signal 2006:pe14–pe14
92. Aillon KL, Xie Y, El-Gendy N, Berkland CJ, Forrest ML (2009) Effects of nanomaterial physicochemical properties on in vivo toxicity. Adv Drug Deliv Rev 61:457–466
93. Walkey CD, Olsen JB, Song F, Liu R, Guo H, Olsen DWH, Cohen Y, Emili A, Chan WCW (2014) Protein corona fingerprinting predicts the cellular interaction of gold and silver nanoparticles. ACS Nano 8:2439–2455

Part IV
Case Studies

Chapter 9
An Integrated Data-Driven Strategy for Safe-by-Design Nanoparticles: The FP7 MODERN Project

Martin Brehm, Alexander Kafka, Markus Bamler, Ralph Kühne, Gerrit Schüürmann, Lauri Sikk, Jaanus Burk, Peeter Burk, Tarmo Tamm, Kaido Tämm, Suman Pokhrel, Lutz Mädler, Anne Kahru, Villem Aruoja, Mariliis Sihtmäe, Janeck Scott-Fordsmand, Peter B. Sorensen, Laura Escorihuela, Carlos P. Roca, Alberto Fernández, Francesc Giralt, and Robert Rallo

Abstract The development and implementation of safe-by-design strategies is key for the safe development of future generations of nanotechnology enabled products. The safety testing of the huge variety of nanomaterials that can be synthetized is unfeasible due to time and cost constraints. Computational modeling facilitates the implementation of alternative testing strategies in a time and cost effective way. The development of predictive nanotoxicology models requires the use of high quality experimental data on the structure, physicochemical properties and bioactivity of nanomaterials. The FP7 Project MODERN has developed and evaluated the main components of a computational framework for the evaluation of the environmental and health impacts of nanoparticles. This chapter describes each of the elements of the framework including aspects related to data generation,

M. Brehm • R. Kühne
UFZ Department of Ecological Chemistry, Helmholtz Centre for Environmental Research, Permoserstrasse 15, 04318 Leipzig, Germany

A. Kafka
UFZ Department of Ecological Chemistry, Helmholtz Centre for Environmental Research, Permoserstrasse 15, 04318 Leipzig, Germany

Faculty for Chemistry and Mineralogy, University of Leipzig, Johannisallee 29, 04103 Leipzig, Germany

M. Bamler • G. Schüürmann
UFZ Department of Ecological Chemistry, Helmholtz Centre for Environmental Research, Permoserstrasse 15, 04318 Leipzig, Germany

Institute for Organic Chemistry, Technical University Bergakademie Freiberg, Leipziger Strasse 29, 09596 Freiberg, Germany

L. Tran et al. (eds.), *Modelling the Toxicity of Nanoparticles*, Advances in Experimental Medicine and Biology, Vol. 947, DOI 10.1007/978-3-319-47754-1_9

management and integration; development of nanodescriptors; establishment of nanostructure-activity relationships; identification of nanoparticle categories; hazard ranking and risk assessment.

Keywords Nanotoxicology • Nanoinformatics • Nanodescriptors • QNAR • Risk assessment

9.1 Introduction

Nanotechnology, through the development of novel materials, nano-enabled products and advanced processes, will impact significantly many areas of research, industry and society. In economic terms, the nanomaterial market is expected to reach $11.8 billion by 2020[1]. At an industrial scale, the safe application of nanotechnology requires the comprehensive analysis of the potential environmental and

[1] http://www.researchandmarkets.com/research/4sxcfc/global.

L. Sikk
Institute of Chemistry, University of Tartu, Ravila 14a, Tartu 50411, Estonia

Institut de Chimie de Nice (UMR CNRS 7272), Université Nice Sophia Antipolis, 06108 Nice, France

J. Burk • P. Burk • K. Tämm
Institute of Chemistry, University of Tartu, Ravila 14a, Tartu 50411, Estonia

T. Tamm
Institute of Technology, University of Tartu, Nooruse 1, Tartu 50411, Estonia

S. Pokhrel • L. Mädler
Foundation Institute of Materials Science (IWT), Department of Production Engineering, University of Bremen, Bremen, Germany

A. Kahru • V. Aruoja • M. Sihtmäe
Laboratory of Environmental Toxicology, National Institute of Chemical Physics and Biophysics, Akadeemia tee 23, Tallinn 12618, Estonia

J. Scott-Fordsmand • P.B. Sorensen
Department of Bioscience, Aarhus Universit,
Vejlsovej 25, PO BOX 314, DK 8600 Silkeborg, Denmark

L. Escorihuela • C.P. Roca • A. Fernández • F. Giralt
Departament d'Enginyeria Química, Universitat Rovira i Virgili,
Av. Paisos Catalans, 26, 43007 Tarragona, Spain

R. Rallo (✉)
Departament d'Enginyeria Informatica i Matematiques, Universitat Rovira i Virgili,
Av. Paisos Catalans, 26, 43007 Tarragona, Spain
e-mail: robert.rallo@urv.cat

human health impacts after the introduction of engineered nanoparticles (eNP) and nanomaterials (NMs) in the environment.

The environmental risk assessment of eNPs requires data on emission sources, physicochemical properties, fate, transformations and persistence, and on their potential toxicity. However, nanosafety data are uncertain and with significant knowledge gaps. Since the goal of nanosafety assessment is to provide meaningful information in relatively short time and using a reasonable amount of economic resources, the whole assessment process and methodology must be simplified. The development of computational models suitable to predict properties constitutes a key element to achieve cost-effectiveness in nanosafety assessment. Specifically, the importance of developing Quantitative nanostructure-activity relationships (QNAR) for hazard ranking, risk assessment and safe-by-design nanoparticles has been recognized as one of the key objective by the EU Nanosafety Cluster[2] in its Strategic Research Agenda for 2015–2025 [68].

MODERN[3] (**MOD**elling the **E**nvi**R**onmental and human health effects of **N**anomaterials) is an FP7 Project funded by the European Commission whose main objective was the development of a robust computational framework suitable for evaluating the environmental and health impact of eNPs. In environmental conditions, nanoparticle effects depend on bioavailability and toxicity mechanisms, which in turn are influenced by physicochemical and structural properties. The development of predictive nanotoxicity models requires the collection and integration of the experimental data available in the literature and in public data repositories into a well-characterized nanoparticle knowledge base. It is also essential to develop new nanodescriptors suitable to relate the eNP structure with the corresponding property and bioactivity profiles. Data on property profiles, together with the in vitro/in vivo screening of toxicity, can be summarized as eNP signatures that can be subsequently used to establish nanoparticle categories that will increase the confidence in nanosafety assessments. Fig. 9.1 summarizes the integration of all the above elements within a consistent framework to facilitate the development of predictive models for the safety assessment of nanomaterials.

This chapter introduces the components of the integrative framework developed in MODERN. Section 2 focuses on data-related issues including data generation and strategies and tools for data management and integration. Section 3 discusses the development of nanodescriptors based on the electronic structure of nanoparticles and size-dependent descriptors based on molecular modelling. Section 4 introduces the development of structure-activity relationships for nanomaterials. Section 5 develops nanoparticle categories and hazard ranking schemes. As conclusion, Sect. 6 provides a perspective on the use of computational modelling for developing a new generation of safe-by-design nanoparticles.

[2] http://www.nanosafetycluster.eu.

[3] http://modern-fp7.biocenit.cat.

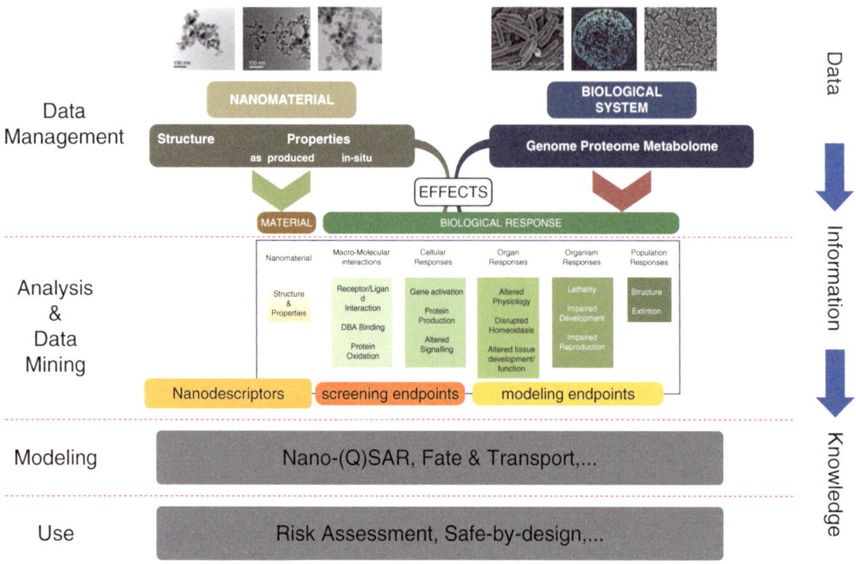

Fig. 9.1 MODERN's computational framework for safe-by-design nanoparticles

9.2 Data Generation and Data Management

The first step in the implementation of a computational strategy for transforming nanoparticle data into relevant knowledge for nanosafety assessment is the collection of high quality information. Literature data collection/curation in MODERN has been complemented with the development of a well characterized, both in terms of physicochemical properties and biological activity, library of nanoparticles. The project focus on inorganic nanoparticles, specifically metals and metal oxides. This section provides an overview of the synthesis and physicochemical characterization process, together with the characterization of the impact of nanoparticles in environmentally relevant organisms. The final part of this section provides an overview of the data management system developed in MODERN and outlines strategies for integrating biological (e.g., omics) data into the model development process.

9.2.1 Synthesis and Physicochemical Characterization of Nanoparticles

Fig. 9.2 shows a schematic of the main elements in a Flame spray pyrolysis (FSP) process, which is a versatile technique to produce new and functional nanoparticles [34]. Compared to the other aerosol techniques, the variety of materials that can be produced is much wider because the process utilizes liquid precursors that are directly atomized and ignited forming a spray flame [73]. During the flame spray process, the liquid precursors (carrying all the energy into the flame) in the form of

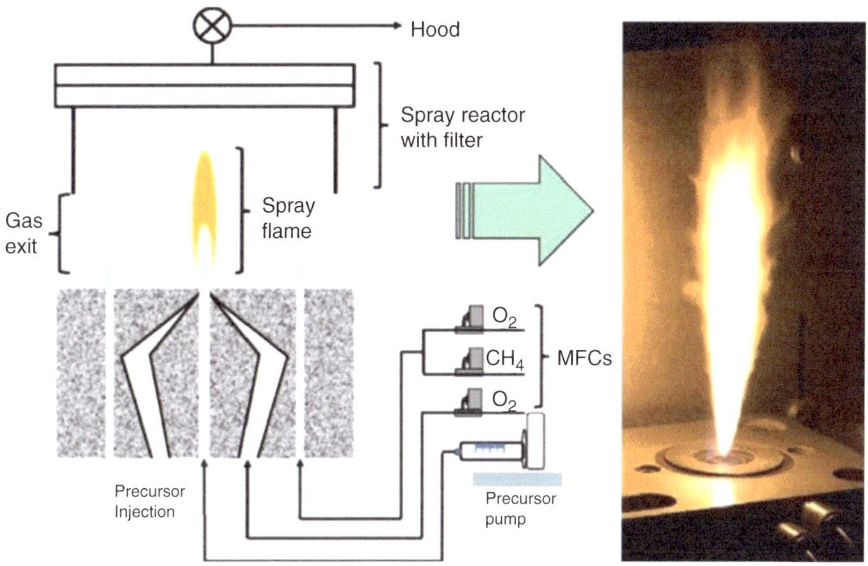

Fig. 9.2 Flame Spray Pyrolysis (FSP) technique for NPs synthesis. *Left* panel: the schematic diagram of the FSP reactor. *Right* panel: a photograph of a roaring flame during particle production

droplets are introduced to the flame subsequently vaporizing followed by atomization and gas phase chemical reactions producing cluster seeds of the NPs [72]. The liquid precursor atomization is achieved with a two fluid nozzle that utilizes high gas velocities to disintegrate the liquid into fine droplets [47].

Each droplet approaching the flame through the spray nozzles essentially serves as a micro-reactor with the exact stoichiometry required for the reaction, producing materials with homogeneous chemical composition in a single step [47, 71]. The resulting cluster seeds grow rapidly by collision and coalescence mechanisms and/or surface growth to produce NPs [34]. The subsequent nanoparticle production is a function of the thermal energy flux that directly influences the particle formation parameters such as temperature profile and residence time within the high temperature environment ($\geq$ 2000 °C). Eventually, the NP aerosol is quenched to room temperature and NPs are collected on a filter unit or directly deposited as porous film on the substrates [39]. Due to the enormously broad range of liquid precursors available, FSP is currently considered as one of the most promising techniques for synthesis of a very large diversity of sophisticated inorganic NPs [73].

9.2.1.1 Physicochemical Characterization of the Nanoparticle Library

A library composed by 11 NPs including ZnO, CuO, Co_3O_4, Fe_3O_4, Mn_3O_4, TiO_2, Sb_2O_3, Al_2O_3, SiO_2, MgO and metallic Pd was obtained using FSP. The specific surface area and primary particle sizes of the particles, determined using BET measurements, were in the range of 53–289 m^2/g and 8–13 nm, respectively. The

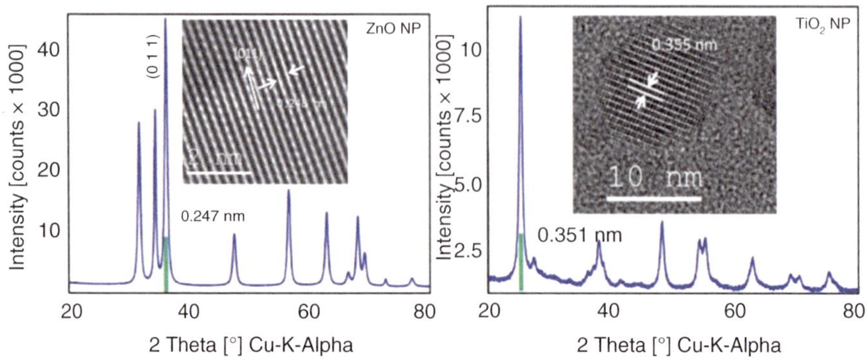

Fig. 9.3 Matching of X-ray diffraction pattern with the lattice distances of (**a**) ZnO NPs (**b**) TiO_2 NPs

determination of the crystal structure of ZnO and TiO_2 showed undetectable changes in the lattice parameters. At constant FSP reactor conditions, TiO_2 had equivalent primary particle size (d_{BET}) and the crystallite size (d_{XRD}) in the range of 12–14 nm and the weight % of anatase and rutile content was found to be 12.4 and 87.6 %, respectively. Similarly, for the other NPs, CuO, Sb_2O_3, Al_2O_3 and MgO, the data for the lattice parameters were in reasonable agreement with the reference data suggesting well-formed single crystalline ultrafine particles. The phase analysis of the spinel type mixed oxides such as Mn_3O_4, Fe_3O_4 and Co_3O_4 showed single phase Mn_3O_4 with very minor changes in the crystal structure relative to the standard model. Similarly, Co_3O_4 was also a single phase spinel crystalizing in cubic system. However, unlike Mn_3O_4 and Co_3O_4, Fe_3O_4 had two phase mixtures (hematite (Fe_2O_3) and magnetite (Fe_3O_4) with a distribution of 64.7 and 35.3 % by weight, respectively. The independent crystallite size analysis showed that within the mixture magnetite (8.2 nm) was slightly smaller than hematite (9.4 nm). In order to estimate the size and morphology of these metal oxide NPs, TEM investigation was carried out. The images of ZnO showed particles with 20 nm (d_{TEM}). The bright field image and the corresponding selected area electron diffraction (SAED) shows highly crystalline nature of the ZnO, Fe_3O_4, MgO and TiO_2 NPs. TEM images of Co_3O_4, TiO_2, CuO and MgO showed spherical nanocrystals and all particles were homogenous except few bigger spheres scattered in the matrix with size distribution ranging 9–15 nm. The morphology of FeO_x and WO_3 were similar with reduced crystallinity observed through low intensity XRD patterns. The primary particle sizes were in the range of 10–15 nm clearly agreeing with the mixed particle sizes of Fe_2O_3 and Fe_3O_4 (9.4 and 8.3 nm, respectively). The NPs of Sb_2O_3 are quite different relative to the other nanoparticles. The long chain agglomerates of the Sb_2O_3 NPs might be due to either insufficient oxygen during combustion or to the mismatch of the precursor/solvent combination.

Selected nanoparticles such as ZnO and/or TiO_2 were also analyzed using high resolution transmission electron microscopy and XRD (Fig. 9.3). For ZnO, the

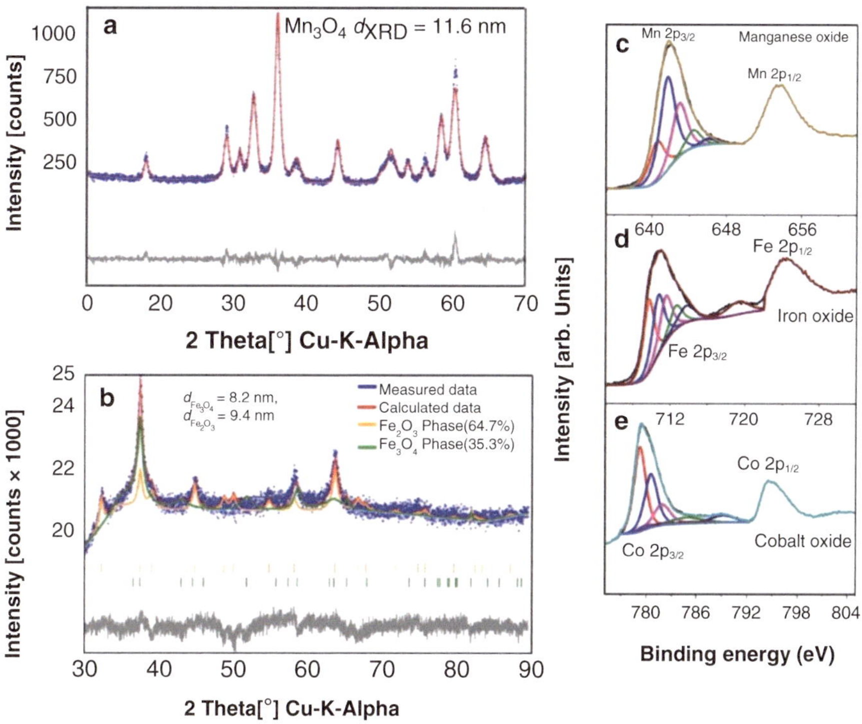

Fig. 9.4 X-ray diffraction patterns and Rietveld analysis of (**a**) manganese oxide and (**b**) iron oxide NPs. The Rietveld analysis showed Mn_3O_4 was a single phase material while iron oxide was a mixture of hematite (Fe_2O_3) and magnetite (Fe_3O_4). (**c–e**) XPS spectra of Mn_3O_4, FeOx and Co_3O_4 NPs

respective *d*-spacing perpendicular to these dense rows of maxima is 0.248 nm which corresponds to d_{011} = 0.247 nm from the X-ray diffraction measurements. From symmetry considerations in space group P 6_3 m c (Laue group *6/mmm*) the [h k l] lattice direction observed in the HRTEM images are symmetrically equivalent to [0 1 1] (see Fig. 9.4a, b).

The chemical composition of Co_3O_4, Fe_3O_4 and Mn_3O_4 was evaluated by XPS analysis (Fig. 9.4c–e). The result showed presence of M^{2+} and M^{3+} coordination. However, with the increase in the atomic number of the transition metal atom (Mn to Fe to Co) the corresponding coupling distances of $2p_{1/2}$ to $2p_{3/2}$ also increased. The peak analysis of the spectra of cobalt, iron and manganese showed that (1) binary oxides are present in the form of Co_3O_4, Mn_2O_3 or Fe_2O_3, (2) only iron oxide (Fe_2O_3) surface is clean without any additional impurities, and (3) Mn_3O_4 has surface MnO whereas Co_3O_4 has surface hydroxylation. Unusual is that the XRD analysis showed two phase mixtures (hematite (Fe_2O_3) and magnetite (Fe_3O_4) in FeO_x but XPS analysis showed 100 % of Fe_2O_3, which is very unlikely. Similarly, XRD analysis of Mn_3O_4 showed single phase material while XPS showed only Mn_2O_3 with 1 % of MnO.

The discrepancies observed using the two different techniques for the spinel characterization are quite striking. Actually, XPS is a surface characterization technique and the sample penetration depth does not exceed 10 nm. If the surface is enriched with hematite during XPS analysis, the signal from the magnetite is apparently lost.

9.2.2 Characterization of the Interaction of Nanoparticles with Biological Systems

Currently, the number of papers published on the biological effects of NPs exceeds 40,000 and it is growing rapidly. Most of this publicly available nanotoxicity data are for metal oxides that has led to the development of QNAR models based mostly on bacterial toxicity endpoints. However, the inconsistencies in the available data related to the complexities inherent to NPs as well as their use in biological experiments hinder the quality and applicability of the models. In MODERN, available literature data for metal and metal oxide eNPs were extended by generating homogenous toxicity data for a number of species relevant in ecotoxicology. The generated data was subsequently used to develop/validate QNAR models based on nanodescriptors obtained from electronic structure and by molecular modelling of eNPs.

9.2.2.1 Selection of Test Species

According to Commission Regulation (EU) No 286/2011 the criterion for classifying and categorizing substances as *hazardous to the aquatic environment* is as follows: *chemicals are considered of acute (short-term) aquatic hazard when their L(E)C50 ≤ 1 mg/L), based on the results of standardized toxicity tests with fish (96 h) and/or crustaceans (48 h) and/or algae (72 or 96 h).*

The selection of test species for mandatory testing of chemicals for regulatory purposes addresses different trophic levels as well as the diverse biological complexity present in the ecosystems. In the case of eNPs, also the ability of test organisms to ingest/internalize particles is an important consideration for species selection. In addition, the composition of the test battery is also influenced by practical considerations such as ease of cultivation in the laboratory and modifications in test protocols needed to study eNPs.

At the base of every ecosystem there are primary producers (autotrophs) that convert CO_2 to organic matter using light or chemical energy, normally consumed by several levels of consumers. In the majority of aquatic environments, the principal primary producers are microalgae, which justifies their inclusion in the test battery. In the current work algae were represented by *Pseudokirchneriella subcapitata*, a freshwater species commonly used in aquatic toxicology, known to be sensitive to metals, including metal-containing eNPs [2, 6, 29]. Algae are considered 'particle-proof' as the algal cell wall should prevent nanoparticles entry into the cells. Indeed, there are only very few reports of particle internalization into algae [33, 79].

The primary consumers in the current test battery were represented by a protozoan *Tetrahymena thermophila*. Protozoa are ecologically widely spread naturally particle-feeding organisms that are also important in biological wastewater treatment [14, 33]. As an example, *T. thermophila* has been shown to ingest single walled carbon nanotubes and bacteria as food with no apparent discrimination [32]. Reported long residence times of eNPs such as CdSe/ZnS quantum dots in *T. thermophila* refer to the increased risks of transfer of these eNPs to higher trophic levels in the ecosystem [51].

We also included bacteria (prokaryotes) in our test suite to represent biodegraders that are of crucial importance in all ecosystems. Naturally luminescent bacterium *Vibrio fischeri* was chosen as the Microtox test (*Vibrio fischeri* bioluminescence inhibition assay) is probably the most used bacterial toxicity test in environmental studies. The kinetic modification of that test (Flash Assay) was used as it is a rapid, simple, cost effective and sensitive method for the evaluation of toxic properties of chemicals, including eNPs [49]. Analogously to unicellular algae bacteria can be classified as organisms that do not internalize nanoparticles.

The selection of *Tetrahymena* as well as *Vibrio fischeri* has additional relevance in terms of QNAR development. Toxicity data obtained using these species have been used extensively for the development of structure-activity relationships [11, 54].

9.2.2.2 Description of Experimental Protocols

Preparation of Nanoparticle Suspensions For all assays the NPs were mixed with deionized (DI) water to yield 25 ml of 200 mg/l stock suspensions that were vortexed and sonicated for 4 min before use (40 W, Branson probe sonicator, USA). For toxicity experiments the NP stock suspensions were diluted 1:1 with 200 % respective test medium to obtain 100 mg/l suspensions in each medium.

Algal 72 h Growth Inhibition Assay with *Pseudokirchneriella Subcapitata* The OECD 201 algal growth inhibition test guidelines [57] were followed as described in detail in Aruoja et al. [2]. Briefly, the exponentially growing *P. subcapitata* cells were exposed to various concentrations of NPs in standard 20 ml glass scintillation vials containing 5 ml of algal growth medium. The vials were shaken on a transparent table constantly illuminated from below with fluorescent tubes in a temperature-controlled environment (24 ± 1 °C) for up to 72 h. Algal biomass was determined at least every 24 h by measuring the fluorescence of algal extract [22]. EC_{50} values (effective concentration that leads to 50 % reduction of biomass) were calculated from dose-response data. By using the vials only once we managed to keep the coefficient of variation of biomass density in replicate control cultures below 5 % throughout the experiments. It was therefore sufficient to have samples in duplicate with four controls distributed evenly on the transparent table.

24 h Cell Viability Assay with the Protozoan *Tetrahymena Thermophile* Protozoan culture (*T. thermophila* strain BIII) was grown as described by [50]. The cells were harvested during the exponential growth phase and washed twice with DI water. For toxicity analysis 100 µl of harvested and washed *T. thermophila* suspension in DI water was added to 100 µl of the suspension of NPs that were previously diluted in DI water in 96-well polystyrene plates. Final cell density in the test was 5 × 10^5 cells/ml. Protozoan suspension and NPs in DI water were used as non-treated and abiotic controls, respectively. The test plates were incubated in 25 °C in the dark. After 24 h viability of the cells was determined based on the ATP content as previously described [50]. A volume of 10 µl of ATP standard (10–5 M) was used for internal calibration. All the luminescence measurements were done using Orion II plate luminometer (Berthold Detection Systems, Germany). The amount of the ATP in each well was calculated according to the following equation:

$$\mathrm{ATP}, \mu mol = \frac{\mathrm{RLU}_{\mathrm{sample}} - \mathrm{RUL}_{\mathrm{background}}}{\mathrm{RULATP}_{\mathrm{standard}}} \times \mathrm{ATP}_{\mathrm{standard}} \tag{9.1}$$

The ATP concentrations in the samples were expressed as percentages of the non-treated controls. The EC_{50} values (effective concentration leading to a 50 % cell death) was calculated from the concentration-effect curves.

***Vibrio Fischeri* Kinetic Bioluminescence Inhibition Assay (a Flash-Assay)** Acute bioluminescence inhibition assay (exposure time 30 min) with *Vibrio fischeri* was carried out in room temperature (~20 °C) in 96-well microplates according to the Flash-Assay protocol [28, 49]. A volume of 100 µl of bacterial suspension was added to 100 µl of test suspension by automatic dispensing in the luminometer testing chamber. The luminescence was recorded for the first 10 s after dispense of the bacteria in each well without additional mixing. After 30 min incubation the luminescence was recorded again. The Microplate Luminometer Orion II (Berthold Detection Systems, Pforzheim, Germany), controlled by Simplicity Version 4.2 Software was used. Reconstituted *V. fischeri* Reagent (Aboatox, Turku, Finland) was used for the bacterial suspension and all chemicals were prepared in 2 % NaCl. The inhibition of bacterial luminescence (INH%) was calculated as follows:

$$INH\% = 100 - \frac{IT_{30}}{KF * IT_0} * 100; \quad KF = \frac{IC_{30}}{IC_0}; \tag{9.2}$$

KF (correction factor) characterizes the natural loss of luminescence of the control (i.e. bacterial suspension in 2 % NaCl). IC_0 and IT_0 are the maximum values of luminescence during the first 5 seconds after dispense of 100 µl of bacteria to 100 µl of control or sample, respectively. IC_{30} and IT_{30} are the respective values after 30 min. EC_{50} is the concentration of a compound reducing the bioluminescence by 50 %.

9.2.2.3 Comparative Analysis of Nanoparticle Toxicity

The results of the toxicity analysis of MODERN's eNP library are presented in Table 9.1. All eNPs were about the same primary size (10–20 nm) and their average hydrodynamic size in DI water was between 65 and 171 nm, except for Mn_3O_4 (395 nm) and MgO (1964 nm) that formed bigger agglomerates. Alga was clearly the most sensitive among the three species used, with EC_{50} values below 100 mg/l for 10 out of 12 eNPs. Algal EC_{50} values lower than 1.0 mg/l were determined for ZnO, Pd and CuO, whereas only ZnO and CuO were consistently the most toxic across all three species (EC_{50}<10 mg/l). The relative sensitivity of algae was expected, based on previous studies that have revealed algae and crustaceans as the most sensitive groups to metallic eNP exposure [32]. Similarly, ZnO and CuO have shown consistent toxicity to various aquatic test species due to highly toxic Zn and Cu ions leaching from these particles [6, 24, 29, 37]. Taken into account the data on solubilization and abiotic reactive oxygen species (ROS) production of the studied eNPs, only ZnO toxicity could be fully explained by Zn solubility whereas the mechanism of CuO toxicity probably involved the generation ROS. At least partially, the observed growth inhibition of algae by eNPs could be due to entrapment of algal cells into eNPs agglomerates creating a physical barrier for nutrients and/or light. Also, the intimate contact between cells and eNPs in cell-nanoparticle agglomerates probably amplified the harmful effects of ROS if generated by eNPs, thus explaining the observed higher sensitivity of the algal assay relative to bacterial and protozoan tests. However, the interpretation of the algal assay is not straightforward since the cells inside the eNP agglomerates can stay viable and resume growth upon dilution. We thus hesitate in assigning the 'acutely hazardous' classification to Co_3O_4, TiO_2, Mn_3O_4 and Fe_3O_4 despite the algal 72 h EC_{50} values near the 1 mg/l threshold. On the other hand, while the majority of the eNPs did not affect the viability of protozoa below the 100 mg/l exposure level, the accumulation of eNPs in their food vacuoles may lead to food-web transfer and bioaccumulation, and therefore to potential harm.

In order to understand the mechanisms of (aquatic) toxicity of eNPs and ultimately move closer to cost-efficient toxicity prediction a quantitative approach to mechanisms is required. This approach should be usable across different organisms or even for the whole aquatic ecosystem and should therefore include data from a battery of test organisms belonging to different trophic levels. One variant of the test

Table 9.1 Toxicity, classification to acute toxic hazard categories and tentative mechanisms of toxic action for 12 metallic nanoparticles tested within FP7 project MODERN [3]

	Trophic level			Primary producer	Consumer	Degrader		
	Cell type			Eukaryote	Eukaryote	Prokaryote		
	Internalizes nanoparticles			No?	Yes	No?		
Nano-particle[¥]	Solubility (%)[§]	ROS (HPF)[$]	ROS (DCF)[+]	Algal 72 h growth inhibition OECD 201	Protozoan 24 h viability	Bacterium 30 min. luminescence ISO 21338	Mechanism of toxicity	Classification[†]
ZnO	56.1	-	-	<1.0	1…2	10...50	Zn ions	
Pd	<0.5	+	++	<1.0	>100	10...50	ROS	Acute aquatic hazard
CuO	5.14	+++*	-	<1.0	1…2	1…2	Cu ions & ROS	
Co_3O_4	1.25	+	+	1…2	>100	>100	ROS	
TiO_2	<0.83	+++	+	1…2	10...50	>100	ROS	
Mn_3O_4	11.1	-	+++	1…2	>100	>100	ROS	Acute aquatic hazard?
Fe_3O_4	<1.38	+*	-	1…2	10...50	>100	ROS	
Al_2O_3	0.40	+	-	10...50	>100	>100		
SiO_2	NA	-	-	10...50	>100	>100		
WO_3	63.2	-	-	10...50	>100	50...100		
MgO	38.1	-	-	>100	>100	>100		
Sb_2O_3	56.3	+	-	>100	>100	50...100		

The multitrophic test battery consisted of alga *Pseudokirchneriella subcapitata*, protozoa *Tetrahymena thermphila* and bacterium *Vibrio fischeri*. Tentative ranking of eNPs for acute aquatic hazard is based on the threshold L(E)C50 ≤1 mg/l of the most sensitive assay. The highest nominal concentration of NPs in the tests was 100 mg/l. EC50 values are based on nominal concentrations of NPs and the range of EC50 values for each toxicity assay is color-coded from red (hazardous) to dark green (not classified). Mechanisms of observed toxicity of NPs are proposed based on the solubility and/or the potential to generate Reactive Oxygen Species (ROS) by studied eNPs in the abiotic conditions

¥ – MODERN library of nanoparticles synthesized using flame spray pyrolysis; primary sizes from 8 to 20 nm, specific surface area from 30 to 290 m^2/g;

§ – soluble metal in 10 % nanoparticle suspension in deionized water determined after ultracentrifugation by atomic adsorption spectroscopy;

$ – formation of Reactive Oxygen Species (ROS) in abiotic conditions as measured using 3'-(p-hydroxyphenyl) florescein assay [17];

+ – formation of ROS in abiotic conditions as measured using 2.7-dichlorodihydrofluorescein diacetate assay [68];

* – in the presence of hydrogen peroxide;

† – adhering to European Commission regulation (EU) No 286/2011

battery is presented in the current study and includes algae as primary producers, protozoa as consumers and bacteria as decomposers. As the current study reveals quite similar toxic effects of metal-based eNPs across species at different trophic levels and in a range of biotests, it is highly probable that QNAR models based on one aquatic species (for example *P. subcapitata*, *T. thermophila* or *V. fischeri*) could at least to some extent predict the toxic effects also to other aquatic organisms paving the way to establish activity-activity relationships for nanomaterials.

9.2.3 Integration of "Omics" Data into Nanosafety Models

To integrate biological data into the models, we propose to use high-throughput data, such as gene expression (transcriptomics) or protein expression (proteomics) data. Their advantages are that they provide comprehensive information about changes in cellular processes, they are expected to detect the early onset of toxic responses, and they are also expected to facilitate the mechanistic understanding of toxicity. Using high-throughput data, however, poses important challenges [27, 43]. The experimental platforms and the methods of data analysis are relatively young. In particular, there is no general agreement on a recommended workflow of data analysis, which causes reproducibility issues. Moreover, the interpretation of results is difficult, because of the enormous level of detail, the amount of noise and the degree of data correlation. We propose therefore to incorporate high-throughput data not in its raw form but aggregated into pathway expression data. This reduces the data dimensionality in a biologically relevant manner. Pathway results are easier to interpret, easier to validate with specific biomarkers, and they can provide hints about the underlying toxicity mechanisms.

To define a workflow of high-throughput data analysis, we divide the analysis in three steps, each one with a well-defined purpose. We exemplify the process for the case of transcriptomics data obtained with microarrays, which is the most commonly used today. The steps are: (1) preprocessing of gene expression, (2) normalization of gene expression, and (3) analysis of differential pathway expression.

Preprocessing of Gene Expression Data This step encompasses all the quality control checks and data correction recommended by the platform builder. The aim is to remove as much technical variation as possible *within each sample*. It is worth noting that gene expression microarrays have different hybridization affinities per probe (each probe tests a small section of a gene), which implies that there is unavoidable technical variation across genes. As a consequence, the analysis pipeline has to take into account the fact that expression levels obtained with this technology are not directly comparable between genes.

Normalization of Gene Expression The objective in this case is to remove the technical variation *between samples*, which emerges from small differences in total mRNA concentration, reagent concentration, hybridization conditions, etc. This step is much less dependent on the specifics of each experimental platform, and thus it can be addressed with general methods. The final result is a GxS matrix of gene expression data, for G genes and S samples.

Analysis of Differential Pathway Expression To aggregate gene expression into pathway expression, it is necessary to define a score for each pathway, which has a clear biological interpretation. Several options have been proposed [40], the two more popular being the number of differentially expressed genes in the pathway, and the rank comparison of genes in the pathway with the complete expression profile. The first approach makes difficult the assessment of statistical significance, using the unrealistic null hypothesis which results from permuting the gene labels [19]. The second approach puts the emphasis on changes of gene expression that are highly correlated in the pathway (many genes over- or under-expressed in a similar amount). This may be the case in the alteration of signaling pathways, but not so much in other cases, like for example regulation pathways.

For these reasons, we propose a method which uses all the expression data available (no preliminary filtering of gene hits) and which does not assume any kind of pattern in gene expression variation when a pathway is altered. The statistic is inspired in the logistic function, used to map continuous variables into a probability estimate in logistic regressions. Let t_i be the t-statistic resulting from comparing expression levels between an experimental treatment and the corresponding control condition. The differential expression g_i for the gene i is defined as:

$$g_i = \frac{1-e^{-t_i^2}}{1+e^{-t_i^2}} = \tanh\left(\frac{t_i^2}{2}\right) \tag{9.3}$$

This definition of g_i treats over- and under-expression changes equally, compares and scales each gene independently (removing differences in hybridization affinities) and takes into consideration the uncertainty in the estimation of differential expression. Afterwards, the differential expression p_j of pathway j is defined as the average differential expression of all the genes P_j annotated to the pathway:

$$s_j = \frac{\sum_{i \in P_j} g_i}{\left|P_j\right|} \tag{9.4}$$

This manner, we obtain a non-negative, normalized TxP matrix, with T rows representing experimental treatments and P columns representing pathways. Each coefficient p_{ij} represents the differential expression of pathway j in treatment i, as a number between 0 and 1. The resulting matrix of biological information can be easily integrated with physicochemical data to develop predictive models.

The above pathway analysis workflow was implemented in MODERN as a web application[4] (Fig. 9.5) that identifies differentially expressed pathways from high-throughput transcriptomics or proteomics assays. It also highlights the important genes in connection to those pathways. The tool makes use of pathway reference information from a curated database. The algorithm is based on network statistics calculated over a genome-wide network, built from the data available for the target species [13].

[4] http://biocenitc-deq.urv.cat/anapath.

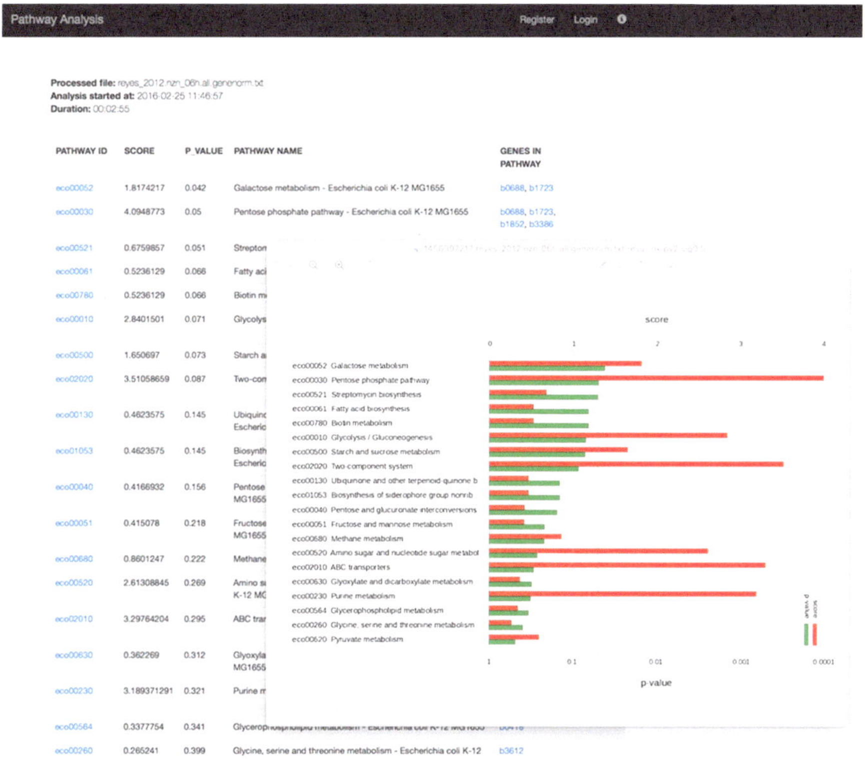

Fig. 9.5 Screenshot of the pathway analysis tool developed in MODERN. Sample output showing pathways identified from gene differential expression data (http://anapath.biocenit.cat)

9.2.4 *Management of Nanosafety Data*

The development of reliable nanosafety assessment strategies requires the compilation of a significant amount of very heterogeneous data (e.g., emission sources, nanoscale properties, intermedia distribution, transformations and persistence, and effects in biological systems). Although these data are rapidly emerging, there is still a critical need for implementing efficient data management protocols aimed to facilitate data retrieval, transparent data sharing and the development of robust structure/property/activity relationships. To be effective, data repositories must provide researchers and regulators with tools for knowledge extraction from annotated data. Data collected from nanotechnology research are fundamental for the identification of correlations between nanomaterial's structure, physicochemical properties and biological activity. Establishing these relationships is of paramount importance to identify mechanisms of toxicity and to guide safe-by-design strategies for new nanomaterials.

Currently, most of the data on nanosafety are widely scattered and remain inaccessible as tables and figures in scientific literature or in non-public databases (e.g., research project results and industrial R&D activities). Data ambiguity, the lack of standardization together with unstructured and heterogeneous data sources makes the impact assessment of nanomaterials a challenging task plagued with uncertainties. The use of ontologies (i.e., controlled vocabularies and relationships that capture knowledge in a specific domain) provides a unifying approach for data structuring and annotation. The use of standards for data annotation allows the integration of heterogeneous data sources, aggregation and presentation in an accessible format and facilitates the computational analysis of the integrated data sets. The combination of controlled vocabularies with database systems enables querying databases not only through their logical schema, but also through the concepts of the ontology and their semantic relationships. Consequently, users can retrieve data in the context of what the data are about, which includes the ontological terms related to the data, their taxonomic parents, related terms and ontology definitions [21]. Although ontologies are common in biology and biomedicine, their use is still scarce in nanotechnology. This can be attributed, in part, to the novelty of the discipline but also to the limited amount of publicly available data.

The *NanoParticle Ontology* (NPO, [74]) was a pioneering work in the development of ontology for nanomaterials and their applications. The ontology, which includes 1903 classes and 81 properties, encompasses knowledge underlying the preparation, chemical composition, physicochemical characterization and in vitro/in vivo characterization of nanomaterials and is intended for top-level modelling of nanomedicine and nanosafety concepts. However, because of its initial development in nanotechnology cancer diagnosis and therapy, the low-level details of NPO are mainly focused on the applications of nanotechnology in cancer research. Complementary approaches to extend ontologies to the whole nanosafety domain have been developed within the FP7 eNanoMapper Project[5] [23]. The eNanoMapper ontology[6] covers the full scope of terminology needed to support research into nanomaterial safety. It builds on multiple pre-existing external ontologies resulting in a total of 6690 classes and 587 properties.

To extract the maximum amount of relevant information, nanomaterial data should not be analysed independently of the overall nanosafety context. For instance, data only on in vitro toxicity effects are not sufficiently relevant/informative to get the complete picture of the potential impact of a nanomaterial. Additional information such as in situ nanomaterial properties, information on synthesis process, exposure conditions and actual dose and distribution of the particles taken up by cells/animals, and characteristics of the biological endpoint are also required. Based on these principles, a data management system with full support for ontology annota-

[5] http://www.enanomapper.net.

[6] http://bioportal.bioontology.org/ontologies/ENM.

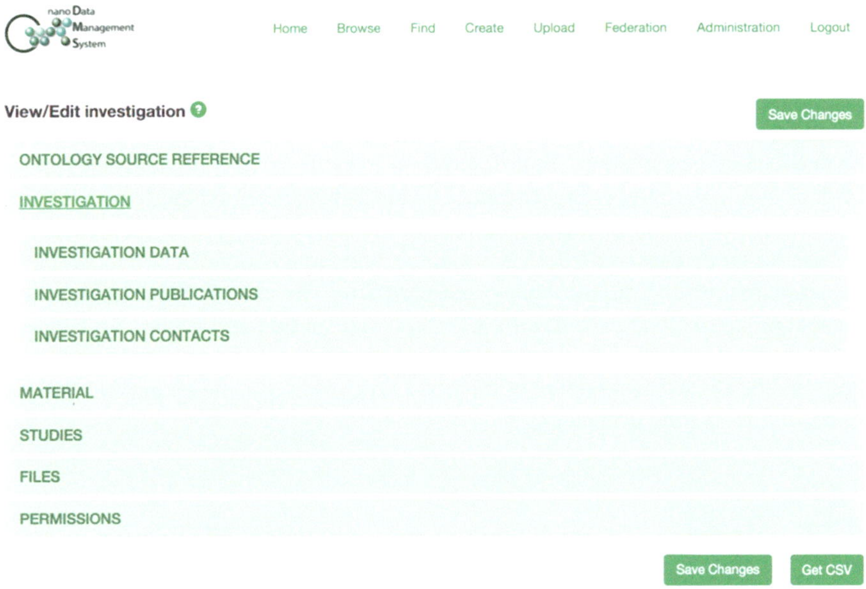

Fig. 9.6 Web interface of the nanoDMS system showing the ISA-TAB_Nano structure (http://nanodms.biocenit.cat)

tion was conceived and implemented in MODERN. The function of the data management system is to store nanosafety data in a semantically consistent manner by leveraging existing ontologies for specific parts of the nanosafety domain. In addition, the system has been designed to be interoperable and to facilitate data sharing. To this end, the data management system implements the ISA-TAB-Nano data exchange format [75].

Following the above principles, the *nano Data Management System*[7] (nanoDMS, see Fig. 9.6) implements a unified and semantically annotated nanosafety data repository based on ISA-TAB-Nano for data sharing. Within MODERN, the use of the data management system has been complemented by the development of procedures for the curation of data so that data quality and provenance information are ensured. The usability of the system has been evaluated in collaboration with other modelling projects (e.g., nanoPUZZLES[8], [48]). The system provides different access control levels to protect data confidentiality, integrity and preserve intellectual property rights. The system also offers support for server federation, allowing the easy deployment of a distributed nanosafety cloud.

[7] http://nanodms.biocenit.cat.

[8] http://www.nanopuzzles.eu/.

9.3 Computational Characterization of eNPs: Development of Nanodescriptors

The development of in silico toxicity models for nanoparticles is mainly hindered by two factors: lack of appropriate descriptors and scarcity of consistent experimental data. While the latter is largely caused by the difficulties in synthesizing nanoparticles with well characterized and narrow size distribution, agglomeration, etc., the former is mainly caused by the very complex and large structures of nanoparticles (as compared to organic molecules) not permitting the use of most atom-explicit computational methods. MODERN has implemented two approaches for nanodescriptor development. The first is based on quantum chemistry calculation to describe the electronic structure of the nanoparticle whereas the second is based on molecular modelling principles.

9.3.1 Nanodescriptors Based on nanoparticle's Electronic Structure

For a chemical compound X containing N electrons, the ionization potential IP and electron affinity EA are defined as difference in energy E for the process of subtracting or adding one electron, which can be approximated through Koopmans theorem as negative energies of the highest occupied and lowest unoccupied molecular orbitals, ε_H (HOMO energy) and ε_L (LUMO energy), respectively [8, 60]:

$$\mathrm{IP} = E_{N-1} - E_N = E\left(\mathrm{X}^+\right) - E\left(\mathrm{X}\right) \approx -\varepsilon_H \tag{9.5}$$

$$\mathrm{EA} = E_N - E_{N+1} = E\left(\mathrm{X}\right) - E\left(\mathrm{X}^-\right) \approx -\varepsilon_L \tag{9.6}$$

In Eq. 9.6, the physicist (and former chemist) sign convention for EA is used that provides a positive value in case the anion is more stable than the neutral compound.

The electronegativity EN (often termed χ) of a molecule is a measure of its initial attraction towards electronic charge. Its rigorous definition as negative partial derivative of energy E on N can be transformed to a finite-difference approximation employing IP and EA, which in turn can be approximately replaced by the HOMO and LUMO energies [8, 60]:

$$\mathrm{EN} = -\frac{\partial E}{\partial N} \approx \frac{1}{2}\left(\mathrm{IP} + \mathrm{EA}\right) \approx -\frac{1}{2}\left(\varepsilon_H + \varepsilon_L\right) \tag{9.7}$$

Correspondingly, the hardness HD (often termed η) characterizing the resistance to charge transfer can be expressed in a simple finite-difference form [8, 59]:

$$\mathrm{HD} = \frac{1}{2}\frac{\partial^2 E}{\partial N^2} \approx \frac{1}{2}(\mathrm{IP} - \mathrm{EA}) \approx -\frac{1}{2}(\varepsilon_{\mathrm{H}} - \varepsilon_{\mathrm{L}}) = \frac{1}{2}(\varepsilon_{\mathrm{L}} - \varepsilon_{\mathrm{H}}) \tag{9.8}$$

In a solid-state conductor with a typical electron density of 5×10^{22} cm^3 (number of conducting electrons N per unit volume V), the probability of occupancy of a given energy level ε_i, f_i (a number between 0 and 1) is governed by the Fermi-Dirac statistics -where k and T denote the Boltzmann constant and absolute temperature, respectively- [69]:

$$f_i = \frac{1}{\exp\left(\frac{\varepsilon_i - \varepsilon_{\mathrm{F}}}{kT}\right) + 1} \tag{9.9}$$

Here, the Fermi level ε_F is associated with the occupation probability of 0.5 at 0 K, and thus can be understood as highest occupied energy level of the respective electronic band (that is only partly filled in case of metals). For a free-electron gas as simple model of a metallic conductor with ε_0 denoting the lowest occupied energy level, the energy difference [69]:

$$E_{\mathrm{F}} = \varepsilon_{\mathrm{F}} - \varepsilon_0 = \frac{h^2}{8m}\left(\frac{3N}{\pi V}\right)^{2/3} \tag{9.10}$$

(with h = Planck's constant, and m = electron mass) quantifies the kinetic energy of the Fermi-level electron, and as such is a positive quantity. In practice, E_F and ε_F are often used synonymously and thus both called Fermi energy. Considering Eq. 9.10, however, we call ε_F (that for bound electrons is a negative quantity) the Fermi level of the system of interest. E_F increases by the 2/3 power with increasing density of conducting electrons (N/V); taking the E_F values of sodium (Na) and silver (Ag) as example, the significantly higher value of the latter (5.5 vs. 3.1 eV) is driven by a more than twofold higher electron density.

Since for conductors ε_F can be approximately taken as highest occupied level, the negative of its energy essentially corresponds to the minimum work required to remove an electron from a solid. The latter is called the work function reflecting the electronic structure of the solid and its surface. On the one hand, the work function corresponds to the (lowest) IP of a molecule (see Eq. 9.5). On the other hand, this correspondence holds only approximately because of the variation in electronic characteristics between bulk and surface-layer atoms, the latter of which are less shielded and thus more prone to ionization.

In a semiconductor, ε_F is located in the forbidden region halfway between the filled valence band and the empty conduction band (referring to 0° K). Denoting ε_v and ε_c as highest filled and lowest unfilled levels of the valence and conduction band, respectively, these levels represent the semiconductor ionization potential and electron affinity according to:

$$\text{IP} \approx -\varepsilon_{\text{V}} \tag{9.11}$$

$$\text{EA} \approx -\varepsilon_{L} \tag{912}$$

(with IP approaching $-\varepsilon_{\text{F}}$ upon metallization of the material), and the solid-state analogues of Eqs. 9.7 and 9.8 provide working equations for the semiconductor electronegativity and hardness:

$$\text{EN} \approx -\frac{1}{2}\left(\varepsilon_{\text{V}} + \varepsilon_{c}\right) = -\varepsilon_{\text{F}} \tag{9.13}$$

$$\text{HD} \approx -\frac{1}{2}\left(\varepsilon_{\text{V}} + \varepsilon_{c}\right) = \frac{1}{2}\left(\varepsilon_{c} - \varepsilon_{\text{V}}\right) \tag{9.14}$$

Considering further that at room temperature, the Fermi-Dirac distribution is smeared out with finite occupation probabilities of energy levels above ε_{F}, the semiconductor work function could be estimated through $-\varepsilon_{\text{F}}$ and thus EN (Eq. 9.13) or $-\varepsilon_{\text{v}}$ (the solid-state IP) or through their arithmetic mean. In any case, Eqs. 9.13 and 9.14 demonstrate that for semiconductors, the band gap reflects the solid-state resistance to charge transfer, and the mean of the highest occupied and lowest unoccupied orbital energies the initial electron attraction of solid-state material, keeping in mind that a more detailed description would require to consider the respective surface-layer modulation of these electronic structure characteristics.

The descriptors developed using the approach discussed above are subsequently used in Sect. 4 to explore structure-activity relationships for the library of metal oxide nanoparticles developed in MODERN.

9.3.2 *Size-Dependent Nanodescriptors Based on Molecular Modelling Approaches*

One of MODERN's goals has been to find a solution to the lack of "true nanodescriptors" capable of distinguishing between the properties of compounds in the bulk and in nanoparticles of different sizes. To address this issue, the approach chosen was to model eNPs as whole-particles since it is the most consistent and size-aware option. Naturally, true quantum chemistry is not applicable to such large systems, therefore, molecular-mechanics/dynamics based methods have been used.

Using this approach, a series of descriptors derived from the full molecular mechanic simulation of metal oxide nanoparticles have been developed. Due to the increasing processing power of computers, it is possible to calculate the energy and structural parameters of nanoparticles in a relatively small timescale using simple interatomic potentials.

The first step of the simulation is the generation of atomic coordinates for a nanoparticle. The thermodynamically most stable crystal structure for each metal oxide is selected and the corresponding unit cell parameters are used to generate a

Table 9.2 Classes of size-dependent nanodescriptors for metal oxide eNPs

Descriptor related to	Basis of descriptors
Chemical composition	Total number of atoms in nanoparticle, in the core and shell regions.
Potential energy	Average potential energy of all atoms in nanoparticle, of metal atoms or oxygen atoms, in electron volts.
Lattice energy	Lattice energy of the whole nanoparticle, relative lattice energy (per diameter or per surface area or as compared to a perfect crystal) of the particle in electron volts.
Topology	Average coordination number of all atoms, metal atoms or oxygen atoms in the nanoparticle.
Size	Diameter, surface area and volume of the nanoparticle in Å, $Å^2$, $Å^3$, respectively.

spherical nanoparticle with the desired diameter. The atoms in this sphere are divided into two groups: core and shell. The atoms in the core are assumed to have similar characteristics to the bulk material while the shell atoms are destabilized. While the positions of the atoms can be optimized according to different schemes, the approach works even without optimization. Already from a single-point calculation, the potential energy and coordination numbers can be extracted. These values are the basis to derive different categories of descriptors for nanoparticles. Table 9.2 describes the basis of each class of nanodescriptor. Constitutional descriptors, reflecting the chemical composition, can be as simple as the number of metal or oxygen atoms in the two respective nanoparticle regions. Topological (i.e., connectivity based) descriptors include the average coordination number of metal and oxygen atoms in the shell group and in the core group. Descriptors based on potential energy can either be derived solely from the nanoparticle (average potential energy of metal atoms in shell regions) or in comparison with bulk material (difference between the lattice energy of nanoparticle and bulk material). Some of the 35 descriptors developed were specific to metal oxide eNPs (parameters related to oxygen atoms or metal atoms) but the concept can be adjusted to pure metal NPs or other types of particles in a straightforward manner.

9.3.2.1 Nanodescriptor Calculation

The calculation of nanodescriptors was performed with the LAMMPS[9] molecular dynamics simulator program [63] using Buckingham potentials to calculate the energies. The most stable form of the metal oxide crystal structure was selected as input for developing the nanodescriptors. Energy of the unit cell was calculated, the optimal cutoff values for Coulombic interactions were calibrated which were later used for nanoparticle calculations. Atomic coordinates of nanoparticle were found by replicating the selected unit cell and cutting out the desired shape of the nanoparticle (sphere in the present case). To ensure the charge neutrality of the nanoparticle,

[9] http://lammps.sandia.gov.

an appropriate number of metal or oxygen atoms were added to random lattice positions on the surface of the nanoparticle. After that, descriptors from classes described above were calculated from the results of a single-point energy calculation.

A large number of the descriptors are derived from the potential energy which is composed of two parts: pairwise energy calculated by the Buckingham potential (Eq. 9.15), and Coulombic interactions, which are calculated by Wolf summation [82]:

$$E_B = A * e^{-r/\rho} - \frac{C}{r^6} \text{ r} < r_c, \qquad (9.15)$$

where A, ρ, C are constants of the Buckingham potential; r is the interatomic distance; r_c is the cut-off radius.

Wolf summation was used as the computationally much more affordable alternative to the traditional Ewald summation [15]. The required cut-off radiuses for the Wolf summation were derived from the modelling of respective infinite crystals by periodic calculation of small clusters of crystal unit cells.

One of the main requirements for these calculations are the constants for Buckingham potentials. For many metal oxides, these values can be found from the literature, but for example Sb_2O_3, these constants had to be derived. Density functional theory (at the level of B3LYP/Def2-TZVDP) was used to calculate the interatomic potential parameters. The ability to derive the interatomic potential fully theoretically based on by ab-initio calculations and the subsequent calculation of descriptors using these parameters is a great advantage since the only experimental parameter required for the calculation of nanodescriptors is the determined structure of the unit cell. As many metal oxides can exist in multiple crystal structures, the thermodynamically most stable crystal structure under standard conditions was used for the calculation of nanodescriptors in all cases.

Geometric descriptors are based on the calculated diameter of the nanoparticle, which is defined as the maximum distance between any two atoms in the nanoparticle. Constitutional descriptors are based on the chemical composition of nanoparticle. Descriptors which are based on the potential energy indicate the stability of the core and shell regions in the nanoparticle, respectively. Topologic descriptors are based on the coordination number of atoms (defined as counting the neighboring atoms which lie inside radius R.

$$R = 1.2 * (R_M + R_O) \qquad (9.16)$$

where R_M, R_O are the ionic radii of metal and oxygen atoms, respectively. The last group of descriptors is based on the lattice energy, the difference of the latter compared to that of a perfect crystal and the proportion to diameter, surface area, and volume.

The main advantages of these nanodescriptors over previously published descriptors are:

1. Current descriptors require only one experimental parameter for calculation, namely the structure of the unit cell of the metal oxide. This information is available for many different metal oxides (and other compounds)

2. Descriptors are size-dependent.
3. Method for calculating descriptors is easily extendable to include solvent effects or to nanoparticles with non-spherical shape.

The unit cell structure is necessary to calculate the required constants for the Buckingham potential and to generate the structure of the nanoparticle. While at the moment the thermodynamically most stable crystal structure is used, in principle any other crystal structure can be used if so desired.

9.3.2.2 Analysis of Size-Dependency

The size-dependency of the new nanodescriptors is paramount, as the nanoparticles with the same chemical composition but different size can have different toxicity/property values. The size-dependency of a Cr_2O_3 nanoparticle descriptor *"Difference between lattice energies of nanoparticle and perfect crystal"* is depicted on Fig. 9.7 based on data generated from molecular dynamic calculations in MODERN.

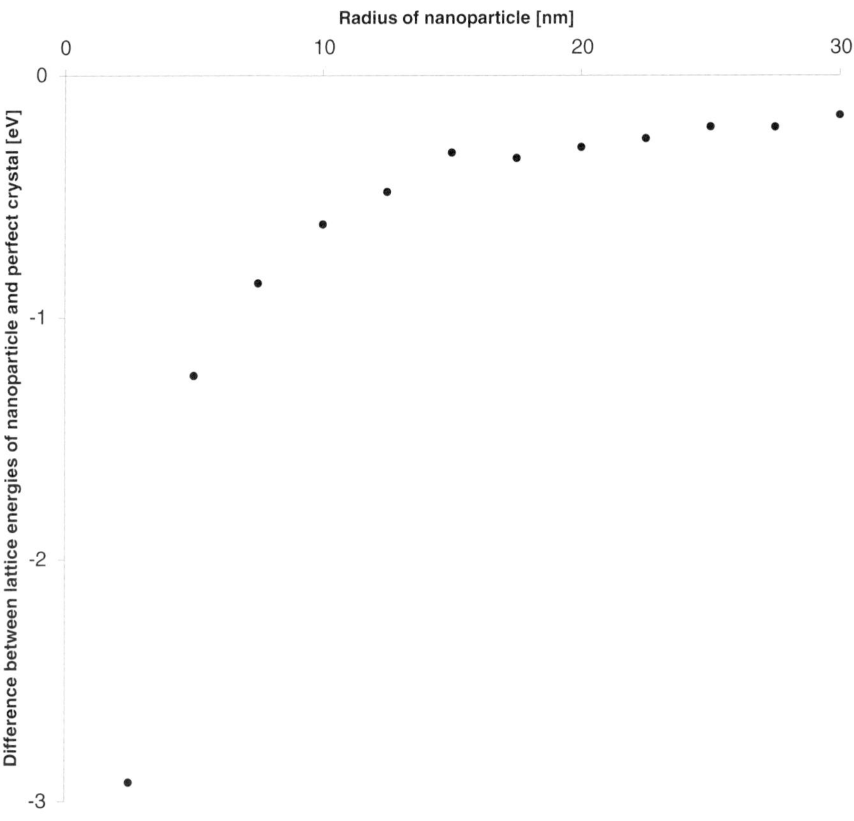

Fig. 9.7 The size-dependency of the descriptor *"Difference between lattice energies of nanoparticle and perfect crystal"* for a Cr_2O_3 nanoparticle

As the radius of nanoparticle increases, the difference between lattice energy of nanoparticle and the corresponding perfect crystal is reduced, approaching a constant. This follows the notion that as the size of nanoparticle approaches macroscopic levels, the properties of the nanoparticle start to become similar to the ones of macroscopic particle/bulk material. The curve presented in Fig. 9.7 agrees well with the commonly accepted position that nanoparticles of size below 20 nm require the most attention as they start to possess significantly different properties as compared to bulk material.

9.4 Structure-Activity Relationships for Nanoparticles

The approach of Quantitative Structure-Activity Relationships (QSAR) is widely used in chemical and biological sciences, attempting to relate the differences in chemical structure represented by a set of descriptive parameters (descriptors) to a physicochemical property or biological activity. Biological activity usually depends on several descriptors and the corresponding multilinear QSAR model has the following form:

$$P = a_0 + a_1 X_1 + \ldots + a_n X_n \tag{9.17}$$

In the above example, the biological activity (P) depends on structural features represented by n descriptors ($X_1 \ldots X_n$) and $n+1$ coefficients ($a_0 \ldots a_n$), which are determined by multilinear regression analysis. Once a valid QSAR model has been determined, the prediction of physicochemical properties or biological activity for related compounds becomes possible. The quality of the predictions of a QSAR model depends on the accuracy of the experimental data, the choice of descriptors and the statistical methods, and on the validation of the model. Naturally, each QSAR model is limited to a certain applicability domain, depending on the structural features present in the source dataset and the range of the experimental values. The same general principles for QSAR development can be applied to eNPs. However, for eNPs one most include additional information in the structure of the models to account for nanoparticle size as well as for changes in its properties in exposure conditions. In addition to data related issues (i.e., scarcity, uncertainty and variability), the lack of detailed mechanistic knowledge of nanoparticle toxicity adds an additional complexity level to model development, validation and use.

9.4.1 Structure-Activity Relationships Based on Nanoparticle's Electronic Structure Descriptors

In recent years it became apparent that besides the often-discussed redox activity of nanoparticles that may induce reactive oxygen species (ROS), other toxicological pathways may play a role [53]. In this context, a particular issue is phagocytosis (Trojan horse) as highly efficient route of uptake into the cell [42], offering the possibility of metal-specific toxicity at very high concentrations upon intracellular dissolution. In the following, algal toxicity in terms of 72-h EC_{50} (growth inhibition 50 %, see Sect. 2.2) values are analyzed from the viewpoint of quantum chemical reactivity parameters calculated for eNP monomers, including trend analyses of the electronic structure characteristics when going from the monomers to clusters with increasing molecular size.

9.4.1.1 Endpoint Selection and Computational Details

Algal toxicity data in terms of 72-h EC_{50} [mg/L] values -effective concentration inhibiting the growth by 50 %- for a subset of metal oxide nanoparticles (Al_2O_3, CuO, Fe_2O_3, MgO, Sb_2O_3, SiO_2, TiO_2 and ZnO) were used as endpoint. For their conversion from the original mass-based unit to mol/L, the molar mass of the respective unit cell was divided by its number of metal atoms, considering the fact that in this way, the toxicity is normalized to the concentration of metal ions. Taking the Al_2O_3 NP as an example, the crystal monomer is Al_2O_3, yielding $AlO_{1.5}$ as formal basis for the NP molar mass. The resultant log EC_{50} [mol/L] values were used for comparison with electronic structure characteristics.

The starting geometries for the metal oxide clusters have been prepared in the following way. For each cluster, a part was cut out of the solid-state crystal structure of the material, such that the number of metal atoms (which we use as a measure for the cluster size) was as desired. The cutting was performed in such a way that the surface of the resulting cluster is completely terminated with oxygen atoms. Subsequently, hydrogen atoms have been added to exposed oxygen atoms in order to obtain a neutral charge, turning some oxygen atoms into hydroxyl groups and others into water molecules. This creates a microsolvation around the cluster, and keeps the coordination number of the metal atoms at the same value as in the bulk phase. To give an example, the cluster containing one aluminum atom was cut out of the corundum lattice as formal "AlO_6", because aluminum has six nearest oxygen neighbors in the lattice. Subsequently, hydrogen atoms were added, turning three of the oxygen atoms into hydroxyl groups and the remaining three into water molecules, thus yielding a final cluster of $Al(OH)_3(H_2O)_3$ that is neutral in charge.

The quantum chemical calculations have been performed with the program package Orca [52], employing density functional theory (DFT) with the PBE functional [62] and Grimme's empirical D3 dispersion correction [20]. Atom-centered basis sets of the type def2-TZVPP [80] have been used for all atoms. The SCF convergence criterion was set to *"VeryTight"*. The starting structures of all metal oxides (monomers and clusters containing an increasing number of monomer units) have been cut from the corresponding solid state lattices, and subsequently the exposed surfaces have been saturated by hydroxide ions to obtain charge-neutral species. Water molecules were added to account for solvation effects ("microsolvation"). Based on these structures, geometry optimizations (convergence criterion "Tight") have been performed to determine equilibrium geometries. IP, EA, ε_F (−EN) and HD as electronic structure characteristics have been calculated from the respective DFT orbital energies based on Koopmans theorem (see Eqs. 9.5–9.8 and their solid-state counterparts Eqs. 9.11–9.14).

9.4.1.2 Trend Analysis

The data distributions of log EC_{50} [mol/L] of the eight eNPs vs. calculated IP, EA, ε_F (= −EN) and HD of the NP monomers (saturated by H atoms) are shown in Fig. 9.8. As can be seen from the top left plot, NP toxicity vs. IP suggests a separation between the main-group metal oxides MgO, Sb_2O_3, SiO_2 and Al_2O_3 on the one hand, and the transition-metal oxides CuO, ZnO, Fe_2O_3 and TiO_2 on the other hand.

Whereas both subsets overlap in their IPs as estimated through Koopmans theorem from the HOMO energies, the high toxicity end of the former (log EC_{50}: from −2.6 for MgO to −3.4 for SiO_2) differs by 1.5 log units from the low toxicity end of the latter (log EC_{50}: from −4.9 for Fe_2O_3 to −5.9 for ZnO). Moreover, eNP toxicity increases with decreasing IP for the transition-metal eNPs, which contrasts with the opposite trend observed for the main-group counterparts. Since the IP is inversely related to the ease of electron donation, the latter appears to play a toxicity-enhancing role only for algal toxicity of the transition-metal oxide eNPs, suggesting a redox-

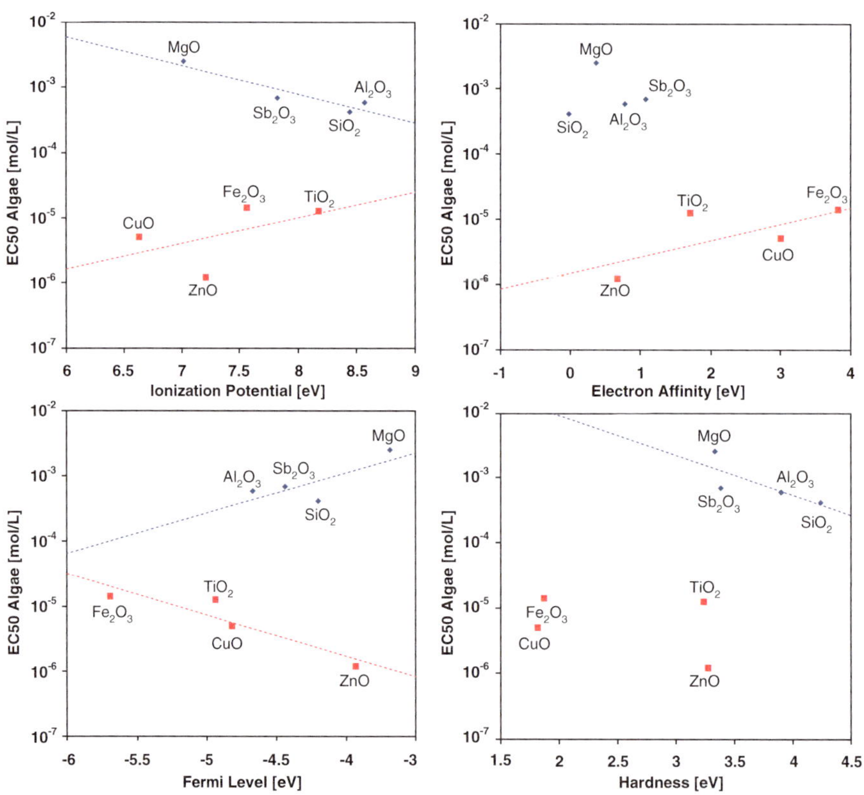

Fig. 9.8 Algal toxicity in terms of logarithmic-scale 72-h EC_{50} of eight NP metal oxides vs. calculated ionization potential (IP, top left), electron affinity (EA, top right), Fermi level (ε_F, bottom left) and hardness (HD, bottom right) from the DFT HOMO and LUMO energies employing Koopmans theorem (see Eqs. 9.5–9.8). Algal toxicity data were taken from [3]

mediated toxicity pathway in contrast to a different toxicological mode of action for the main-group oxide NPs.

In the top right plot of Fig. 9.8, a similar discrimination between the main-group and transition-element metal oxide eNPs is shown with EA as eNP monomer property. In this case, increasing EA corresponds to an increase in algal toxicity for the transition metals, indicating that the latter is enhanced with increasing capability of accepting excess electronic charge (keeping in mind the presently used sign convention for EA as outlined above). Regarding the Fermi level of the eNP monomers (bottom left in Fig. 9.8), eNP toxicity increases with increasing ε_F, suggesting that the NP electron donor strength provides a significant contribution to the algal toxicity. The corresponding decrease in toxicity with increasing EN ($= -\varepsilon_F$) is in line with this interpretation and indicates that in contrast to the main-group metal oxides, the transition-metal eNPs show a decrease in toxicity with increasing initial electron attraction.

When analyzing log EC_{50} from the viewpoint of the electronic hardness (that is inversely related to the polarizability), a similar between-group separation is accompanied by less pronounced within-group trends. In this case, the main-group metal oxides show a weak increase in toxicity with increasing HD, which would also hold with regard to ZnO as compared to the other three transition-metal oxides, but not otherwise. Taking all four plots of Fig. 9.8 together, the level of significant approximation regarding eNP electronic properties should be kept in mind, which concerns both fundamental and methodological issues (eNP monomer vs eNP bulk vs eNP surface, Koopmans theorem, DFT computational chemistry).

In summary, the above quantum chemical analysis of the algal toxicity of metal-oxide eNPs suggests a significant difference in mode of action between the subgroups of transition metals and main-group elements. For the former, toxicity increases with increasing electron-donor capability, indicating that a redox-mediated process involving electron transfer to reactive oxygen species (ROS) plays a crucial toxicological role. Alternatively, and considering the experimental setting with exposure to light and thus radiation energy, the increase in toxicity with decreasing IP may also reflect an increased probability for electronic transitions into the excited state and thus a phototoxic contribution to the observed algal growth inhibition in terms of 72-h EC_{50} values. For the main-group elements, the lack of respective EC_{50} dependencies suggests a non-ROS mode of toxic action.

9.4.2 *QNAR Development from Size-Dependent Nanodescriptors*

For (small) organic molecules, descriptors quantifying geometric, steric or electronic properties of a molecule can be empirically determined or more accurately calculated using quantum chemical methods of computational chemistry, usually at semi-empirical level often based on AM1 parametrization [38].

These methods, however, are not well suited to model inorganic materials, especially nanoparticles due to the large number of atoms and the contribution of the lattice. Moreover, many of the algorithms used to calculate descriptors for organic molecules cannot be directly applied for nanoparticles. One simple approach to overcome this problem is to use external descriptors – known/measured properties of the nanoparticle material (or calculated from small metal oxide clusters) such as lattice energy, electronegativity, size, ionic index etc. [45] have successfully used this type of descriptors to model the probability of toxicity of metal oxide nanoparticles, as the best model has a classification accuracy of 93.74 %. A similar approach was used by [65] to describe the cytotoxicity of 17 metal oxide nanoparticles to bacteria *Escherichia coli*. The resulting QSAR model utilized one descriptor: enthalpy of formation of gaseous metal cation. Descriptors based on the SMILES code of the compound have been also used to model the toxicity and Young modulus of inorganic nanomaterials [64, 76, 78]. The modelling of nanoparticles coated with a layer of organic molecules has been limited to descriptors directly calculated from the structure of the coating material [35, 77]. Unfortunately, none of these approaches is capable of predicting the size-dependency of the toxicity of nanoparticles, a phenomenon observed recently by several experimental studies [9, 36, 61].

A simple example of a QNAR model using size-dependent nanodescriptors was constructed and is presented below. Toxicity data in Table 9.3 were taken from the work of [65].

Using the information in Table 9.3 and the size-dependent **nanodescriptors** developed in Sect. 3.2, a two parameter multilinear QNAR was developed:

$$\log\left(\frac{1}{EC_{50}} \right) = 3.82 + 0.07ND_1 - 0.05ND_2 \tag{9.18}$$

Where ND_1 in Eq. 9.18 corresponds to the nanodescriptor *Average potential energy of atoms in the shell region of the nanoparticle* in electron volts, and ND_2 is the nanodescriptors corresponding to the *Average potential energy of oxygen atoms in the core region of the nanoparticle* in electron volts. The QNAR shows acceptable statistics for performance and stability[10]:

$$\begin{cases} R^2 = 0.87 \\ R^2_{cv} = 0.81 \\ F = 45.26 \\ s^2 = 0.04. \end{cases}$$

Fig. 9.9 shows a plot of the experimental versus predicted endpoint values.

[10] *R^2-*squared correlation coefficient;* R^2_{cv} *–squared cross-validated correlation coefficient; F – Fisher criterion;* s^2 *– squared standard deviation.*

Table 9.3 Experimental and predicted log(1/EC_{50}) values of nanoparticles, calculated nanodescriptors (ND) and size information

	Log(1/EC_{50})		Nanodescriptors		NP size
NP	Exp.	Pred.	ND_1 (eV)	ND_2 (eV)	Diameter (nm)
ZnO	3.45	3.38	−19.36	−19.70	21.0
CuO	3.20	3.34	−21.25	−21.60	48.0
V_2O_3	3.14	2.72	−28.60	−19.47	20.0
Y_2O_3	2.87	2.75	−27.19	−18.00	32.7
Bi_2O_3	2.82	3.01	−17.04	−8.36	51.0
In_2O_3	2.81	2.75	−27.88	−18.94	59.6
Sb_2O_3	2.64	2.74	−28.04	−18.97	20.0
Al_2O_3	2.49	2.49	−32.79	−20.80	31.0
Fe_2O_3	2.29	2.71	−29.64	−20.71	20.0
SiO_2	2.20	1.96	−41.98	−23.11	20.0
ZrO_2	2.15	2.22	−32.16	−14.06	25.0
SnO_2	2.01	2.11	−37.23	−19.27	21.0
TiO_2	1.74	1.77	−43.31	−21.05	15.0
CoO	3.51	3.37	−19.83	−20.14	20.0
NiO	3.45	3.37	−19.92	−20.33	20.0
Cr_2O_3	2.51	2.68	−30.37	−21.09	20.0
La_2O_3	2.87	2.77	−12.12	4.24	24.6

ND_1 is the *Average potential energy of atoms in the shell region of the eNP* and ND_2 is *Average potential energy of oxygen atoms in the core region of the eNP*. Toxicity data were taken from [66]

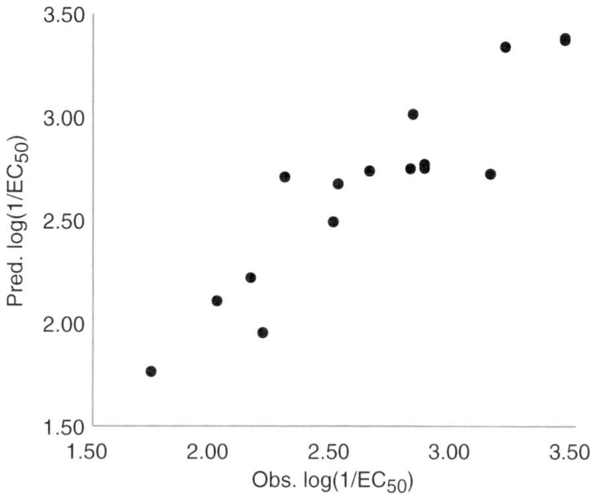

Fig. 9.9 Plot of observed vs. predicted log(1/EC_{50}) values of nanosized metal oxides, based on the two nanodescriptor model (Eq. 9.18) Toxicity data were taken from [66]

9.4.2.1 Model Interpretation

The first nanodescriptor – ND_1 – in Eq. 9.18 is a very obvious choice, since many of the outstanding characteristics of eNPs are related to the uncompensated potential energies of the atoms at or near the surface of the particle. The descriptor is not that much composition dependent, instead it relates to the shape and size of the particle. The second nanodescriptor – ND_2 – is more composition specific, as the potential energy of the oxygen atoms in the unperturbed lattice of an oxide depend both on the lattice structure and the metal atom involved. Therefore, with just two descriptors the model can account for the chemical composition, the lattice structure, the size, and the shape of eNPs.

The computational method described above was designed to be expandable to allow the development of a more accurate representation of nanoparticles and more complex descriptors. While at the moment the descriptors were calculated from single-point energies, it is easily expandable to include full or partial minimization of the geometry of the nanoparticle before the calculation of descriptors or even perform molecular dynamics simulation of these systems to replicate the pyrolysis conditions used in the synthesis of nanoparticles. In this way, even the synthesis conditions can be reflected by the model. Computationally more expensive potential improvements include the use of solvent and coating materials around the nanoparticle. In order to prove the benefits of such more elaborate approaches, however, a very consistent and accurate experimental dataset has to be used with known and controlled particle size distribution and other variables.

9.5 Nanoparticle Grouping and Hazard Ranking

9.5.1 Identification of Nanoparticle Categories.

The large number of nanotechnology enabled products and the multitude of different types of nanomaterials makes impracticable, in terms of time and costs, their exhaustive hazard and risk assessment. Nanomaterial categorization criteria are needed to develop “grouping schemes” that will make nanosafety assessment more efficient. From a computational nanotoxicity perspective, the large number of possible nanoparticle types (e.g., diverse combinations of core composition, surface modifications and functionalization) hinders the development of “universal” models. It is thus fundamental to develop similarity metrics involving different nanomaterial characteristics (e.g., physicochemical property profile and biological activity) across their entire life cycle. The use of appropriate similarity metrics will facilitate the grouping of nanoparticles into homogeneous categories where more accurate and reliable models can be developed and validated. The establishment of eNP categories will also enable the ranking of their environmental and human health impact paving the way to the development of a risk assessment framework for nanomaterials.

Within MODERN, different techniques and algorithms have been used to group the metal and metal oxide nanoparticles into similar groups using different types of

Table 9.4 Toxicities of nanoparticle (eNP) suspensions to algae *Pseudokirchneriella subcapitata*, protozoa *Tetrahymena thermophila* and bacteria *Vibrio fischeri*

eNP	Algae z EC_{50}	Protozoa EC_{50}	Bacteria EC_{50}	BET size (BET)	Hydrodinamic size (nm, DI water)	Zeta potential (mV, DI water)	Oxidation level
ZnO	0.1	1.84	11.52	20.4	171	16.4	2
CuO	0.43	2	1.78	13.1	130	17	2
TiO_2	1.26	52.6	100	12.2	171	–13.6	4
Fe_3O_4	1.93	26.03	100	9.7	128	22.2	2.67
Co_3O_4	1.11	100	100	11.5	99	23	2.67
Mn_3O_4	1.34	100	100	15.2	395	–14.4	2.67
Pd	0.41	100	55.42	15.1	127	–27.8	0
SiO_2	35.58	100	100	7.8	148	–33.2	4
$Al2O_3$	30.8	100	100	11.4	95	39.2	3
WO_3	57.8	100	87.07	10.6	63	–45.2	6
MgO	100	100	100	13.6	1964	6.9	2
Sb_2O_3	100	100	73.74	20.5	125	–24.3	3

Data were taken from [3]

information. The different category schemes developed in the current analysis are discussed in terms of the *soluble*, *active* and *passive* groups proposed by [1] in their *DF4nanoGrouping* framework. In what follows, a given eNP is considered as active if it has EC_{50} values lower than 100 mg/L for at least one of the tested species.

The data set in Table 9.4, covering material characters and hazard estimates will be used in the following to illustrate the application of the hierarchical analysis.

9.5.1.1 Categories Derived from Physicochemical Properties

Categories can be developed from measured physicochemical properties of the eNPs. Fig. 9.10 depicts the categories obtained using the Self-Organizing Map (SOM) algorithm [41] to group the 11 metal oxide nanoparticles in MODERN's library [3]. Each SOM map unit ((i.e., circles in Fig. 9.10) can be interpreted as a cluster. Accordingly, nanoparticles assigned to the same unit form a category. The features used for grouping include BET size, hydrodynamic diameter in DI water, zeta potential in DI water and the oxidation level (see Table 9.4). Prior to SOM development, data were centered and scaled. Map topology was defined as toroidal (i.e., periodic boundaries) to avoid border effects and the map grid was rectangular with a dimension of 3 × 2 units. The position of eNPs within each unit reflects their similarity (i.e., the closer the labels the more similar the eNPs). In addition, the distance of a nanoparticle to the unit center (i.e., center of each circle) is related to the ability of a given nanoparticle to act as a representative element for the group of eNPs assigned to the unit. The SOM analysis identifies six different groups of nanoparticles.

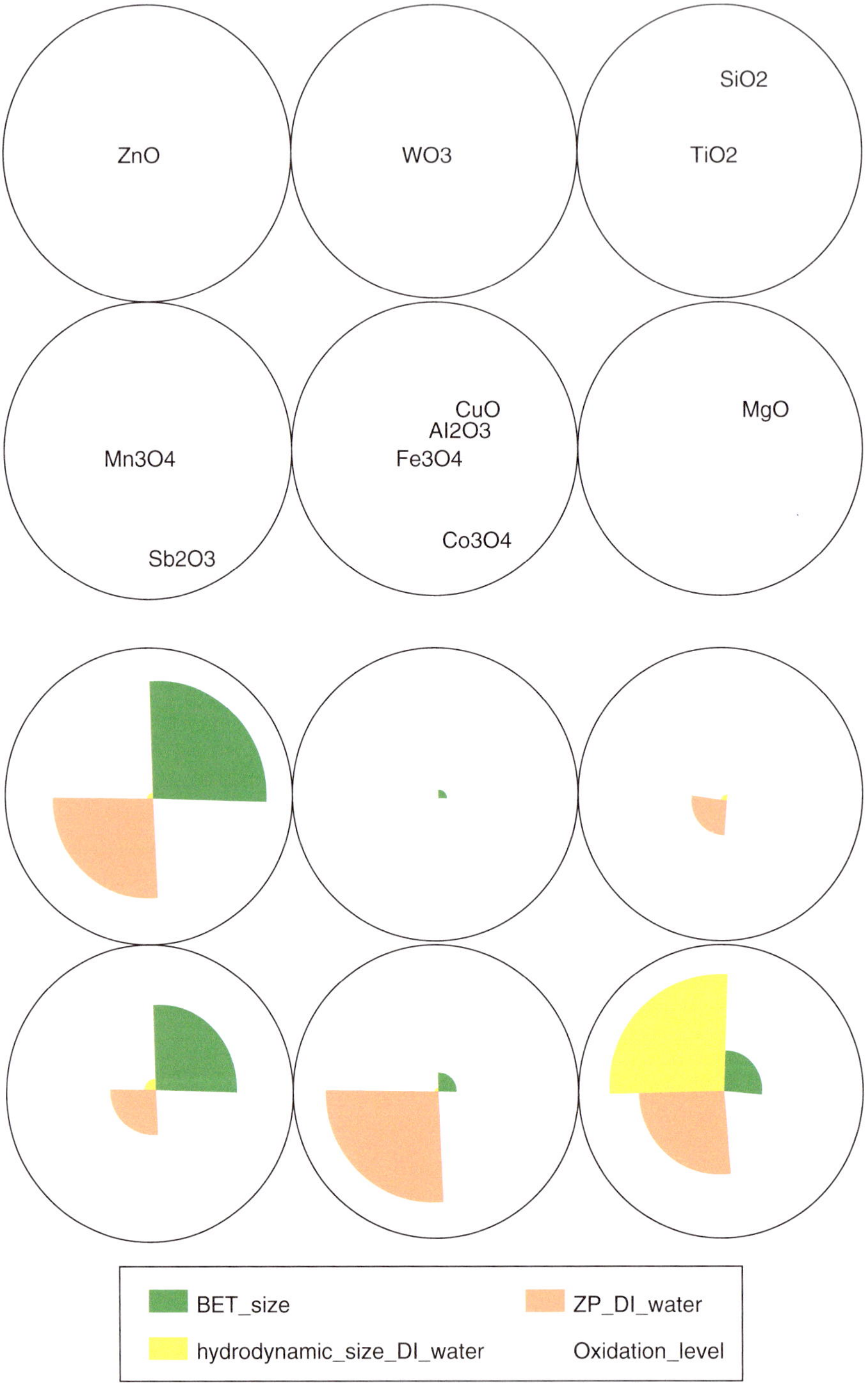

Fig. 9.10 Self-Organizing Map (SOM) clustering of metal oxide nanoparticles in MODERN's library. Clustering was performed based on the group of physicochemical properties reported in Table 9.5. (*Top*) Identification of eNPs in each category. (*Bottom*) Physicochemical property profile of each category

Data in Fig. 9.10 provides the basis for the interpretation of the categories obtained from SOM analysis. Category 1 is formed only by ZnO and its main characteristics are large BET size and a moderate positive ZP in DI water. The distinctive features of Category 2 {WO_3} are the oxidation level and a highly negative surface charge. Category 3 comprises {SiO_2, TiO_2}, the main characteristics of this group are the oxidation level and negative surface charge. The representative eNP of this category is TiO_2. The fourth category includes {Mn_3O_4, Sb_2O_3} and its main features are similar to category 1. However, BET sizes are smaller and the surface charge in DI water is negative. The representative eNP of this category is Mn_3O_4. Category 5 is the most populated {CuO, Al_2O_3, Fe_3O_4, Co_3O_4} and contains positively charged eNPs. The category representative is Fe_3O_4. Finally, category 6 is formed by MgO and its distinctive property is the hydrodynamic size with large values.

Categories 1, 2, 3 and 5 include eNPs which are either *soluble* or *active* to the species tested ($EC_{50} < 100$ mg/L). Category 6 includes a *passive* eNP whereas category 4 mixes *active* with *passive* eNPs.

9.5.1.2 Categories Derived from the Ecotoxicity Profile

Grouping can also be derived from the observed nanoparticles' effects on different test species using different assays. The toxic effects of the 12 eNPs in MODERN's library were compared by grouping the different eNPs according to type of assay and the corresponding EC_{50} values (Table 9.5). Despite the fact that different species and assays were used, similarities in terms of toxicity emerged. Soluble metal oxide eNPs (CuO and ZnO) were the most toxic to all the species. Similarly, Pd and Co_3O_4 were toxic to alga and bacteria relatively low concentrations. The only non-toxic (i.e., passive) eNP to all species in all testing conditions was MgO. The remaining eNPs (Al_2O_3, Co_3O_4, Fe_3O_4, Mn_3O_4, SiO_2, TiO_2) showed no toxic effects at concentrations below 100 mg/l. The algal growth inhibition assay was the most sensitive. According to this assay, MgO and Sb_2O_3 did not show any toxicity even at very high concentrations (100 mg/l). In contrast, CuO, ZnO and Pd showed growth inhibition at very low concentrations (<1 mg/L).

Using a data-driven grouping approach based on community detection on complex networks three categories of eNPs were automatically detected. Complex networks are graphs containing a set of nodes, representing nanoparticles, and a set of edges connecting pairs of nodes [55]. Data used to build the network were standardized by subtracting their mean and dividing by their standard deviation, and then the Euclidean distance between each pair of nanoparticles was computed. Distances were finally converted into weights in a range between 0 and 1, in order to obtain a completely connected network with weighted edges.

The grouping of nanoparticles inside the network was done looking for what it is called the analysis of the community structure of the networks [18], i.e. a partition of the network into communities, which are subsets of nodes more strongly connected between them than with the rest of the nodes in the network.

The drawing of the network in Fig. 9.11 as well as the analysis of their community structure was done using the software *Gephi* [5]. *Gephi* looks for community

Table 9.5 Expert criteria categorization of eNPs based on the toxicity values (EC_{50} or MBC, mg compound/L) to bacteria, protozoa and algae. All NPs were tested in nominal concentrations from 0.01 up to 100 mg/L

EC_{50} mg compound/L	**72 h EC_{50}**	**24 h EC_{50}**	**30 min EC_{50}**
Organisms:	Algae	Protozoa	Bacteria
Species:	*Pseudokirchneriella subcapitata*	*Tetrahymena thermophila*	*Vibrio fischeri* (*G-*)
Exposure medium:	Mineral medium	DI water	2 % NaCl
0.1-1	**CuO, ZnO, Pd**	**None**	**None**
>1-10	**Co_3O_4, Fe_3O_4, Mn_3O_4, TiO_2**	**CuO, ZnO**	**CuO**
>10-100	**Al_2O_3, SiO_2, WO_3**	**Fe_3O_4, TiO_2**	**ZnO, Pd, WO_3, Sb_2O_3**
>100	**MgO, Sb_2O_3**	**Al_2O_3, Co_3O_4, MgO, Mn_3O_4, Pd, Sb_2O_3, SiO_2, WO_3**	**Al_2O_3, Co_3O_4, Fe_3O_4, MgO, Mn_3O_4, SiO_2, TiO_2**

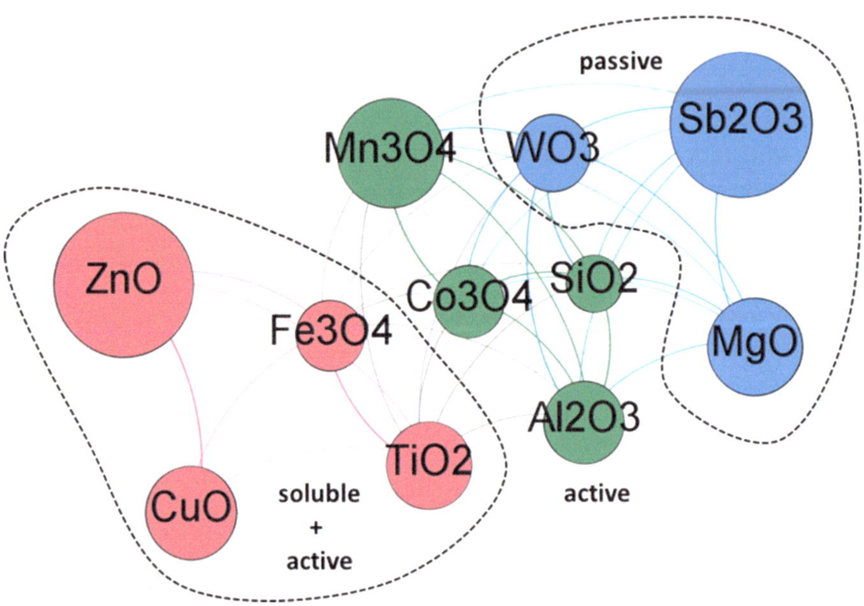

Fig. 9.11 Categories identified from the integrated (algae + protozoa + bacteria) ecotoxicity profile. Color codes correspond to different eNP categories

structure maximizing one of the most successful quality functions for community detection, the modularity function [56], which evaluates the quality of a network's partition into communities.

Interestingly, the category that contains *soluble* and *active* eNPs, formed by {ZnO, CuO, Fe_3O_4 and TiO_2}, is consistent with the grouping observed in the trend analysis of electronic structure descriptors (Fig. 9.8). The second category that comprises {Mn_3O_4, Co_3O_4, SiO_2, and Al_2O_3 also corresponds to *active* eNPs. Finally, the third category that includes {WO_3, MgO and Sb_2O_3} corresponds to eNPs which are *passive* from the ecotoxicity viewpoint (i.e., EC_{50} > 100 mg/L for at least two species).

9.5.1.3 Categories Identified from Integrated Structure-Physicochemical and Ecotoxicological Data

Relevant eNP categories can be identified from data by integrating heterogeneous information related to multiple aspects of the nanomaterial. Fig. 9.12 depicts the categories identified using complex network analysis after integrating three

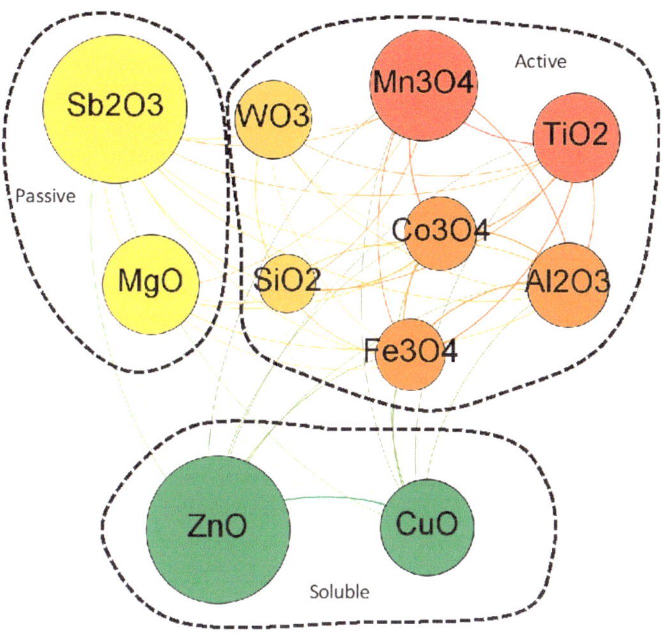

Fig. 9.12 Grouping obtained using complex network analysis techniques. Data used for grouping includes size-dependent nanodescriptors, physicochemical properties (BET size, ZP in DI water, oxidation state) and ecotoxicity profile for algae, bacteria and protozoa. The size of each node is proportional to nanoparticle's BET size. Colors correspond to communities (groups) of eNPs identified using modularity

different types of information including structure, physicochemical properties and ecotoxicity. The *soluble* category includes ZnO and CuO whose toxicity is driven by solubility. The category corresponding to the *active* nanoparticles is formed by three subgroups with different activity levels and includes Fe_3O_4, TiO_2, Mn_3O_4, Co_3O_4, SiO_2, Al_2O_3 and WO_3. Finally, the *passive* category is formed by Sb_2O_3 and MgO.

9.5.2 Integrated Hazard Ranking for Risk Assessment.

Current nano(eco)toxicological knowledge is by far not sufficient to tackle all the issues involved in the hazard evaluation of eNPs. Compared with single-test organism systems, which are already quite complicated covering a myriad of interactions of nanoparticle-organism-environmental factors, ecosystems are by far more complicated involving thousands of species exposed to natural and engineered nanoparticles in various combinations and varying environmental conditions. The integration within a hazard-ranking framework of the information gathered from reliable experimental data and from reliable models should provide cost-effective tools to evaluate the environmental and human health impact of eNPs.

For eNPs, only few MCDA approaches have been reported [26, 44]. However, these previous approaches do not build on the robust order-preserving foundation of partial order theory. Order-preserving should be interpreted in this context in the sense that if new information is entered, the order of the original information will be preserved. This order preserving approach has good potential for hazard assessment, as it allows maintaining previous decisions when new data arrive, in this way refining the decision.

In regard to hazard assessment, the results of tests with eNPs – on characterization, exposure and toxicity – are almost always partial, e.g. either do not have a full set of data for each eNP, the data come from different sources, or the data are not directly comparable because it is unknown how to extrapolate data between eNPs. The problem is complex and the information in form of data tends to be limited, so it is highly important to be aware of uncertainties and to select proper methods that are suited for handling uncertain and ill-defined problems. Here, we will focus on the partial and total order attribute value models to provide a robust framework, where order (ranks/groups) is preserved even when more complex approaches are enforced.

9.5.2.1 Partial Order (POAV) and Total Order (TOAV) Attribute Value Models

As mentioned in the introduction, transfer of knowledge between eNPs is important, partly because there may be insufficient knowledge in regard to a hazard characterization for the individual eNPs, partly because transferring knowledge between eNPs (when relevant) will save resources and, hence, optimize the hazard assessment efforts for example by prioritizing materials. One way to transfer knowledge is by being able to compare information between eNPs by ranking or grouping this information [58].

It should be mentioned that other approaches relevant to hazard decision making (and grouping and ranking) have used more refined tools based on various linear based methods, e.g. [10, 65]. Common for these approaches is that they favor tools which provide advanced answers, but also have a number of pre-requisites and constraints that may inhibit their usefulness in this context when using multiple source data.

Given the values assigned to the attributes, a ranking can be performed. Since there is no preference between attributes, i.e. one attribute is not of different importance than another attribute (see, [7]), the rankings that can be done in the partial ordering are robust rankings because they are independent of weighting factors and ranking functions (e.g. using an analytical hierarchy process). This partial order model assumes a known monotonic relation between the attribute value and the importance for the sub-problem. Thus, an increasing attribute value is assumed to give either a higher or lower hazard level for the whole value interval of the attribute. The POAV model fails if, for the same attribute, there is an increasing hazard level for increasing attribute level for some values and a decreasing hazard level for increasing attribute values for other attribute values.

Total Order Attribute Value models assume a known weighting and functional relation between all attribute values that can yield a final rank of hazard (e.g. using outranking). The rankings generated by a POAV model will be reproduced if the same attribute values are used in a TOAV model. But where the POAV will leave conflicting ranks as an open question, the TOAV will rank all eNPs using Multi Attribute Utility Theory (MAUT). This analysis needs to assume multi attribute preference functions to be valid. The challenge is that these functions are difficult to define and they will control the result. If there is sufficient information to calibrate the ranking model using the relationship between attribute values and "true" rank, then the functions can be found and evaluated using e.g. the UTA method [30]. However, it is difficult to establish so much information in case of eNPs. Basically, there are two main types of preference functions: *additive functions*, where the contribution from one attribute (preference) is independent of the contribution from other attributes, and *multiplicative functions*, where the contribution from an attribute depends on the contribution from other attributes.

The partial order rank can be combined with a categorization to investigate whether one category tends to be ranked above/below another category [66]. The difference between categories and attributes for the ranking is that categories can be any type of labelling, while attributes need to have an order relation. Any kind of category can be applied to look for patterns between the ranking and the category.

9.5.2.2 Example: Ranking of eNPs Using the POAV Model

The eNPs presented in Table 9.4 can be hazard ranked through their attribute values, i.e. by assuming that a lower EC_{50} value poses a higher hazard than a higher EC_{50} value. Each attribute can be ranked as show in yellow rankings in Fig. 9.13, and a joint rank, that only makes rankings between two eNPs that have the same rank for all three attributes, is shown as a Hasse-diagram. For example, the rank ZnO > TiO_2 is true for all three attributes (cell tests), while both the rank ZnO > CuO and CuO < ZnO are present for the single attributes, so ZnO and CuO are not ranked in the Hasse-diagram (Fig. 9.13).

The rankings in the Hasse-diagram are "robust" because they are independent on any preference between the attributes. In the Hasse-diagram, there are 45 robust eNP rankings out of the total $12 \times 11/2 = 66$ possible pairs of eNPs to be ranked.

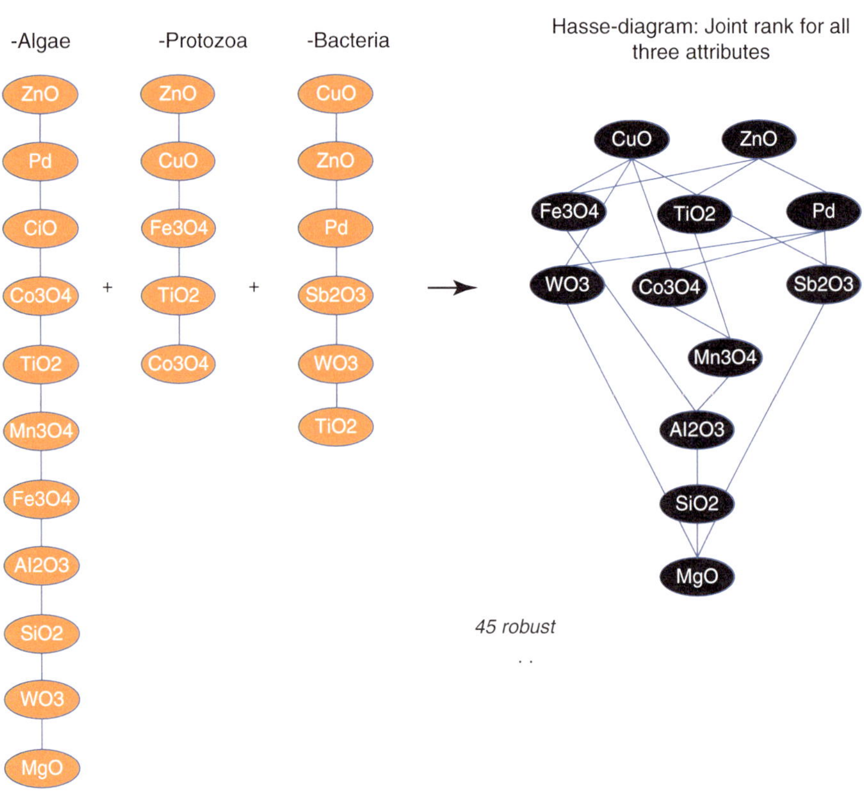

Fig. 9.13 Rank of each of three sets of toxicity data and the combined rank of the same data. Some NMs are equal ranked for respectively Algae as (MgO, Sb_2O_3), Protozoa as (Co_3O_4, Sb_2O_3, Mn_3O_4, Pb, SiO_2, Al_2O_3, WO_3, MgO, Sb_2O_3), Bacteria as (TiO_2, Fe_3O_4, Co_3O_4, Mn_3O_4, SiO_2, Al_2O_3, MgO)

Thus, 68 % of all rankings between the eNPs can be made without any assumed weighting of the attributes in relation to each other or any preference function that aggregate the attributes to one rank.

9.5.2.3 Risk Ranking

Following analyses of the material and hazard attributes, it is possible to proceed to the actual risk analysis, obviously only if risk is associated with the material or hazard in a known way. Apart from ranking of the known risk among materials, it is also possible to some extent to rank an unknown material. The ranking-grouping aspect can be used to assess possible risks if a set of benchmarking materials are available. A benchmark is an eNP for which there exists knowledge about descriptors (e.g. material characteristics) and the related risk. By ranking "unknown" eNPs material-characteristics together with benchmarks, knowledge from benchmarks can be used to interpret the unknown eNPs. In this case, the unknown eNPs will be ranked in a bounded risk banding interval, i.e. depending on the benchmarking materials the top boundary condition (known) can be identified and all unknown eNPs that have the combination of descriptors below this are not likely to cause risk.

In summary, the ranking approach can be used to derive a risk banding level and support risk decision, even when a continuous model cannot be applied to the data. This is a key point in the grouping and read across for risk assessment of nanomaterials. The ranking approach may further support a priori or posterior grouping of the materials and in this way optimize the risk characterization. It will likely be especially useful to identify materials of different mode of action, defined as different concentration-response curve parameters.

9.6 Towards Safe-by-Design eNPs

The production of safe nanomaterials requires controlling the property profile of new nanoparticles during their synthesis and the understanding of the interactions between nanoparticle structure, properties and biological activity. Safe-by-design strategies for nanoparticles can be implemented by introducing changes in their structure, which in turn modify their intrinsic properties and biological effects.

It is well known that ZnO nanoparticles dissolve in the biological medium. Accordingly, the toxicological effects of ZnO NPs have an important relationship with particle dissolution, which start in the tissue culture or environmental medium and, making its way inside the cells, to the different organs of living organisms [4, 31, 81, 83]. The demonstration of Zn^{2+} dissolution leading to an increased cytotoxicity points out the necessity for re-engineering ZnO through surface modification such as Fe doping that might slow down the dissolution effect due to its stronger coordination with neighboring Zn in the crystal structure consequently improving the cytotoxicity profile [12, 16, 84].

Similar to the use of the inherent dissolution property of ZnO as a toxicity paradigm, TiO_2, an insoluble nanoparticle, is prone to light activation and has a potential to induce toxicity under UV exposure conditions [25]. The electrons are excited to the conduction band of TiO_2 creating a hole in the valence band via UV light irradiation. Materials that are capable of separating an e^-/h^+ pair in the electronic bands are technologically important however are critically hazardous to the environment since the e^-/h^+ pair is able to interact with surrounding H_2O and molecular oxygen to generate ROS ($HO^•$ radical and/or superoxide) [25, 46, 70]. To study and control bioactivity under light irradiation, TiO_2 could be doped so that band gap values could be tuned in such a way that the electronic excitation is possible at low energy wavelength.

A third toxicity paradigm could be the overlap of the conduction band energy of metal oxides with the oxidative redox potential of reactions within the cell. Re-engineering the conduction band energy would allow safer designing of the particles.

The development of in silico models, relating the structure and properties of nanomaterials with their biological activity, will be key for the implementation of safe-by-design strategies in future generations of nanomaterials [85]. From data analysis and structure/property-activity modelling, toxicity can be related to a number of physicochemical properties of nanoparticles. The combination of in vitro toxicity assessment with the application of in silico models has the potential to contribute to establish a new paradigm for development and testing of nanomaterials. The integrative approach implemented in MODERN has demonstrated the feasibility of developing nanoparticle descriptors that can be subsequently used for the establishment of predictive nanotoxicity models, which in turn can inform hazard ranking and nanosafety assessment

Acknowledgments Authors acknowledge the financial support received from the European Commission through the FP7 MODERN Project (Contract No. 309314). RR also acknowledges the support received from Generalitat de Catalunya (2014SGR 1352).

References

1. Arts JHE, Hadi M, Irfan M-A, Keene AM, Kreiling R, Lyon D, Maier M, Michel K, Petry T, Sauer UG, Warheit D, Wiench K, Wohlleben W, Landsiedel R (2015) A decision-making framework for the grouping and testing of nanomaterials (DF4nanoGrouping). Regul Toxicol Pharmacol 71:S1–27. doi:10.1016/j.yrtph.2015.03.007
2. Aruoja V, Dubourguier H-C, Kasemets K, Kahru A (2009) Toxicity of nanoparticles of CuO, ZnO and TiO2 to microalgae Pseudokirchneriella subcapitata. Sci Total Environ 407:1461–1468. doi:10.1016/j.scitotenv.2008.10.053

3. Aruoja V, Pokhrel S, Sihtmäe M, Mortimer M, Mädler L, Kahru A (2015) Toxicity of 12 metal-based nanoparticles to algae, bacteria and protozoa. Environ Sci Nano 2:630–644. doi:10.1039/C5EN00057B
4. Bai W, Zhang Z, Tian W, He X, Ma Y, Zhao Y, Chai Z (2009) Toxicity of zinc oxide nanoparticles to zebrafish embryo: a physicochemical study of toxicity mechanism. J Nanopart Res 12:1645–1654. doi:10.1007/s11051-009-9740-9
5. Bastian M, Heymann S, Jacomy M (2009) Gephi: an open source software for exploring and manipulating networks. ICWSM 8:361–362
6. Bondarenko O, Juganson K, Ivask A, Kasemets K, Mortimer M, Kahru A (2013) Toxicity of Ag, CuO and ZnO nanoparticles to selected environmentally relevant test organisms and mammalian cells in vitro: a critical review. Arch Toxicol 87:1181–1200. doi:10.1007/s00204-013-1079-4
7. Brüggemann R, Patil G (2011) Ranking and prioritization for multsi-indicator systems: Introduction to partial order applications. Springer, New York
8. Chattaraj PK, Giri S, Duley S (2011) Update 2 of: electrophilicity index. Chem Rev 111:PR43–PR75. doi:10.1021/cr100149p
9. Choi O, Hu Z (2008) Size dependent and reactive oxygen species related nanosilver toxicity to nitrifying bacteria. Environ Sci Technol 42:4583–4588. doi:10.1021/es703238h
10. Cohen Y, Rallo R, Liu R, Liu HH (2013) In silico analysis of nanomaterials hazard and risk. Acc Chem Res 46:802–812. doi:10.1021/ar300049e
11. Cronin MTD, Schultz TW (1997) Validation of Vibrio fisheri acute toxicity data: mechanism of action-based QSARs for non-polar narcotics and polar narcotic phenols. Sci Total Environ 204:75–88. doi:10.1016/S0048-9697(97)00179-4
12. Damoiseaux R, George S, Li M, Pokhrel S, Ji Z, France B, Xia T, Suarez E, Rallo R, Mädler L, Cohen Y, Hoek EMV, Nel A (2011) No time to lose – high throughput screening to assess nanomaterial safety. Nanoscale 3:1345–1360. doi:10.1039/c0nr00618a
13. Eom HJ, Roca CP, Roh JY, Chatterjee N, Jeong JS, Shim I, Kim HM, Kim PJ, Choi K, Giralt F, Choi J (2015) A systems toxicology approach on the mechanism of uptake and toxicity of MWCNT in *Caenorhabditis elegans*. Chem Biol Interact 239:153–163. doi:10.1016/j.cbi.2015.06.031
14. Esteban G, Tellez C, Bautista L (1992) The indicator value of Tetrahymena thermophila populations in the activated sludge process. Acta Protozool 31:129–132
15. Ewald PP (1921) Die Berechnung optischer und elektrostatischer Gitterpotentiale. Ann Phys 369:253–287. doi:10.1002/andp.19213690304
16. George S, Pokhrel S, Xia T, Gilbert B, Ji Z, Schowalter M, Rosenauer A, Damoiseaux R, Bradley KA, Mädler L, Nel AE (2010) Use of a rapid cytotoxicity screening approach to engineer a safer zinc oxide nanoparticle through iron doping. ACS Nano 4:15–29. doi:10.1021/nn901503q
17. George S, Pokhrel S, Ji Z, Henderson BL, Xia T, Li L, Zink JI, Nel AE, Mädler L (2011) Role of Fe doping in tuning the band gap of TiO2 for the photo-oxidation-induced cytotoxicity paradigm. J Am Chem Soc 133:11270–11278. doi:10.1021/ja202836s
18. Girvan M, Newman MEJ (2002) Community structure in social and biological networks. Proc Natl Acad Sci U S A 99:7821–7826. doi:10.1073/pnas.122653799
19. Goeman JJ, Bühlmann P (2007) Analyzing gene expression data in terms of gene sets: methodological issues. Bioinformatics 23:980–987. doi:10.1093/bioinformatics/btm051
20. Grimme S, Antony J, Ehrlich S, Krieg H (2010) A consistent and accurate ab initio parametrization of density functional dispersion correction (DFT-D) for the 94 elements H-Pu. J Chem Phys 132:154104. doi:10.1063/1.3382344
21. Gupta A, Condit C, Qian X (2010) BioDB: an ontology-enhanced information system for heterogeneous biological information. Data Knowl Eng 69:1084–1102
22. Hartmann NB, Engelbrekt C, Zhang J, Ulstrup J, Kusk KO, Baun A (2012) The challenges of testing metal and metal oxide nanoparticles in algal bioassays: titanium dioxide and gold nanoparticles as case studies. Nanotoxicology 7:1082–1094

23. Hastings J, Jeliazkova N, Owen G, Tsiliki G, Munteanu CR, Steinbeck C, Willighagen E (2015) eNanoMapper: harnessing ontologies to enable data integration for nanomaterial risk assessment. J Biomed Semantics 6:10. doi:10.1186/s13326-015-0005-5
24. Heinlaan M, Ivask A, Blinova I, Dubourguier H-C, Kahru A (2008) Toxicity of nanosized and bulk ZnO, CuO and TiO2 to bacteria Vibrio fischeri and crustaceans *Daphnia magna* and *Thamnocephalus platyurus*. Chemosphere 71:1308–1316. doi:10.1016/j.chemosphere.2007.11.047
25. Hoffmann MR, Martin ST, Choi W, Bahnemann DW (1995) Environmental applications of semiconductor photocatalysis. Chem Rev 95:69–96. doi:10.1021/cr00033a004
26. Hristozov DR, Gottardo S, Cinelli M, Isigonis P, Zabeo A, Critto A, Van Tongeren M, Tran L, Marcomini A (2014) Application of a quantitative weight of evidence approach for ranking and prioritising occupational exposure scenarios for titanium dioxide and carbon nanomaterials. Nanotoxicology 8:117–131. doi:10.3109/17435390.2012.760013
27. Ioannidis JPA, Khoury MJ (2011) Improving validation practices in “omics” research. Science 334:1230–1232. doi:10.1126/science.1211811
28. ISO 21338:2010 – Water quality – Kinetic determination of the inhibitory effects of sediment, other solids and coloured samples on the light emission of Vibrio fischeri (kinetic luminescent bacteria test) [WWW Document], n.d. URL http://www.iso.org/iso/catalogue_detail.htm?csnumber=44880. Accessed 17 Feb 2016
29. Ivask A, Kurvet I, Kasemets K, Blinova I, Aruoja V, Suppi S, Vija H, Käkinen A, Titma T, Heinlaan M, Visnapuu M, Koller D, Kisand V, Kahru A (2014) Size-dependent toxicity of silver nanoparticles to bacteria, yeast, algae, crustaceans and mammalian cells in vitro. PLoS One 9:e102108. doi:10.1371/journal.pone.0102108
30. Jacquet-Lagreze E, Siskos J (1982) Assessing a set of additive utility functions for multicriteria decision-making, the UTA method. Eur J Oper Res 10:151–164. doi:10.1016/0377-2217(82)90155-2
31. Ji Z, Jin X, George S, Xia T, Meng H, Wang X, Suarez E, Zhang H, Hoek EMV, Godwin H, Nel AE, Zink JI (2010) Dispersion and stability optimization of TiO2 nanoparticles in cell culture media. Environ Sci Technol 44:7309–7314. doi:10.1021/es100417s
32. Kahru A, Dubourguier H-C (2010) From ecotoxicology to nanoecotoxicology. Toxicology 269:105–119. doi:10.1016/j.tox.2009.08.016
33. Kahru A, Dubourguier H, Blinova I, Ivask A, Kasemets K (2008) Biotests and biosensors for ecotoxicology of metal oxide nanoparticles: a minireview. Sensors 8:5153–5170
34. Kammler HK, Mädler L, Pratsinis SE (2001) Flame synthesis of nanoparticles. Chem Eng Technol 24:583–596. doi:10.1002/1521-4125(200106)24:6<583::AID-CEAT583>3.0.CO;2-H
35. Kar S, Gajewicz A, Puzyn T, Roy K (2014) Nano-quantitative structure-activity relationship modeling using easily computable and interpretable descriptors for uptake of magnetofluorescent engineered nanoparticles in pancreatic cancer cells. Toxicol In Vitro 28:600–606. doi:10.1016/j.tiv.2013.12.018
36. Karlsson HL, Gustafsson J, Cronholm P, Moller L (2009) Size-dependent toxicity of metal oxide particles-A comparison between nano- and micrometer size. Toxicol Lett 188:112–118
37. Kasemets K, Ivask A, Dubourguier H-C, Kahru A (2009) Toxicity of nanoparticles of ZnO, CuO and TiO2 to yeast *Saccharomyces cerevisiae*. Toxicol In Vitro 23:1116–1122. doi:10.1016/j.tiv.2009.05.015
38. Katritzky AR, Lobanov VS, Karelson M (1995) QSPR: the correlation and quantitative prediction of chemical and physical properties from structure. Chem Soc Rev 24:279. doi:10.1039/cs9952400279
39. Kemmler JA, Pokhrel S, Birkenstock J, Schowalter M, Rosenauer A, Bârsan N, Weimar U, Mädler L (2012) Quenched, nanocrystalline In4Sn3O12 high temperature phase for gas sensing applications. Sens Actuators B 161:740–747. doi:10.1016/j.snb.2011.11.026
40. Khatri P, Sirota M, Butte AJ (2012) Ten years of pathway analysis: current approaches and outstanding challenges. PLoS Comput Biol 8:e1002375. doi:10.1371/journal.pcbi.1002375
41. Kohonen T (1990) The self-organizing map. Proc IEEE 78:1464–1480

42. Krug HF, Wick P (2011) Nanotoxicology: an interdisciplinary challenge. Angew Chem Int Ed Engl 50:1260–1278. doi:10.1002/anie.201001037
43. Lay JO, Liyanage R, Borgmann S, Wilkins CL (2006) Problems with the "omics". TrAC Trends Anal Chem 25:1046–1056. doi:10.1016/j.trac.2006.10.007
44. Linkov I, Satterstrom F, Steevens J, Ferguson E, Pleus R (2007) Multi-criteria decision analysis and environmental risk assessment for nanomaterials. J Nanopart Res 9:543–554
45. Liu R, Zhang HY, Ji ZX, Rallo R, Xia T, Chang CH, Nel A, Cohen Y (2013) Development of structure-activity relationship for metal oxide nanoparticles. Nanoscale 5:5644–5653. doi:10.1039/c3nr01533e
46. Long TC, Saleh N, Tilton RD, Lowry GV, Veronesi B (2006) Titanium dioxide (P25) produces reactive oxygen species in immortalized brain microglia (BV2): implications for nanoparticle neurotoxicity †. Environ Sci Technol 40:4346–4352. doi:10.1021/es060589n
47. Mädler L (2004) Liquid-fed aerosol reactors for one-step synthesis of nano-structured particles. KONA Powder Part J 22:107–120. doi:10.14356/kona.2004014
48. Marchese Robinson RL, Cronin MTD, Richarz A-N, Rallo R (2015) An ISA-TAB-Nano based data collection framework to support data-driven modelling of nanotoxicology. Beilstein J Nanotechnol 6:1978–1999. doi:10.3762/bjnano.6.202
49. Mortimer M, Kasemets K, Heinlaan M, Kurvet I, Kahru A (2008) High throughput kinetic Vibrio fischeri bioluminescence inhibition assay for study of toxic effects of nanoparticles. Toxicol In Vitro 22:1412–1417. doi:10.1016/j.tiv.2008.02.011
50. Mortimer M, Kasemets K, Kahru A (2010) Toxicity of ZnO and CuO nanoparticles to ciliated protozoa Tetrahymena thermophila. Toxicology 269:182–189. doi:10.1016/j.tox.2009.07.007
51. Mortimer M, Kahru A, Slaveykova VI (2014) Uptake, localization and clearance of quantum dots in ciliated protozoa Tetrahymena thermophila. Environ Pollut 190:58–64. doi:10.1016/j.envpol.2014.03.021
52. Neese F (2012) The ORCA program system. Wiley Interdiscip Rev Comput Mol Sci 2:73–78. doi:10.1002/wcms.81
53. Nel A, Mädler L, Velegol D, Xia T, Hoek E, Somasundaran P, Klaessig F, Castranova V, Thompson M (2009) Understanding biophysicochemical interactions at the nano–bio interface. Nat Mater 8:543–557
54. Netzeva TI, Schultz TW (2005) QSARs for the aquatic toxicity of aromatic aldehydes from Tetrahymena data. Chemosphere 61:1632–1643. doi:10.1016/j.chemosphere.2005.04.040
55. Newman M (2010) Networks: an introduction. Oxford University Press Inc, New York
56. Newman M, Girvan M (2004) Finding and evaluating community structure in networks. Phys Rev E 69:026113. doi:10.1103/PhysRevE.69.026113
57. OECD (2006) OECD Guidelines for the Testing of Chemicals, Section 2, Test No. 201: Freshwater Alga and Cyanobacteria, Growth Inhibition Test. Organization for Economic Cooperation and Development, Paris
58. Oomen AG, Bleeker EAJ, Bos PMJ, van Broekhuizen F, Gottardo S, Groenewold M, Hristozov D, Hund-Rinke K, Irfan M-A, Marcomini A, Peijnenburg WJGM, Rasmussen K, Jiménez AS, Scott-Fordsmand JJ, van Tongeren M, Wiench K, Wohlleben W, Landsiedel R (2015) Grouping and read-across approaches for risk assessment of nanomaterials. Int J Environ Res Public Health 12:13415–13434. doi:10.3390/ijerph121013415
59. Parr RG, Pearson RG (1983) Absolute hardness: companion parameter to absolute electronegativity. J Am Chem Soc 105:7512–7516. doi:10.1021/ja00364a005
60. Parr RG, Donnelly RA, Levy M, Palke WE (1978) Electronegativity: the density functional viewpoint. J Chem Phys 68:3801. doi:10.1063/1.436185
61. Passagne I, Morille M, Rousset M, Pujalté I, L'azou B (2012) Implication of oxidative stress in size-dependent toxicity of silica nanoparticles in kidney cells. Toxicology 299:112–124. doi:10.1016/j.tox.2012.05.010
62. Perdew JP, Burke K, Ernzerhof M (1996) Generalized gradient approximation made simple. Phys Rev Lett 77:3865–3868. doi:10.1103/PhysRevLett.77.3865

63. Plimpton S (1995) Fast parallel algorithms for short-range molecular dynamics. J Comput Phys 117:1–19. doi:10.1006/jcph.1995.1039
64. Puzyn T, Leszczynska D, Leszczynski J (2009) Toward the development of "Nano-QSARs": advances and challenges. Small 5:2494–2509
65. Puzyn T, Rasulev B, Gajewicz A, Hu X, Dasari TP, Michalkova A, Hwang H-M, Toropov A, Leszczynska D, Leszczynski J (2011) Using nano-QSAR to predict the cytotoxicity of metal oxide nanoparticles. Nat Nanotechnol 6:175–178. doi:10.1038/nnano.2011.10
66. Restrepo G, Weckert M, Brüggemann R, Gerstmann S, Frank H (2008) Ranking of refrigerants. Environ Sci Technol 42:2925–2930. doi:10.1021/es7026289
67. Rushton EK, Jiang J, Leonard SS, Eberly S, Castranova V, Biswas P, Elder A, Han X, Gelein R, Finkelstein J, Oberdörster G (2010) Concept of assessing nanoparticle hazards considering nanoparticle dosemetric and chemical/biological response metrics. J Toxicol Environ Heal Part A 3:445–461
68. Savolainen K, Backman U, Brouwer D, Fadeel B, Fernandes T, Kuhlbusch T, Landsiedel R, Lynch I, Pylkkänen L (2013) Nanosafety in Europe 2015–2025: Towards Safe and Sustainable Nanomaterials and Nanotechnology Innovations. Helsinki, Finish Institute of Occupational Health
69. Singh J (2001) Semiconductor devices. Basic principles. Wiley, New York
70. Sinha RP, Häder D-P (2002) UV-induced DNA damage and repair: a review. Photochem Photobiol Sci 1:225–236. doi:10.1039/b201230h
71. Suzuki R, Shimodaira H (2006) Pvclust: an R package for assessing the uncertainty in hierarchical clustering. Bioinformatics 22:1540–1542
72. Tani, T. 2003. Flame Spray Pyrolysis of Zinc Oxide/silica Particles. PhD Thesis, Swiss Federal Institute of Technology, Zurich. Dissertation ETHNo. 15266, 1-116
73. Tani T, Mädler L, Pratsinis SE (2002) Homogeneous ZnO nanoparticles by flame spray pyrolysis. J Nanopart Res 4:337–343. doi:10.1023/A:1021153419671
74. Teoh WY, Amal R, Mädler L (2010) Flame spray pyrolysis: an enabling technology for nanoparticles design and fabrication. Nanoscale 2:1324–1347. doi:10.1039/c0nr00017e
75. Thomas DG, Pappu RV, Baker NA (2011) NanoParticle Ontology for cancer nanotechnology research. J Biomed Inform 44:59–74. doi:10.1016/j.jbi.2010.03.001
76. Thomas DG, Gaheen S, Harper SL, Fritts M, Klaessig F, Hahn-Dantona E, Paik D, Pan S, Stafford GA, Freund ET, Klemm JD, Baker NA (2013) ISA-TAB-Nano: a specification for sharing nanomaterial research data in spreadsheet-based format. BMC Biotechnol 13:2. doi:10.1186/1472-6750-13-2
77. Toropov AA, Toropova AP, Puzyn T, Benfenati E, Gini G, Leszczynska D, Leszczynski J (2013) QSAR as a random event: modeling of nanoparticles uptake in PaCa2 cancer cells. Chemosphere 92:31–37. doi:10.1016/j.chemosphere.2013.03.012
78. Toropova AP, Toropov AA, Rallo R, Leszczynska D, Leszczynski J (2015) Optimal descriptor as a translator of eclectic data into prediction of cytotoxicity for metal oxide nanoparticles under different conditions. Ecotoxicol Environ Saf 112:39–45. doi:10.1016/j.ecoenv.2014.10.003
79. von Moos N, Slaveykova VI (2014) Oxidative stress induced by inorganic nanoparticles in bacteria and aquatic microalgae – state of the art and knowledge gaps. Nanotoxicology 8:605–630. doi:10.3109/17435390.2013.809810
80. Weigend F, Ahlrichs R (2005) Balanced basis sets of split valence, triple zeta valence and quadruple zeta valence quality for H to Rn: design and assessment of accuracy. Phys Chem Chem Phys 7:3297–3305. doi:10.1039/b508541a
81. Wesselkamper SC, Chen LC, Gordon T (2001) Development of pulmonary tolerance in mice exposed to zinc oxide fumes. Toxicol Sci 60:144–151. doi:10.1093/toxsci/60.1.144
82. Wolf D, Keblinski P, Phillpot SR, Eggebrecht J (1999) Exact method for the simulation of Coulombic systems by spherically truncated, pairwise r[sup −1] summation. J Chem Phys 110:8254. doi:10.1063/1.478738
83. Xia T, Kovochich M, Liong M, Mädler L, Gilbert B, Shi H, Yeh J, Zink J, Nel A (2008) Comparison of the mechanism of toxicity of zinc oxide and cerium oxide nanoparticles based on dissolution and oxidative stress properties. ACS Nano 2:2121–2134

84. Xia T, Zhao Y, Sager T, George S, Pokhrel S, Li N, Schoenfeld D, Meng H, Lin S, Wang X, Wang M, Ji Z, Zink JI, Mädler L, Castranova V, Lin S, Nel AE (2011) Decreased dissolution of ZnO by iron doping yields nanoparticles with reduced toxicity in the rodent lung and zebrafish embryos. ACS Nano 5:1223–1235. doi:10.1021/nn1028482
85. Zhang H, Ji Z, Xia T, Meng H, Low-Kam C, Liu R, Pokhrel S, Lin S, Wang X, Liao Y-P, Wang M, Li L, Rallo R, Damoiseaux R, Telesca D, Mädler L, Cohen Y, Zink JI, Nel AE (2012) Use of metal oxide nanoparticle band gap to develop a predictive paradigm for oxidative stress and acute pulmonary inflammation. ACS Nano 6:4349–4368. doi:10.1021/nn3010087

Chapter 10
Compilation of Data and Modelling of Nanoparticle Interactions and Toxicity in the NanoPUZZLES Project

Andrea-Nicole Richarz, Aggelos Avramopoulos, Emilio Benfenati, Agnieszka Gajewicz, Nazanin Golbamaki Bakhtyari, Georgios Leonis, Richard L Marchese Robinson, Manthos G Papadopoulos, Mark TD Cronin, and Tomasz Puzyn

Abstract The particular properties of nanomaterials have led to their rapidly increasing use in diverse fields of application. However, safety assessment is not keeping pace and there are still gaps in the understanding of their hazards. Computational models predicting nanotoxicity, such as (quantitative) structure-activity relationships ((Q)SARs), can contribute to safety evaluation, in line with general efforts to apply alternative methods in chemical risk assessment. Their development is highly dependent on the availability of reliable and high quality experimental data, both regarding the compounds' properties as well as the measured toxic effects. In particular, "nano-QSARs" should take the nano-specific characteristics into account. The information compiled needs to be well organized, quality con-

A.-N. Richarz (✉) • R.L. Marchese Robinson • M.T.D. Cronin
School of Pharmacy and Biomolecular Sciences, Liverpool John Moores University, Liverpool, UK
e-mail: andrea.richarz.nano@gmail.com

A. Avramopoulos • G. Leonis • M.G. Papadopoulos
Institute of Biology, Pharmaceutical Chemistry and Biotechnology, National Hellenic Research Foundation, Athens, Greece

E. Benfenati • N. Golbamaki Bakhtyari
Laboratory of Environmental Chemistry and Toxicology, Istituto di Ricerche Farmacologiche Mario Negri, Milan, Italy

A. Gajewicz • T. Puzyn
Laboratory of Environmental Chemometrics, Institute for Environmental and Human Health Protection, Faculty of Chemistry, University of Gdańsk, Gdańsk, Poland

L. Tran et al. (eds.), *Modelling the Toxicity of Nanoparticles*, Advances in Experimental Medicine and Biology, Vol. 947, DOI 10.1007/978-3-319-47754-1_10

trolled and standardized. Integrating the data in an overarching, structured data collection aims to (a) organize the data in a way to support modelling, (b) make (meta) data necessary for modelling available, and (c) add value by making a comparison between data from different sources possible.

Based on the available data, specific descriptors can be derived to parameterize the nanomaterial-specific structure and physico-chemical properties appropriately. Furthermore, the interactions between nanoparticles and biological systems as well as small molecules, which can lead to modifications of the structure of the active nanoparticles, need to be described and taken into account in the development of models to predict the biological activity and toxicity of nanoparticles. The EU NanoPUZZLES project was part of a global cooperative effort to advance data availability and modelling approaches supporting the characterization and evaluation of nanomaterials.

Keywords Nanoparticle • Nanomaterial • Toxicity • Interactions • Data compilation • Data quality • Data standardization • Nano-descriptors • Nano-QSAR

10.1 Introduction

The number of nanomaterials and their applications in nanotechnology is increasing rapidly and will grow considerably in the foreseeable future. It is the particular properties of the nanomaterials, based on specific characteristics such as the particle shape and size distribution, surface area and chemistry, solubility and porosity as well as the state of agglomeration and dispersion, which have led to their increased use in diverse fields of application such as medicine, cosmetics, food and textile products, water treatment, catalysis or in construction and electronics.

However, safety assessment of nanoparticles is not keeping pace with their rapid development and there are still gaps in the understanding of the hazards posed by nanomaterials. The basic requirement for establishing a nanoparticle-specific hazard assessment is the availability of reliable and consistent nanotoxicity data, associated with high quality characterization of the identity and structural/physico-chemical properties of the materials investigated. There are many efforts worldwide to elucidate mechanisms of action for nanoparticles and generate toxicity data for diverse endpoints. However, the publication of the data is disperse and thus makes their comparability and uptake for safety assessment difficult.

Risk assessment of all chemicals including nanoparticles is increasingly being supported by computational methods, such as (quantitative) structure–activity relationships ((Q)SARs), based on the understanding that the biological activity of a substance is related to its chemical structure and physico-chemical properties. The development of these in silico approaches is highly dependent on the availability of reliable and high quality experimental data. QSAR models are well established for different endpoints for organic molecules, however, the same techniques cannot necessarily be applied directly to nanoparticles. As such, so called "nano-QSARs" have to take the nano-specific characteristics of nanoparticles into account. Therefore the compilation of data of high quality, both regarding the compound characteristics, particular for nanomaterials, as well as the measured toxic effects, is imperative as the basis for the modelling of nanotoxicity. Ideally these data should be made available in a unified form and in an open, electronic, easily accessible format.

Within the 3-year European NanoPUZZLES project (European Commission 7th Framework Programme, FP7-NMP-2012-SMALL-6, grant agreement no. 309837), from January 2013 to December 2015, the partners University of Gdańsk (Gdańsk, Poland), Istituto di Ricerche Farmacologiche Mario Negri (Milan, Italy), National Hellenic Research Foundation (Athens, Greece), Liverpool John Moores University (Liverpool, England) and BioBaltica (Gdańsk, Poland) have collaborated to advance knowledge and computational methods for modelling the relationships between the structure, properties, molecular interactions and toxicity of nanoparticles. The work has been organised into four complementary thematic areas (the "puzzles"), namely: NanoDATA, collation and evaluation of available physico-chemical and toxicological data for nanoparticles; NanoDESC, descriptors specific to nanoparticles and their characteristic features; NanoINTER, interactions of nanoparticles with biological systems; NanoQSAR, using the knowledge gained from within the first three puzzles to understand the relationships between experimental toxicity data and nanoparticle properties and develop nano-QSAR models. The project has focussed on two groups of nanomaterials: inorganic nanoparticles such as metal oxides and carbon nanomaterials such as fullerenes and fullerene derivatives.

A general approach to data driven modelling of nanomaterial toxicity, starting from the review of available data, is summarized in Fig. 10.1 [25]. The data collection entails consideration of the data quality and suitability of the data for modelling (e.g. availability of key endpoint data for a suitably large number of nanomaterials obtained from a single source) as well as standardization of the format and terminology, essential for nanotoxicity modelling, also in view of the broader application of the data by integration into overarching databases, supporting predictive model development and safety evaluation of nanomaterials.

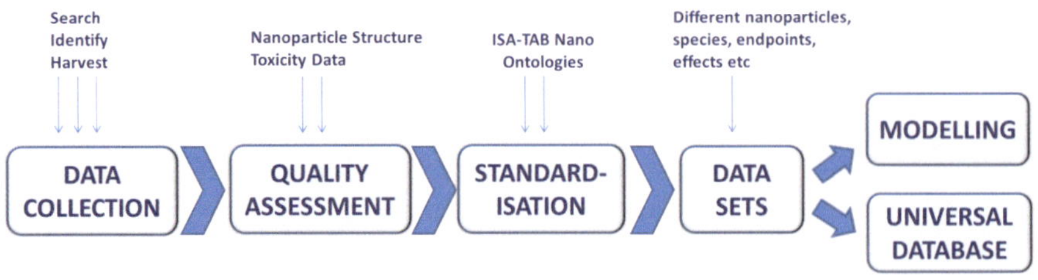

Fig. 10.1 A general approach to data collection to model nanoparticle toxicity

10.2 Data Collection

Reliable information on the structural characteristics and physico-chemical properties of nanomaterials, as well as available experimental data on biological effects, is the basis for the development of nano-QSARs. Therefore, the first step for modelling is to identify and collate available data, from the scientific literature or, for example, other research initiatives generating data on nanomaterials.

Compiling the data in an overarching, even inter-project or global, data collection has the following aims:

(a) to organize the data in a form to optimally support the modelling,
(b) to ensure all (meta)data necessary for modelling are available,
(c) to add value to the available data by making comparison of data from different sources possible and thus to enable the integration of these data from diverse efforts to characterize nanomaterial hazard for the development of models to support nanosafety assessment.

To achieve these aims, particular efforts have been made in the NanoPUZZLES project to identify available data through a comprehensive literature review and these data were evaluated according to the requirements for building a data collection useful for nano-QSAR modelling. In order to merge heterogeneous datasets and make them comparable, a suitable standardised data format and vocabulary is needed.

Within the NanoPUZZLES project, data from more than 300 publications were identified, including data obtained through the NanoBridges project (FP7-PEOPLE-2011-IRSES, grant agreement no. 295128), principally focussing on the endpoints cytotoxicity, genotoxicity and aquatic toxicity for fullerenes, carbon nanotubes, metal/metal oxides and silica nanoparticles. Physico-chemical and structural information which were identified also included size characterisation measurements from multiple techniques such as transmission electron microscopy (TEM) and dynamic light scattering (DLS).

The specific procedures and suggestions for the standardisation and quality assessment of the information collected are described below.

10.2.1 Data Quality and Availability

The quality of an in silico prediction model depends on the quality of the data used for the model development. Therefore the consideration and evaluation of the data quality is a crucial step for modelling.

Data suitability considerations comprise three different areas, which include evaluation of some nanomaterial specific properties, as follows:

- the amount of data and their quality sufficient or necessary for modelling
- the quality and completeness of the nanomaterial structural characterization and physico-chemical properties data,
- the quality of the nanomaterial toxicity data.

Lubinski et al. [14] suggested that at least 10 data points, from a single source, are required for (Q)SAR modelling. A literature survey was performed in NanoPUZZLES and the results confirmed that it is a challenge, in practice, to achieve this criterion. Out of a sample of 363 primary experimental articles considered in 2014, with cytotoxicity and genotoxicity data for nanomaterials, most reported on too few nanomaterials/data points to develop models based on data from a single source. The use of data from different sources would improve the data availability situation, if they were sufficiently comparable. Therefore the recording of (physico-chemical and toxicity) data in a standardized form is the prerequisite to compare data from different sources and possibly allow the development of models based on multiple sources. The standardization of data recording is discussed in the next section.

Lubinski et al. [14] proposed a general set of quality criteria for any kind of nanomaterial experimental data, based on Klimisch scores [9]. The Nanomaterial Registry proposes a set of "compliance levels" [18, 20] for evaluating the curated physico-chemical data. These approaches were considered when developing proposals for data quality and completeness assessment within NanoPUZZLES. In addition, published proposals regarding priority physico-chemical characterization parameters [31, 32] as well as the need to consider the potential for artefacts and misinterpretations in data obtained from nanotoxicology tests [21] was taken into account to develop provisional data quality/completeness schemes for assessing the curated physico-chemical and toxicity data.

The importance of transparency of this evaluation has to be emphasized, i.e., as also recommended for the ToxRTool [29], all outcomes obtained from evaluation of individual checklist criteria as well as the final quality assessment should be reported to allow for critical evaluation of the quality assessment by end-users. The most appropriate manner in which the completeness and quality of curated nanomaterial data should be evaluated remains a critical question (see Marchese Robinson et al. [16]).

10.2.2 Standardization of Data Recording

To make it possible to integrate data(sets) from different sources and experimental measurements into one database, there is a need for standardization of the data format, the vocabulary and metadata used. This also provides the necessary basis to allow for the possible comparison of heterogeneous experimental data.

ISA-TAB-Nano [6, 15, 34], an extension of the ISA-TAB specification for reporting of data from biological experiments based upon a standardized representation of (meta)data [26, 27], was proposed as a global standard for nanomaterial data sharing and has been adopted as such for the nanotoxicity modelling projects. Specific data collection templates, designed to support collection of specific nanosafety data and metadata, were developed within NanoPUZZLES [15]. Columns and rows were pre-defined based upon relevant items of (meta)data capturing key physico-chemical measurements, toxicological data and experimental conditions.

The ISA-TAB-Nano specification is based upon a set of four linked spreadsheet-like tab-delimited text file types, designed to capture different kinds of nanoaterial (meta)data from experimental measurements, the Investigation, Study, Assay and Material files, with a pre-defined file structure and syntax for (meta)data. The Study and Assay files describe the samples and the corresponding assays respectively, including assay conditions and measured values. A Material file records chemical composition information, associated with the original nanomaterial sample, and (as of ISA-TAB-Nano version 1.2) nominal/vendor supplied characteristics. There are two types of studies: "physico-chemical characterization" or "*in vitro* and *in vivo* characterization" studies. For the former, the (derivative) nanomaterial sample is considered "the sample" which is being tested in an assay whilst, for the latter, the (derivative sample of the) biospecimen is considered "the sample" and the nanomaterial is considered an experimental variable whose effect on the assay outcome is being evaluated – or, in the nomenclature of the ISA-TAB-Nano specification, a "factor". The Investigation file links all of the previous three file types together and allows for links to additional metadata, such as primary literature references and links to ontologies. Figure 10.2 shows the concept of the links between different ISA-TAB-Nano file types schematically.

These file types are named according to the following conventions: "i_xxxx.txt" (Investigation files), "s_xxxx.txt" (Study files), "a_xxxx.txt" (Assay files), "m_xxxx.txt" (Material files) where "xxxx" denotes some unique file identifier. Figure 10.3 shows a simplified example of how the linked ISA-TAB-Nano files were used to record data from a literature source in the NanoPUZZLES project. In addition, ISA-TAB-Nano also specifies how to create links between nanomaterial samples and any kind of external (raw) data files as well as images, e.g. from transmission electron microscopy.

In addition to the standardization of the "technical" data format, it is essential to define a) a consistent vocabulary to describe the information and, b) which information on the nanomaterial characterization and toxicity assays needs to be recorded

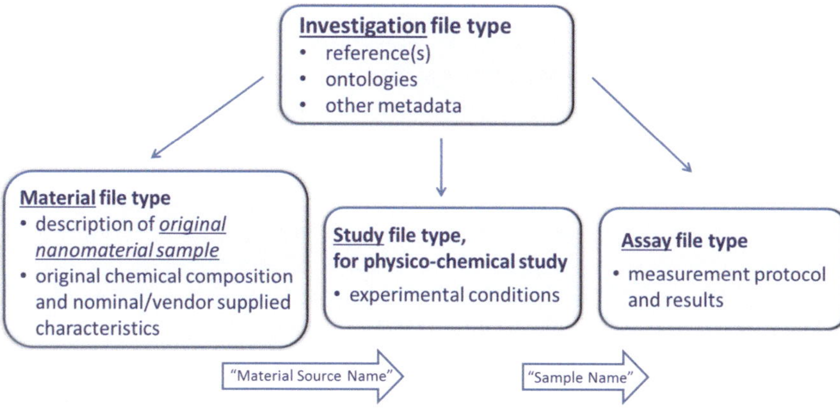

Fig. 10.2 ISA-TAB-Nano defines a set of linked spreadsheet files (Investigation, Material, Study, and Assay), with a pre-defined file structure and syntax for (meta)data. Capture of physico-chemical (meta)data is illustrated

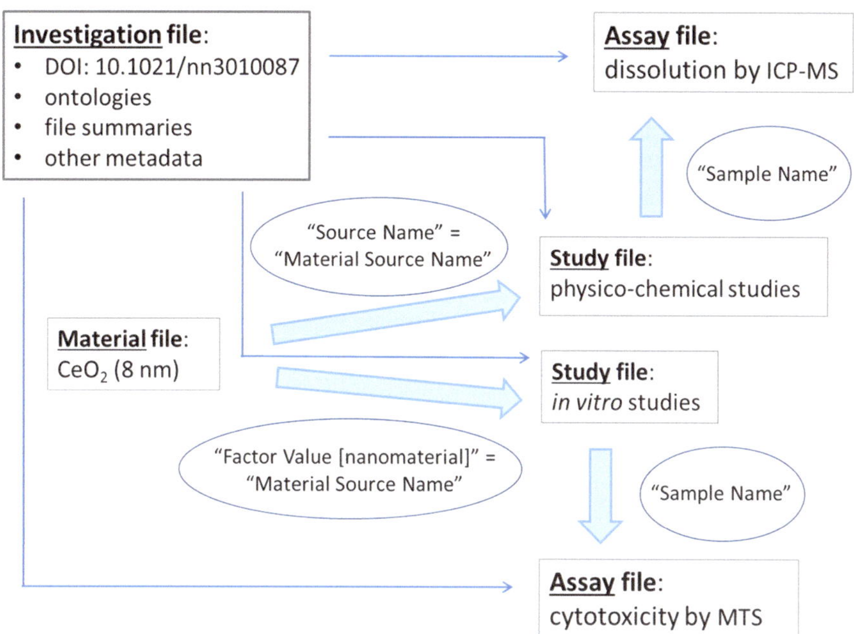

Fig. 10.3 A simplified illustration of how linked ISA-TAB-Nano files are used to record data derived from physico-chemical characterization as well as biological characterization assessments of nanomaterials reported in a single primary literature reference (Adapted from Marchese Robinson et al. [16]. Creative Commons Attribution License, http://creativecommons.org/licenses/by/2.0)

to ensure possible comparison and valid deduction of models (minimum information requirements). While the generic ISA-TAB-Nano specification pre-defines file structure and syntax, it does not specify all (meta)data to be recorded.

Recording data in a unified, consistent terminology is required to avoid inconsistencies, for example due to synonyms. Therefore, for data extracted from the literature, links to terms from ontologies, retrieved using BioPortal [44], for example the NanoParticle Ontology [33], were identified. Links to those ontologies are established via the Investigation file. The selection of appropriate terms is still a challenge due to inconsistent definitions of terms and incompleteness of existing ontologies. This standardized labeling facilitates the possibility to filter the compiled data collection, for example to find all data associated with the same specific toxic effect, and allows comparison of data measured in different experiments.

The adequate characterization of the nanomaterial and description of the assay performed are crucial to allow exploration of intra-and inter-assay and -laboratory comparison and thus to clearly define a uniform basis for the development of nano-QSAR models. For example, an adequate description of the structure is needed to characterize a nanomaterial. A comprehensive and universal format for uniquely representing nanomaterial structures, however, is not yet established, due to the complex nature of nanomaterials. Therefore chemical structures of the nanomaterials were recorded where possible, e.g. for fullerenes and their derivatives, using the SMILES (Simplified Molecular-Input Line-Entry System) line notation. However, the inherent complexity of the nanomaterial structures makes this notation inapplicable for most nanoparticles. In some cases, partial representations based on Crystallographic Information Files (CIF) are available. The ISA-TAB-Nano Material and Assay files are used to record key structural/physico-chemical data such as crystal phase and size information or link to transmission electron microscopy image files.

In general, to ensure a certain quality standard in terms of completeness, minimum information requirements need to be determined regarding which (meta)data should be recorded, i.e. to make certain that all necessary information for modelling the biological activity/hazard of nanomaterials are made available and to allow the evaluation of whether measurements were obtained under sufficiently similar experimental conditions to allow the combination in one single modelling dataset. Proposals for nanomaterial data Minimum Information Standards have been discussed in the literature, with an emphasis on the parameters required to adequately define the structural and physico-chemical nanomaterial properties, e.g. by the Minimum Information on Nanoparticle Characterization (MINChar) initiative [16, 31, 32].

The ISA-TAB-Nano specification, as proposed by its developers, does not specify exactly which nanomaterial characteristics, experimental details and measurements should be recorded. Therefore extended data collection templates were designed within NanoPUZZLES to capture important toxicity and physico-chemical/structural measurements and key experimental details [15]. Within NanoPUZZLES it was suggested that the minimum information criteria should sufficiently define the exact parameters to be recorded. For example, the list of parameters proposed by the MINChar initiative [32] to be reported for nanoparticles in toxicological studies only states "particle size/size distribution" as an essential parameter. It would be useful to further specify how to record this, e.g. as the z-average hydrodynamic diameter and polydispersity index determined from dynamic light scattering.

10.2.3 Contributions to a Universal Database

The physico-chemical and toxicity data of nanomaterials are collected with the aim of supporting the modelling of nanomaterial properties and toxicity as well as to build an increasing data basis to support hazard assessment of nanomaterials. Ideally all data should be integrated in one universal database, or be searchable through interoperable databases.

The assessment of the data quality and adherence to minimum information requirements as well as their preparation in a standardized format contributes to the effort of organizing the diverse experimental data for inclusion in a database. The aim of the data collection, quality evaluation and standardization is to organize the data and to also allow for a comparison between data from different sources and thus give added value to the data collection. This possible comparison of different data collected for the same nanomaterials will bring together the dispersed efforts to characterize nanomaterials to support safety assessment. Therefore the characterization of the sample and adequate description of the experimental measurements are crucial to allow intra-and inter-assay and -laboratory comparison.

For this overall aim, the general intention is to feed the collected, evaluated and standardized data into an interoperable, searchable nanotoxicity database, which may be consulted on its own to identify hazards and support nanomaterial risk assessment, or used as the data basis for the development of predictive models. The preparation of datasets based on the ISA-TAB-Nano specification was done in view of integration of these data within external databases. For example, NanoPUZZLES ISA-TAB-Nano datasets were submitted to the nanoDMS online database, developed within the EU MODERN project [19].

10.3 Modelling Interactions and Toxicity of Nanoparticles

In silico models such as (quantitative) structure-activity relationships ((Q)SARs), are developed from experimental data and are based on the premise that the biological activity of a substance, e.g. an observed effect on human health or environmental species, can be related to the chemical structure and physico-chemical properties of the compound. This relationship can be expressed as mathematical equations including descriptors of the relevant physico-chemical properties.

Although traditional QSARs have been successfully applied to nanomaterials in some cases, more specific approaches are needed for the model development for nanoparticles [45].

10.3.1 Need for Specific Nano-descriptors

In order to develop QSARs for nanoparticles, suitable descriptors need to be found taking into account the specific characteristics of nanoparticles and their relationship to the activities to be modeled, i.e. in particular to describe the structure and nanoparticle physico-chemical properties appropriately [3, 5, 23].

The development of novel descriptors for nanoparticle structures was the aim of the NanoPUZZLES NanoDESC work package. The "nano-descriptors" developed should reflect the essential properties of nanomaterials such as parameters characterizing the particular nanoparticle structure and electronic states resulting from quantum effects of the nano-size. An overview of the type of the general characteristics described by 0D to 4D descriptors and examples of specific nanoparticle properties are given in Table 10.1.

While the chemical structure of a substance can be generally represented by, e.g., an orbital graph, a 2D SMILES string or the unique InChI identifier, a novel concept is needed for the majority of nanomaterials to express the specific features of the "nano structure": The molecular architecture of nanoparticles is very large and complex, with inorganic and organic elements and exact stoichiometry varying between materials [2], and it is not always possible to represent the specific interactions between different parts of the nanosystems topologically and/or by means of molecular and quantum mechanics.

A possible approach for the creation of a model for nanomaterial(s), when traditional descriptors such as 2D topological indexes, 3D stereochemical and quantum mechanical descriptors are not adequate, is to consider the measured endpoint as a mathematical function of all available eclectic information. The traditional QSAR paradigm 'Endpoint = *F*(molecular structure)' will therefore be replaced with

Table 10.1 Molecular descriptors: simple mathematical representation of a molecule, used to encode significant features of molecules

Dimensions	Characterization
0D descriptors	Constitutional
1D descriptors	Physico-chemical
2D descriptors	Topology of a molecule
3D descriptors	Based on a space representation of a molecule
4D descriptors	Various conformers of the same compound

Properties	Experimental measurements
Shape and aggregation	High resolution microscopy
Composition, purity and surface chemistry	Spectroscopy and chromatography methods
Surface charge	Zeta potential analysis
Crystal structure	X-ray diffraction
Particle number and size distribution	e.g. SEM, TEM, DLS
Surface area	Brunauer-Emmett-Teller adsorption

SEM scanning electron microscopy, *TEM* transmission electron microscopy, *DLS* dynamic light scattering

'Endpoint = *F*(eclectic information)' for nano-QSARs [40]. This information includes atomic composition, conditions of synthesis or preparation of the nanomaterial, as well parameters such as particle size and its distribution, agglomeration state, porosity, particle shape, symmetry, surface area and charge, coating, metal content, dissolution, crystal structure, electronic properties (reactivity, conductivity, interaction energies), chiral vectors of nanotubes, number of walls in the nanotubes [4]. Table 10.2 shows examples of different descriptors applied to the modelling of nanomaterials to predict the listed properties.

It should be noted that (most of) the nanospecific properties cannot be calculated but have to be provided experimentally for the specific nanomaterial considered (see Table 10.1). In consequence the nano-QSAR models are dependent on experimental measurements, rather than calculated descriptors from the molecule representation, and the experimental procedure to determine the respective values may influence the model quality. Moreover, the characterization of the applicability domain of a nano-QSAR model has to be extended consequently according to the new types of descriptors (see for example [41]).

Further examples of descriptors include the metal electronegativity (χ), the charge of the metal cation corresponding to a given oxide (χox), the atomic number and valence electron number of the metal, which were used as simple molecular periodic table-based descriptors for QSAR model development, for example to predict the cytotoxicity of metal oxide nanoparticles towards *E. coli* [7].

In collaboration with the NanoBridges project, scanning electron microscope (SEM) and transmission electron microscope (TEM) images of nanoparticles were computationally processed and analyzed by means of a customized algorithm implemented in the ImageJ software to obtain new descriptors. Morphological (large scale) features were analyzed in relation to shape and size and expressed in the form of ten key descriptors: area, perimeter, major axis, minor axis, aspect ratio, maximum Feret's diameter, minimum Feret's diameter, roundness, circularity, and solidity.

The "Liquid Drop" Model (LDM) descriptors were also developed with the NanoBridges project. They describe the geometric and volume features of studied metal oxide nanoparticles. The model is based on the representation of the nanoparticle as a spherical drop, the elementary particles are densely packed and the density is equal to the density of bulk [30].

10.3.2 Modelling Interactions of Nanoparticles with Biological Systems and Small Molecules

The knowledge of the interactions between nanoparticles and biological systems, such as DNA, proteins and membranes is limited. However, these interactions can lead to the formation of protein coronas, particle wrapping, intercellular uptake and biocatalytic processes, and thus modify the structure of the active nanoparticle, which has an impact on model development, e.g., the descriptors to be calculated.

Table 10.2 Examples of descriptors used to predict nanoparticle properties

Nanomaterials	Descriptors	Predicted property	Reference
24 metaloxides	Conduction band energy levels	Ability to induce oxygen radicals, oxidative stress, and inflammation (cellular redox potential)	[47]
17 metaloxides	$\Delta H_{Me+,}$ representing the enthalpy of formation of a gaseous cation having the same oxidation state as that in the MOx structure	Cytotoxicity (EC_{50}: the effective concentration of a compound resulting in a 50 % reduction in bacteria viability)	[24]
29 (metaloxides, nitrites, silicon carbide)	Product of the correlation weight (DCW) DCW = ΠkCW(Ik)	Prediction of Young's modulus	[35]
Fullerene C_{60}	RNCG-relative negative charge (Zefirov's PC), ASIC-average structural information content, Eminee(C-C)-min: exchange energy for the C-C bound	Solubility in n-heptane	[17]
Fullerene C_{60}	Correlation of Highest Occupied Molecular Orbital (HOMO), certain heteroatom fragments, and geometrical parameters with solubility	Solubility in organic solvents	[22]
Carbon nanotubes	Chiral vector components (the components of chiral vector contain information about rolling up graphite layer in formation of carbon nanotubes)	Water solubility (log S, S in mol/L) and octanol water partition coefficient (log P)	[37]
Carbon nanotubes, r-graphene oxide, Ag-silica, Ag-carbon, C_{70}TGA, $nC_{60}(OH)_{20}$, $nC_{60}(OH)_{32}$, SiO_2, TiO_2)	Hydrophobicity, hydrogen bond, polarity/polarizability, and lone-pair electrons	Surface adsorption forces (the nanomaterial interaction with biological components)	[46]
Carbon nanotubes	C(N2)/C(T) (number of non-sp^2 hybridized carbons per total carbons) and chiral angle identified as critical descriptors for both Young's modulus and Poisson's ratio	Mechanical property prediction (for Young's modulus and Poisson's ratio)	[1]

Therefore, the aim of the NanoINTER part of the NanoPUZZLES project was to develop methods to simulate and predict interactions between engineered nanoparticles and biomolecules, their environment (e.g., solvents) as well as small molecules.

For the development of a computational protocol for the reliable calculation of the properties of large interacting systems, extensive calculations have been performed, which involved: a large array of techniques (e.g., semi-empirical-PM6-, HF, DFT-B97D, wB97XD, M062X, B3LYP, PBE0-, MP2, molecular dynamics, molecular mechanics Poisson-Boltzmann surface area (MM-PBSA)), and basis sets (e.g., 6-31G*, 6–31+G*, 6-311G*); ab initio techniques for the calculation of the intermolecular interaction energy (e.g., the Su and Li approach, Kitaura and Morokuma (KM) method, effective fragment potential approach (EFP), a variational perturbation scheme); the atoms-in-molecules (AIM) technique for the computation of the hydrogen bond energy; geometry optimization techniques (e.g., B97D/6-31G*, PM6). Several properties, such as the interaction energy, the energies of the Highest Occupied Molecular Orbital (E_{HOMO}) and Lowest Unoccupied Molecular Orbital (E_{LUMO}), ionization potential, electron affinity, dipole moment, polarizabilities and first hyperpolarizabilities, binding free energies, hydrogen bond energies, and structural features were computed. A variety of nanoparticles were used to evaluate their structure and interaction properties, for example: $C_{24}H_{12}$, $C_{84}H_{20}$, $C_{24}H_{12}$, $C_{114}H_{30}$, $C_{222}H_{42}$, $C_{366}H_{54}$; C_{60}, $C_{60}F_n$, $(TiO_2)_n$, single wall carbon nanotube (SWCNT, C_{360}). The calculations were tested on a variety of biomolecules, such as proteins or protein fragments (e.g., HIV-1PR, renin, a G protein-coupled receptor (GPCR), human serum albumin, and human DNA topoisomerase II-alpha).

In conclusion, the following computational methods were recommended:

- A DFT technique for computing the interactions of nanoparticle-nanoparticle systems.
- Ab initio techniques for calculating the interaction properties of biomolecule-biomolecule systems: the AIM method for the computation of the hydrogen bond energy and the Su and Li method for the calculation of the interaction energy.
- Molecular dynamics for studying large biomolecule-nanoparticle or biomolecule-biomolecule systems: the AIM method for the computation of the hydrogen bond energy and MM-PBSA for the calculation of the interaction energy.

The interaction energy of a series of fullerene derivatives with human serum albumin (HSA), or appropriately defined models of HSA, has been computed and resolved into a variety of meaningful contributions by employing ab initio methods. MM-PBSA for the resolution of the binding free energy has also been used. After a detailed analysis of the large body of the computed data, the resolution of the interaction energy yielded the following results:

- For small or average size systems, the use of the Su and Li method in connection with DFT may provide valuable information on the interaction mechanism.
- The fragment molecular orbital method gives very useful information for the whole interacting system, but it is computationally demanding.

- For large interacting systems, the employment of MM-PBSA is recommended.

Furthermore, the interaction with the solvent environment was investigated. Water soluble fullerene derivatives are essential for many biomedical applications. Several studies have examined the relation between solubility and toxicity [28]. Pilot computations on the solubility of several fullerenes were undertaken. It was shown that the solubilization energy (ΔE) depends on the dielectric constant (ε) and the functional groups attached to the C_{60} core, while an increase in ε leads to an increase in |ΔE|. Also, it was shown that the solvent may have a significant effect on some properties (e.g., the average polarizability) of the interacting system.

Three different geometry conformations were used to study the interaction between a fullerene derivative and a model of graphene oxide, i.e. between two different nanoparticles, in aqueous solution. The observed energy difference between the two dominant conformations may be attributed to the number of hydrogen bonds formed between the fullerene and the surface of graphene oxide.

Additionally, molecular aggregation of fullerene systems is obtained spontaneously in aqueous solution, as represented by stable hydrophobic clusters throughout the molecular dynamics simulations. Conversely, water-soluble fullerenes do not form agglomerates in water regardless of their concentration and initial member separation. Moreover, to study the relationship between aggregation and toxicity, (e.g., [13]), pilot computations were performed on the aggregation of fullerenes, $(C_{60})n$, n = 1–6, where it was shown that:

- Aggregation has a negligible effect on the considered properties.
- The solvent appears to have a minor effect on the properties of (C60)n.

Another objective was to connect the genotoxicity/mutagenicity of a series of fullerenes, $C_{60}F_n$, n = 10, 12, 14, 16 and 18, with their solubility. It is known that E_{HOMO} increases with mutagenicity. This knowledge allowed for the relationship between mutagenicity and solubility of $C_{60}F_n$ to be investigated since it was shown that an increase in n was followed by an increase in solubility and a decrease in mutagenicity. It was also demonstrated that mutagenicity increases with lipophilicity (logP), Q_F (the sum of the charges on the fluorine atoms of $C_{60}F_n$), E(S-T), the dipole moment, the average polarizability, and the first hyperpolarizability. Therefore, a relationship connecting mutagenicity with solubility and a set of important electronic descriptors may be established. The interaction of $C_{60}F_n$ with DNA and human DNA topoisomerase II-alpha (HT2a) was also studied by performing molecular dynamics and MM-PBSA free energy calculations. Significant interactions have been found between $C_{60}F_n$ and both systems, particularly between $C_{60}F_n$ and HT2a. These interactions may induce an undesirable effect on the DNA function.

Finally, the effect of: (i) chemical composition, (ii) size and shape, (iii) aggregation, (iv) surface charge, and (v) contamination of nanoparticles upon a series of properties, such as the structure, the binding free energy (ΔE_{bind}), the interaction energy (E_{int}), E_{HOMO} and E_{LUMO} of several nanoparticle-biomolecule systems was investigated. The nanoparticles investigated were: a) molecular graphene ($C_{84}H_{24}$,

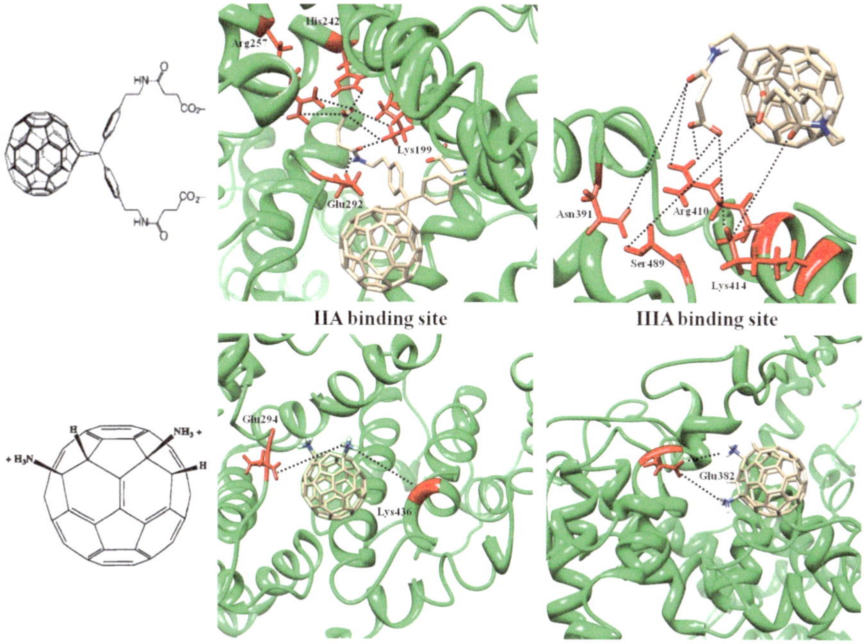

Fig. 10.4 Hydrogen bond interactions between fullerenes and human serum albumin. Main hydrogen bonds between fullerene derivatives and the binding cavities of human serum albumin are shown as dotted lines. Residues that are involved in interactions with the compounds are highlighted (Adapted with permission from Leonis et al. [13], Copyright (2014) American Chemical Society)

$C_{60}B_{12}N_{12}H_{24}$, $C_{78}H_{24}FeN_4$), b) C_{60} and $(C_{60})_4$, c) two single wall carbon nanotubes ($C_{360}H_{40}$, $C_{180}H_{20}$), d) a series of functionalized C_{60} derivatives, and e) $C_{60}F_n$, n = 10, 12, 14, 16, and 18. For biomolecules, the following systems were used: a) the five nucleobases (NBs), thymine (T), cytosine (C), guanine (G), adenine (A) and uracil (U), b) the guanine tetramer (G4), c) HSA, d) HT2a, and e) a DNA sequence were considered. The ab initio study on the effect of size and shape of the nanoparticle was performed on complexes: $C_{84}H_{24}$-X/Y, C_{60}-X, $C_{360}H_{40}$-X/Y, $C_{180}H_{20}$-X, where X=guanine and Y=adenine. It was found that E_{int} is mostly affected by the size and shape and in a lesser extent by $E_{HOMO/LUMO}$. Molecular dynamics runs for fullerenes bound to HSA showed that compounds with longer groups attached to the fullerene core strengthen HSA binding, while compounds with shorter groups diminish protein binding (Fig. 10.4). For the study of the aggregation effect, the $(C_{60})_4$-G4 complex was used. A significant change of the G4 geometry was observed and a high E_{int} (−18.4 kcal/mol) was computed. The effect of the surface charge was studied by computing the E_{int} of C_{60}Q-G, where $Q = 0, -2, -4$. It was found that the charge of C_{60} has a remarkable impact on E_{int} and on the geometry of guanine. Regarding HSA complexes, the calculations showed that negatively charged groups on the fullerenes are necessary to produce profound binding effects [12]. Also, fullerene binding to the IIA site of HSA was associated with allosteric modulation of IIIA and heme binding sites. The above findings may be particularly useful in future nanoparticle design for biological applications.

The influence of contamination was studied with E_{int} and $E_{HOMO/LUMO}$ calculations on $C_{84}H_{24}$-NB, $C_{60}B_{12}N_{12}H_{24}$-NB and $C_{78}H_{24}FeN_4$-NB, where changes were observed in the presence of the contaminant ($B_{12}N_{12}$ and FeN_4). Significant binding for functionalized fullerenes was observed in HSA, HT2a and DNA systems, with the van der Waals and nonpolar terms being the dominant contributions. The detailed description of each investigation performed under NanoPUZZLES, including those which are mentioned above, can be found in [8, 10, 11, 42, 43].

10.3.3 Examples of Nano-QSAR Modelling

The NanoPUZZLES NanoQSAR work package developed nano-QSAR models based on the data and information discussed above, with the aim of extending the understanding of toxicity and behavior of nanoparticles by establishing relationships between experimental and computational properties.

To encode the cytotoxicity profile of metal oxide nanoparticles to the bacterium *E. coli*, periodic table-based descriptors were employed to construct robust interpretable quantitative structure-toxicity relationship (QSTR) models. Ten random splits of the data into the training and test set were examined with extensive validation techniques employing the Organization for Economic Co-operation and Development (OECD) recommendations. The results obtained have demonstrated that simple periodic table derived descriptors (metal electronegativity and the charge of the metal cation corresponding to a given oxide) have a significant influence on the cytotoxicity of metal oxide nanoparticles to bacteria *E. coli*. Both descriptors were obtained from the molecular formula and information derived from the periodic table. Moreover, both descriptors are independent of the nanoparticles' size range, thus avoiding the modelling difficulties arising with the variation of various physical nanoparticle properties with the size range. In addition, the results obtained confirmed that the toxicity of metal oxide nanoparticles is associated with the reductive potential, i.e., the detachment of the electron from the metal oxides [7].

To investigate the differences in the mechanisms of toxicity of metal oxide nanoparticles to the bacterium *E. coli* (prokaryotic system) and a human keratinocyte cell line (eukaryotic system), different computational approaches were utilized. To reflect the nanoparticles' structure for the different levels of organization, i.e. from single molecule to supramolecular ensemble of molecules, a Simplex Representation of Molecular Structure (SiRMS) and the "liquid drop" model (LDM) were used. Classification models were developed based on the combination of descriptors calculated within the NanoPUZZLES NanoDESC work package. Based on these simple fragmentary 2D descriptors (forming two latent variables) a consensus model was developed. The Nano-QSAR models obtained provided reliable pre-

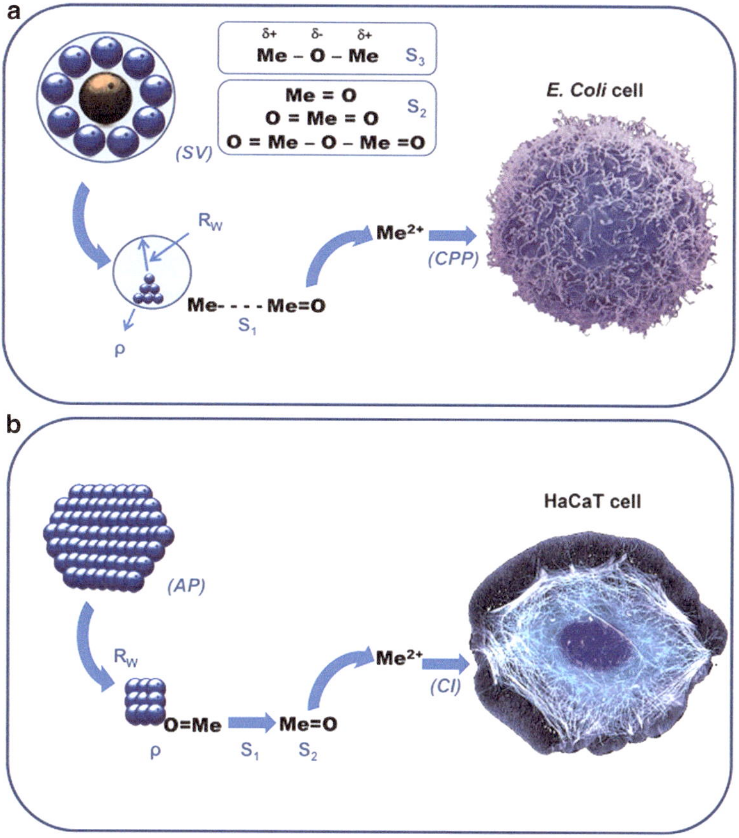

Fig. 10.5 Schematic representation of the mechanism of metal oxide nanoparticle toxicity for: (**a**) bacterium *E. coli* and (**b**) human keratinocyte cell line (HaCaT) (Reproduced from Sizochenko et al. [32], http://dx.doi.org/10.1039/c4nr03487b with permission of The Royal Society of Chemistry)

dictions for all metal oxide nanoparticles studied. It needs to be highlighted, however, that the developed nano-QSAR models refer to different mechanisms of toxicity of nanoparticles towards the bacterium *E. coli* and the human keratinocyte cell line (HaCaT) [30]. The schematic representation of suggested mechanisms is shown in Fig. 10.5.

As further examples, toxicological data collected and novel nanodescriptors calculated were used to model the cellular uptake of 109 magnetofluorescent nanoparticles modified with small organic molecules into $PaCa_2$ pancreatic cancer cells [38] and the membrane damage (units/L) caused by TiO_2 nanoparticles. The latter model is based on the use of eclectic information, in particular the following physico-chemical parameters of the TiO_2 nanoparticles have been adopted: engineered size, size in water suspension, size in phosphate buffered saline, concentration, and zeta potential [39].

With nanomaterials there is the problem of the lack of standardization and harmonization, in addition to the related issue of finding suitable descriptors. The novelty of the CORAL approach, as applied to nano-QSAR, is in its simplicity [40]. It is highly flexible and involves a broad series of parameters of different nature, including experimental values arising from the different protocols. Thus, the model can adapt the features of interest depending on the case, covering a wide set of situations. CORAL is shown to be a general system and that several specific models can be produced through CORAL using the particular parameters relevant in each case. The disadvantage of this, which however is related to the lack of general protocols in many of the experiments with nanomaterials, is that the specific model requires as inputs the values of the specific experimental case used to train the model, and thus the specific model is very local.

Optimal descriptors calculated within NanoPUZZLES have also been used to develop a mathematical model for the mutagenicity of fullerenes. Experimental data for the bacterial reverse mutation test on C_{60} nanoparticles for TA100, and WP2uvrA/pkM101 were employed as dependent variables, i.e. endpoints. The models obtained were the mathematical function of all available eclectic data, such as: dose, metabolic activation (i.e. with or without S9 mix) and illumination (i.e. darkness or irradiation), in the role of logical and digital basis [36].

Moreover, optimal descriptors have been applied to predict thermal conductivity of micro-electro-mechanical systems (MEMS), largely used in nanotechnology. The decimal logarithm of thermal conductivity of MEMS has been modelled as a mathematical function of temperature and physicochemical status of MEMS, such as Ceramic (code = 1), Single crystal (code = 2), Cubic (code = 3), Chemical Vapour Deposition (CVD, code = 4), and Glass (code = 5) [41].

In summary, within NanoPUZZLES, physico-chemical properties as well as toxicity of a series of nanoparticles were predicted using various approaches. It was demonstrated that all models provide reliable predictions, though are characterized by different accuracy as compared to experimental data. The models developed allow for the prediction of physico-chemical properties and toxicity of new nanoparticles, not studied yet experimentally.

10.4 Nanoparticle Evaluation to Be Set in a Global Context

The evaluation of nanomaterials, toxicity tests and development of predictive models cannot be singular efforts but should be coordinated within ongoing global initiatives in order to make valid conclusions on nanoparticle toxicity. The presented work was integrated in international research and discussions, as shown in Fig. 10.6.

The NanoPUZZLES project coordinated approaches with other EU research projects modelling nanotoxicity (ModNanoTox, NanoTransKinetics, ModEnpTox, MembraneNanoPart, PreNanoTox, MODERN). At the European level, all projects addressing different aspects of nanosafety are joined under the overarching umbrella of the NanoSafety Cluster (http://www.nanosafetycluster.eu), with cross-cluster working groups on cross-cutting topics such as databases, hazard, risk and modeling.

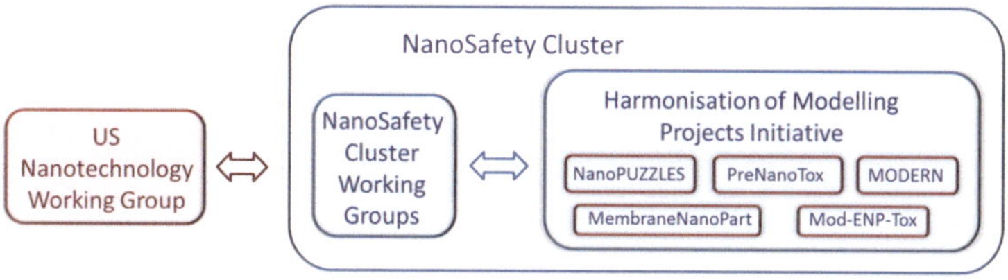

Fig. 10.6 Collaboration of the NanoPUZZLES project with other nanotoxicity modelling projects and setting within global nano-research networks

Furthermore, an EU-US dialogue (http://us-eu.org) framework is coordinating ongoing efforts and discussions on both sides of the Atlantic. In addition, the US Nanotechnology Working Group (https://wiki.nci.nih.gov/display/ICR/Nanotechnology+Working+Group), including the developers of the ISA-TAB-Nano specification, is very active for example in discussing data sharing standards, ontologies and nanomaterial characterisation.

10.5 Conclusions

Computational models predicting nanotoxicity could make an important contribution to safety evaluations, in line with general efforts in applying alternative methods in chemicals risk assessment, as promoted by recent legislation such as REACH and the European Cosmetics Regulation. This includes in silico models such as nano-QSARs for properties and toxicities.

Compiled, well organized, quality controlled and standardized information allows for the optimal use of existing data to support nanomaterial hazard and risk assessment. Integrating the data in an overarching, structured data collection pursues the following aims: a) organizing the data in a way to support the modelling, b) making certain that all (meta)data necessary for modelling are available, c) adding value to the available data by making a comparison between data from different sources possible. Databases of nanoparticle physico-chemical and toxicological properties thus allow for the use and comparison of all available data in integrated risk assessment approaches, avoiding new in vivo tests.

Based on the available data, specific descriptors can be derived to describe the nanomaterial-specific structure and physico-chemical properties appropriately. Furthermore, the interactions between nanoparticles and biological systems as well as small molecules, which can lead to modifications of the structure of the active nanoparticles, need to be described and taken into account in the development of models to predict the biological activity and toxicity of nanoparticles.

Overall, a global effort is required and is being built up to overcome the fragmentation of efforts to create, characterize and assess the toxicity data needed for the nanomaterial hazard and risk assessment. It aims to bridge the gap between experimental, computational as well as regulatory efforts.

Acknowledgments Funding through the European Commission 7th Framework Programme NanoPUZZLES project (FP7-NMP-2012-SMALL-6, grant agreement no.309837) is gratefully acknowledged.

References

1. Borders TL, Fonseca AF, Zhang H, Cho K, Rusinko A (2013) Developing descriptors to predict mechanical properties of nanotubes. J Chem Inf Model 53(4):773–782
2. Burello E, Worth AP (2011) QSAR modeling of nanomaterials. Wiley Interdiscip Rev Nanomed Nanobiotechnol 3(3):298–306
3. Fubini B, Ghiazza M, Fenoglio I (2010) Physico-chemical features of engineered nanoparticles relevant to their toxicity. Nanotoxicology 4:347–363
4. Gajewicz A, Rasulev B, Dinadayalane TC, Urbaszek P, Puzyn T, Leszczynska D, Leszczynski J (2012) Advancing risk assessment of engineered nanomaterials: application of computational approaches. Adv Drug Deliv Rev 64:1663–1693
5. Gallegos Saliner A, Poater A, Worth AP (2008) Toward in silico approaches for investigating the activity of nanoparticles in therapeutic development. IDrugs 11:728–732
6. ISA-TAB-Nano Wiki (2015) https://wiki.nci.nih.gov/display/ICR/ISA-TAB-Nano. Accessed 5 May 2015
7. Kar S, Gajewicz A, Puzyn T, Roy K (2014) Periodic table-based descriptors to encode cytotoxicity profile of metal oxide nanoparticles: a mechanistic QSTR approach. Ecotoxicol Environ Saf 107C:162–169
8. Kellici TF, Ntountaniotis D, Leonis G, Chatziathanasiadou M, Chatzikonstantinou AV, Becker-Baldus J, Glaubitz C, Tzakos AG, Viras K, Chatzigeorgiou P, Tzimas S, Kefala E, Valsami G, Archontaki H, Papadopoulos MG, Mavromoustakos T (2015) Investigation of the interactions of silibinin with 2-hydroxypropyl-β-cyclodextrin through biophysical techniques and computational methods. Mol Pharm 12(3):954–965
9. Klimisch HJ, Andreae M, Tillmann U (1997) A systematic approach for evaluating the quality of experimental toxicological and ecotoxicological data. Regul Toxicol Pharmacol 25:1–5
10. Leonis G, Steinbrecher T, Papadopoulos MG (2013) A contribution to the drug resistance mechanism of darunavir, amprenavir, indinavir, and saquinavir complexes with HIV-1 protease due to flap mutation I50V: a Systematic MM–PBSA and thermodynamic integration study. J Chem Inf Model 53:2141–2153
11. Leonis G, Avramopoulos A, Salmas RE, Durdagi S, Yurtsever M, Papadopoulos MG (2014) Elucidation of conformational states, dynamics and mechanism of binding in human κ-opioid receptor complexes. J Chem Inf Model 54(8):2294–2308
12. Leonis G, Avramopoulos A, Papavasileiou KD, Reis H, Steinbrecher T, Papadopoulos MG (2015) A comprehensive computational study of the interaction between human serum albumin and fullerenes. J Phys Chem B 119:14971–14985
13. Lesar A, Milosev I (2009) Density functional study of the corrosion inhibition properties of 1,2,4-triazole and its amino derivatives. Chem Phys Lett 483:198–203
14. Lubiński L, Urbaszek P, Gajewicz A, Cronin MTD, Enoch SJ, Madden JC, Leszczynska D, Leszczynski J, Puzyn T (2013) Evaluation criteria for the quality of published experimental data on nanomaterials and their usefulness for QSAR modelling. SAR QSAR Environ Res 24(12):995–1008
15. Marchese Robinson RL, Cronin MTD, Richarz A-N, Rallo R (2015) An ISA-TAB-Nano based data collection framework to support data driven modelling of nanotoxicology. Beilstein J Nanotechnol 6:1978–1999
16. Marchese Robinson RL, Lynch I, Peijnenburg W, Rumble J, Klaessig F, Marquardt C, Rauscher H, Puzyn T, Purian R, Åberg C, Karcher S, Vriens H, Hoet P, Hoover MD, Ogilvie Hendren C,

Harper SL (2016) How should the completeness and quality of curated nanomaterial data be evaluated? Nanoscale 8:9919–9943
17. Martin D, Maran U, Sild S, Karelson M (2007) QSPR modeling of solubility of polyaromatic hydrocarbons and fullerene in 1-octanol and n-heptane. J Phys Chem B 111(33):9853–9857
18. Mills KC, Murry D, Guzan KA, Ostraat ML (2014) Nanomaterial registry: database that captures the minimal information about nanomaterial physico-chemical characteristics. J Nanopart Res 16(2):1–9
19. nanoDMS: Nanomaterial Data Management System (2015). http://biocenitc-deq.urv.cat/nanodms. Accessed 11 Sept 2015.
20. Ostraat ML, Mills KC, Guzan KA, Murry D (2013) The Nanomaterial Registry: facilitating the sharing and analysis of data in the diverse nanomaterial community. Int J Nanomedicine 8(Suppl 1):7–13
21. Petersen EJ, Henry TB, Zhao J, MacCuspie RI, Kirschling TL, Dobrovolskaia MA, Hackley V, Xing B, White JC (2014) Identification and avoidance of potential artefacts and misinterpretations in nanomaterial ecotoxicity measurements. Environ Sci Technol 48(8):4226–4246
22. Petrova T, Rasulev BF, Toropov AA, Leszczynska D, Leszczynski J (2011) Improved model for fullerene C_{60} solubility in organic solvents based on quantum-chemical and topological descriptors. J Nanopart Res 13(8):3235–3247
23. Puzyn T, Leszczynska D, Leszczynski J (2009) Toward the development of 'Nano-QSAR': advances and challenges. Small 5:2494–2509
24. Puzyn T, Rasulev B, Gajewicz A, Hu X, Dasari TP, Michalkova A, Hwang HM, Toropov A, Leszczynska D, Leszczynski J (2011) Using nano-QSAR to predict the cytotoxicity of metal oxide nanoparticles. Nat Nanotechnol 6:175–178
25. Richarz A-N, Madden JC, Marchese Robinson RL, Lubiński Ł, Mokshina UP, Kuz'min VE, Puzyn T, Cronin MTD (2015) Development of computational models for the prediction of the toxicity of nanomaterials. Perspect Sci 3:27–29
26. Rocca-Serra P, Brandizi M, Maguire E, Sklyar N, Taylor C, Begley K, Field D, Harris S, Hide W, Hofmann O, Neumann S, Sterk P, Tong W, Sansone S-A (2010) ISA software suite: supporting standards-compliant experimental annotation and enabling curation at the community level. Bioinformatics 26(18):2354–2356
27. Sansone S-A, Rocca-Serra P, Field D, Maguire E, Taylor C, Hofmann O, Fang H, Neumann S, Tong W, Amaral-Zettler L, Begley K, Booth T, Bougueleret L, Burns G, Chapman B, Clark T, Coleman L-A, Copeland J, Das S, de Daruvar A, de Matos P, Dix I, Edmunds S, Evelo CT, Forster MJ, Gaudet P, Gilbert J, Goble C, Griffin JL, Jacob D, Kleinjans J, Harland L, Haug K, Hermjakob H, Sui SJH, Laederach A, Liang S, Marshall S, McGrath A, Merrill E, Reilly D, Roux M, Shamu CE, Shang CA, Steinbeck C, Trefethen A, Williams-Jones B, Wolstencroft K, Xenarios I, Hide W (2012) Toward interoperable bioscience data. Nat Genet 44(2):121–126
28. Sayes CM, Fortner JD, Guo W, Lyon D, Boyd AM, Ausman KD, Tao YJ, Sitharaman B, Wilson LJ, Hughes JB, West JL, Colvin VL (2004) The differential cytotoxicity of water-soluble fullerenes. Nano Lett 4(10):1881–1887
29. Schneider K, Schwarz M, Burkholder I, Kopp-Schneider A, Edler L, Kinsner-Ovaskainen A, Hartung T, Hoffmann S (2009) ToxRTool', a new tool to assess the reliability of toxicological data. Toxicol Lett 189(2):138–144
30. Sizochenko N, Rasulev B, Gajewicz A, Kuzmin VE, Puzyn T, Leszczynski J (2014) From basic physics to mechanisms of toxicity: liquid drop approach applied to develop predictive classification models for toxicity of metal oxide nanoparticles. Nanoscale 6:13986–13993
31. Stefaniak AB, Hackley VA, Roebben G, Ehara K, Hankin S, Postek MT, Lynch I, Fu W-E, Linsinger TPJ, Thünemann AF (2013) Nanoscale reference materials for environmental, health and safety measurements: needs, gaps and opportunities. Nanotoxicology 7(8):1325–1337
32. Thomas DG, Klaessig F, Harper SL, Fritts M, Hoover MD, Gaheen S, Stokes TH, Reznik-Zellen R, Freund ET, Klemm JD, Paik DS, Baker NA (2011a) Informatics and standards for nanomedicine technology. Wiley Interdiscip Rev Nanomed Nanobiotechnol 3(5):511–532
33. Thomas DG, Pappu RV, Baker NA (2011b) NanoParticle ontology for cancer nanotechnology research. J Biomed Inform 44(1):59–74

34. Thomas DG, Gaheen S, Harper SL, Fritts M, Klaessig F, Hahn-Dantona E, Paik D, Pan S, Stafford GA, Freund ET, Klemm JD, Baker NA (2013) ISA-TAB-Nano: a specification for sharing nanomaterial research data in spreadsheet-based format. BMC Biotechnol 13:2
35. Toropov AA, Leszczynski J (2007) A new approach to the characterization of nanomaterials: predicting Young's modulus by correlation weighting of nanomaterials codes. Chem Phys Lett 433(1–3):125–129
36. Toropov AA, Toropova AP (2013) Optimal descriptor as a translator of eclectic data into endpoint prediction: mutagenicity of fullerene as a mathematical function of conditions. Chemosphere 104:262–264
37. Toropov AA, Leszczynska D, Leszczynski J (2007) Predicting water solubility and octanol water partition coefficient for carbon nanotubes based on the chiral vector. Comput Biol Chem 31(2):127–128
38. Toropov AA, Toropova AP, Puzyn T, Benfenati E, Gini G, Leszczynska D, Leszczynski J (2013) QSAR as a random event: modeling of nanoparticles uptake in $PaCa_2$ cancer cells. Chemosphere 92:31–37
39. Toropova AP, Toropov AA (2013) Optimal descriptor as a translator of eclectic information into the prediction of membrane damage by means of various TiO_2 nanoparticles. Chemosphere 93(10):2650–2655
40. Toropova AP, Toropov AA (2015) Mutagenicity: QSAR – quasi-QSAR – nano-QSAR. Mini Rev Med Chem 15(8):608–621
41. Toropova AP, Toropov AA, Puzyn T, Benfenati E, Leszczynska D, Leszczynski J (2013) Optimal descriptor as a translator of eclectic information into the prediction of thermal conductivity of micro-electro-mechanical systems. J Math Chem 51:2230–2237
42. Tzoupis H, Leonis G, Avramopoulos A, Mavromoustakos T, Papadopoulos MG (2014) Systematic molecular dynamics, MM–PBSA and ab initio approaches to the Saquinavir resistance mechanism in HIV-1 PR due to 11 double and multiple mutations. J Phys Chem B 118(32):9538–9552
43. Vrontaki E, Leonis G, Avramopoulos A, Papadopoulos MG, Simčič M, Grdadolnik SG, Afantitis A, Melagraki G, Hadjikakou SK, Mavromoustakos T (2015) Stability and binding effects of silver(I) complexes at lipoxygenase-1. J Enzyme Inhib Med Chem 30:539–549
44. Whetzel PL, Noy NF, Shah NH, Alexander PR, Nyulas C, Tudorache T, Musen MA (2011) BioPortal: enhanced functionality via new web services from the National Center for Biomedical Ontology to access and use ontologies in software applications. Nucleic Acids Res 39(Web Server issue):W541–W545
45. Winkler DA, Mombelli E, Pietroiusti A, Tran L, Worth A, Fadeel B, McCall MJ (2013) Applying quantitative structure-activity relationship approaches to nanotoxicology: current status and future potential. Toxicology 313(1):15–23
46. Xia XR, Monteiro-Riviere NA, Mathur S, Song X, Xiao L, Oldenberg SJ, Fadeel B, Riviere JE (2011) Mapping the surface adsorption forces of nanomaterials in biological systems. ACS Nano 5(11):9074–9081
47. Zhang H, Ji Z, Xia T, Meng H, Low-Kam C, Liu R, Pokhrel S, Lin S, Wang X, Liao Y-P, Wang M, Li L, Rallo R, Damoiseaux R, Telesca D, Mädler L, Cohen Y, Zink JI, Nel AE (2012) Use of metal oxide nanoparticle band gap to develop a predictive paradigm for oxidative stress and acute pulmonary inflammation. ACS Nano 6(5):4349–4368

Chapter 11
Case Study III: The Construction of a Nanotoxicity Database – The MOD-ENP-TOX Experience

Hanne Vriens, Dominik Mertens, Renaud Regret, Pinpin Lin, Jean-Pierre Locquet, and Peter Hoet

Abstract The amount of experimental studies on the toxicity of nanomaterials is growing fast. Interpretation and comparison of these studies is a complex issue due to the high amount of variables possibly determining the toxicity of nanomaterials.

Qualitative databases providing a structured combination, integration and quality evaluation of the existing data could reveal insights that cannot be seen from different studies alone. A few database initiatives are under development but in practice very little data is publicly available and collaboration between physicists, toxicologists, computer scientists and modellers is needed to further develop databases, standards and analysis tools.

In this case study the process of building a database on the in vitro toxicity of amorphous silica nanoparticles (NPs) is described in detail. Experimental data were systematically collected from peer reviewed papers, manually curated and stored in a standardised format. The result is a database in ISA-Tab-Nano including 68 peer

H. Vriens
Center for Environment and Health, KU Leuven, Leuven, Belgium

D. Mertens
Genedata AG, Basel, Switzerland

R. Regret
Rhenovia Pharma, Mulhouse, France

P. Lin
National Health Research Institutes, Zhunan Town, Taiwan

J.-P. Locquet
Solid State Physics and Magnetism Section, KU Leuven, Leuven, Belgium

P. Hoet (✉)
Center for Environment and Health, KU Leuven, Leuven, Belgium

KU Leuven, Department of Public Health and Primary Care, Center for Environment and Health, Herestraat 49 mailbox 706, B-3000 Leuven, Belgium
e-mail: peter.hoet@med.kuleuven.be

L. Tran et al. (eds.), *Modelling the Toxicity of Nanoparticles*, Advances in Experimental Medicine and Biology, Vol. 947, DOI 10.1007/978-3-319-47754-1_11

reviewed papers on the toxicity of 148 amorphous silica NPs. Both the physicochemical characterization of the particles and their biological effect (described in 230 in vitro assays) were stored in the database. A scoring system was elaborated in order to evaluate the reliability of the stored data.

Keywords Amorphous silica nanoparticles • ISA-Tab-Nano • Data reliability • *In vitro* • Nano-informatics

11.1 Introduction

Nanotechnology results in the presence of a variety of different engineered NPs (NPs) in our environment. Due to their small size NPs behave differently from their larger counterparts of the same composition. Therefore adjusted safety assessment (hazard identification, hazard characterisation and exposure assessment) is needed. The last 10 years hazard identification and characterisation has mainly focused on finding the physicochemical properties of NPs that determine their interaction with biological systems and the underlying pathways causing these interactions, resulting in a large number of experimental data published. Drawing conclusions from these data is however difficult because of the large amount of variables possibly determining the toxicity outcome. Variables are associated with the nanomaterial itself and the exposure conditions, the biological test system (in vitro/in vivo) and the toxicological assay.

A structured combination and integration of the existing data could reveal variables which are important determinants of NPs toxicity and can eventually lead to predictive models and QSARs for nanotoxicity. Database initiatives relevant for nanotoxicology are under development: CaNanoLab [1], the Nanomaterial Registry [2], the NP Information Library [3] and the Nanomaterial-Biological Interactions Knowledge Base [4]. In practice very few data is publicly available and collaboration between physicists, toxicologists, computer scientists and modellers is needed to further develop databases, standards and analysis tools.

This case-study describes the construction of a database on the in vitro toxicity of amorphous silica NPs (including particles with a silica shell). A search of PubMed was performed to collect 68 peer-reviewed papers which were manually curated. Both the physicochemical characteristics of the NPs and their interaction with cellular systems were stored in an ISA-Tab-Nano compatible format. To assess the reliability of the stored data a scoring system was elaborated: variables associated with the nanomaterials, toxicity assay and biological system were scored to assess the reliability of the data.

First the process of data collection (including criteria and literature search) (11.2.1), data storage (11.2.2) and data evaluation (11.2.3) is described in order to build a qualitative database. In paragraph 5 the database itself is described: the articles, in vitro assays and NPs. And some results of the reliability scoring system are presented. In paragraph 6 several issues that came up during the construction of the database and some future suggestions to overcome these issues are discussed.

This project is a subproject of the "Modelling Assays Platform "MAP" for hazard ranking of engineered metal-based NPs (MOD-ENP-TOX) " project of the Seventh Framework Program (FP7) funded by the European Union.

11.2 Construction of the Database

11.2.1 Data Collection

11.2.1.1 Including Criteria

Data were collected from peer-reviewed papers which investigate the in vitro toxicity of amorphous silica NPs. This type of nanomaterial is well studied; a preliminary search gave more than 600 hits. And experiments have shown that different silica NPs, although it is one chemical identity, can give different biological effects [5].

Selected papers:

1. study the toxicity of amorphous silica particles with a defined shape (aspect ratio <3), composition (all silica or a silica shell), crystallinity (amorphous) and primary size;
2. study the effect of the particles on cell lines or primary cells (experimental in vitro studies),
3. and include information on cell viability, apoptosis/necrosis, genotoxicity, oxidative stress and pro-inflammation.

The dose of NPs effectively reaching the cells depends on the exposure route/method (dry state, suspension or aerosol). In order to avoid dose-response curves which are difficult to compare, we only included experimental studies that administer particles to the cell in suspension.

11.2.1.2 Literature Search

A sensitive search strategy, to retrieve as many relevant papers as possible, was performed using PubMed. Searching "amorphous silica OR silica NPs AND toxicity NOT review" with a filter of 10 years and full text availability gave 624 hits (13th of May 2014).

Secondly the 624 papers were sorted by relevance. The relevance sort option is based on an algorithm that analyses each PubMed citation that includes the search terms. A "weight" is calculated for citations depending on how many search terms are found and in which fields they are found. In addition, recently-published articles are given a somewhat higher weight for sorting [6]. Only the titles of the 450 most relevant papers were further screened for retrieval (cfr. including criteria 11.2.1.1). The 117 retrieved articles underwent a second

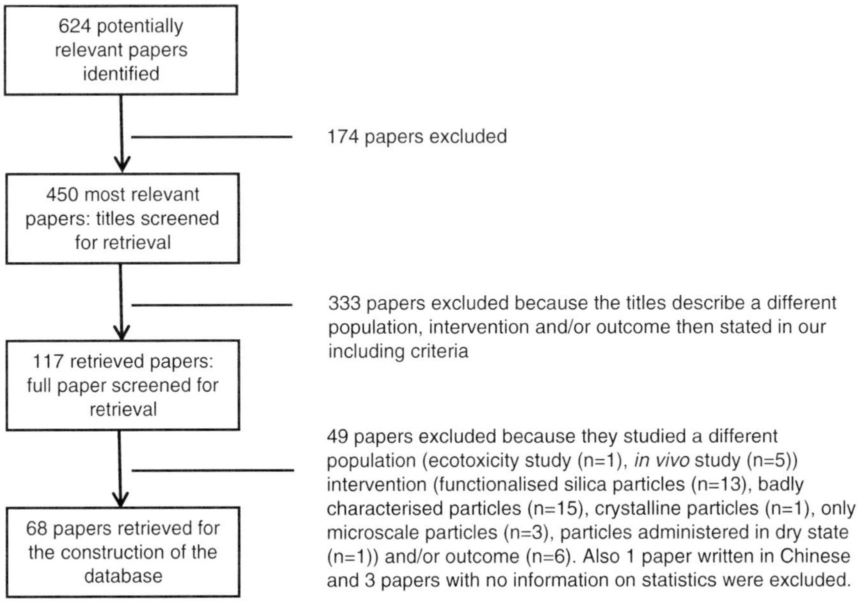

Fig. 11.1 Selection procedure for peer reviewed papers included in the database

evaluation for which the including criteria were applied to the full article. Another 47 papers were excluded because they studied a different population (ecotoxicity study (n = 1), in vivo study (n = 5)), different intervention (used functionalised silica particles (n = 13), badly characterised particles (n = 15), crystalline particles (n = 1), microscale particles (n = 3), particles in dry state (n = 1)) and/or a different outcome (n = 6). Also one paper written in Chinese and three papers with no information on the statistics were excluded. Eventually 68 papers were retrieved for the construction of the database.

The selection procedure is depicted in Fig. 11.1.

11.2.2 Data Storage

Data are stored in ISA-Tab-Nano; an emerging standard format for sharing nanomaterial research data. The format supports the use of ontology terms to promote standardized descriptions, and facilitate search and integration of data. Four types of files are provided to store different types of data:

1. The investigation file contains descriptive information (principle investigators, sponsor, link to full text paper,…) which lays the foundation for the other ISA-Tab-Nano files and links them together.
2. The material file describes the materials used; nanomaterials but also other materials such as positive controls.

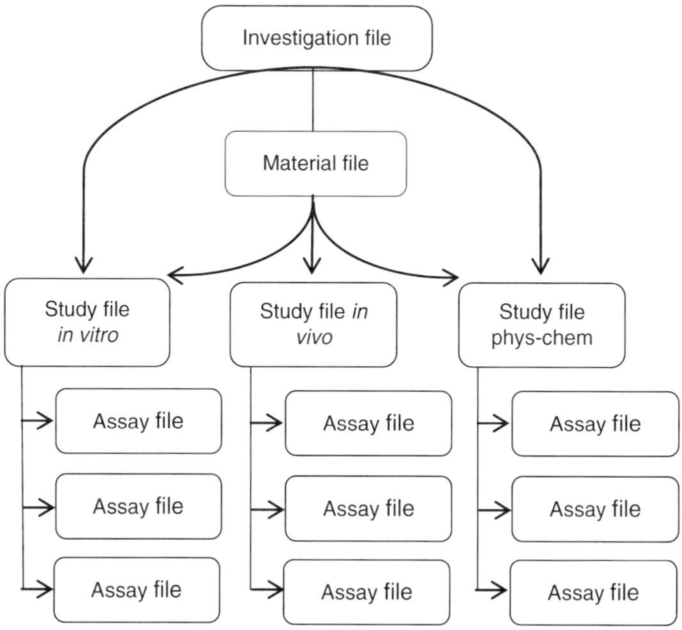

Fig. 11.2 Linkage of the different ISA-Tab-Nano files

3. The study file describes how samples (material and biological samples) are prepared for analysis (physicochemical, in vitro and in vivo characterisation).
4. The assay file is designed to store the measured endpoints of the physicochemical, in vitro or in vivo characterisation of the nanomaterials [7].

Figure 11.2 depicts how the ISA-Tab-Nano files are linked to each other.

ISA-Tab-Nano recommends using the material file only to store the nominal characteristics and chemical composition of the NPs. The experimentally measured physicochemical characteristics are stored in the assay files. We chose to store all the physicochemical characteristics in the material file to make analysis and data integration more convenient afterwards in perspective of the MOD-ENP-TOX project.

The different fieldnames used in the material, investigation, assay and study file can be found in Table 11.1. The data of each article is stored in a set of these four types of files.

11.2.3 Data Reliability Evaluation

Klimisch et al. define the reliability of data as an evaluation of the inherent quality of a test report or publication relating to preferably standardized methodology and the way the experimental procedure and results are described to give evidence of the clarity and plausibility of the findings [8].

Table 11.1 Overview of the fieldnames used in the four ISA-Tab-Nano files used to describe data from one paper

INVESTIGATION FILE: pubmed ID, DOI, title and authors of publication
MATERIAL FILE: material source name, material description, material synthesis, material manufacturer
Primary size: diameter, min feret diameter, max feret diameter assay name
Shape: shape, aspect ratio, assay name
Crystallinity: major crystalline fraction, major crystalline fraction proportion, minor crystalline fraction, minor crystalline fraction proportion, amorphous fraction proportion, assay name
Composition: composition core, composition shell, composition coating, assay name
Purity: purity core/shell, assay name
Surface area: specific surface area, external surface area, assay name
Porosity: porosity, pore volume, pore size, assay name
Surface charge: zeta-potential in water/saline/saline with serum/medium/ medium with serum, assay name
Agglomeration/aggregation: hydrodynamic diameter in water/saline/saline with serum/ medium/ medium with serum, assay name
Solubility: solubility in water, assay name
STUDY FILE (physicochemical characterisation):
STUDY FILE (in vitro assays): source name, material type, cell species, cell organ, cell type, NP sample, particle concentration, exposure route/medium/duration, serum concentration exposure medium, hydrodynamic diameter exposure medium
ASSAY FILE (viability): viability relative to ctrl, LDH medium, cytotoxicity, ATP level, cell death, number of cells
ASSAY FILE (genotoxicity): tail DNA, tail DNA (treatment-control), micronucleated cells, cytokinesis block proliferation index
ASSAY FILE (pro-inflammation): (mRNA) IL-6, (mRNA) IL-8, (mRNA) TNF-α, (mRNA) IL1-β, (mRNA) COX-2, (mRNA) MMP-9, (mRNA) MIP-1 alpha/beta, F3, ICAM1, VCAM1, SELE, NRF-2,cytc, MCP-1, iNOS
ASSAY FILE (oxidative stress): dichlorofluorescin, hydroethidium, hydroxyphenyl fluorescein, glutathione, oxidized/reduced, glutathione, malondialdehyde, hydroxyl-alkenals, HO-1, superoxide dismutase, superoxide dismutase activity, hydrogen peroxide
ASSAY FILE (apoptosis/necrosis): live cells, early apoptotic rate, late apoptotic and necrotic rate, total apoptotic and necrotic rate, caspase-3/9, caspase-3/7/9 activity, bax, bcl-2, p53, p21, cell morphology changes, SubG1/G1/S/G2 phase
STUDY FILE (in vivo assays):

General guidelines on reliability assessment of toxicological data are described in literature. The following is a selected overview of the existing schemes relevant for the data reliability assessment within the MOD-ENP-TOX project. Both general schemes designed for the reliability assessment of chemicals and more specific schemes for nanomaterials are discussed (11.2.3.1). Afterwards a scheme is proposed to assess the reliability of the data in our database (11.2.3.2).

11.2.3.1 Schemes for Assessment of the Reliability of Toxicological Data

Klimisch et al. published a categorisation scheme to assign toxicological data of chemicals to one of four reliability categories: reliable without restrictions, reliable with restrictions, not reliable or not assignable. Distinction between these reliability categories are based on the amount of information provided on the testing procedure and analysis. Tests conducted and reported in accordance to international standards have the highest grade of reliability [8]. Unfortunately, the differentiation between reliability classes is not always clear. To make the decision process of assigning reliability categories more transparent and harmonised the **'ToxRTool'** (Toxicological data Reliability Assessment Tool – [9]) was developed. The tool provides a detailed list of yes-no questions about the identification of the test substance, characterization of the biological system, description of the study design, documentation of the study results and the plausibility of the study design and data [10]. The answers to the yes-no questions are used to attribute the data to one of the Klimisch categories.

The evaluation criteria for toxicity need to be reconsidered for toxicological data of **nanomaterials**, taking into account the following complications:

1. The toxicity of nanomaterials not only depends on the dose and the composition but also on the size, shape, specific surface area, surface coating, porosity, surface charge and the solubility of the nanomaterials [2, 11].
2. No international standardized test procedures exist for both the physicochemical characterisation of the nanomaterials and their toxicity assessment. The OECD initiated a large programme on the safety of manufactured nanomaterials which will result in a set of guidelines supporting standardization [12]. Guidelines already published are: 'Report of the OECD expert meeting on the physical chemical properties of manufactured nanomaterials and test guidelines', 'Guidance on Sample Preparation and Dosimetry for the Safety Testing of Manufactured Nanomaterials',...
3. Nanomaterials can interfere with toxicity assays resulting in false positive and false negative results [13, 14].

The **Nanomaterial Registry** provides a metric, the compliance level (CL), of the quality and quantity of characterization for each nanomaterial entry. In order to be compliant a nanomaterial characterization should include: the synthesis method or processing details, the DOI citation of the synthesis procedure, the manufacturer or synthesis laboratory name, the product name and lot number and the nanomaterial's physical state. For each measurement the following should be reported: the technique, technique protocols and parameters and the best practice information. Compliance levels are only developed for data associated with the nanomaterial, not for the data on the interactions with biological systems [2].

Lubinski et al. published a scheme to assess the quality of nanotoxicity data in the context of developing QSPR's/QSAR's. A checklist of yes-no questions supports the user to assign data to one of five reliability classes. The checklist comprises questions related to the extent to which the nanomaterials were characterised, the degree to which the experimental assays and methods were described and the use of standardised protocols [15].

11.2.3.2 Scheme for Reliability Assessment of In Vitro Nanotoxicity Data

We developed a scoring system to assess the reliability of in vitro nanotoxicity data extracted from papers. Reliability scoring is based on the amount of information available on the different variables influencing in vitro toxicity outcome of the nanomaterials (which depends on the way of reporting, amount of detail given) and on how this information was obtained (methodology). The variables influencing in vitro nanotoxicity were recently listed by Krug [16]. He makes a differentiation between variables associated with the nanomaterial, with the toxicity assay and the biological system. The following is the list of variables copied from Krug and a proposition for a scoring system to assess the reliability of data of in vitro nanotoxicity data.

In the end a score will be assigned to each particle, assay and test system to give an indication of the reliability of the data.

Variables associated with the nanomaterial:

A. sample purification for the removal of biologically relevant trace elements
B. sample characterization of the raw material: composition and purity size shape agglomeration status etc.
C. sample characterization regarding biological impurities: endotoxins etc.
D. dispersion in biological media under relevant conditions: temperature humidity gas concentrations (O_2, CO_2) salinity etc.
E. sample characterization in biological media: size and shape agglomeration status protein corona etc.
F. the measurement device used for the characterisation *(not listed by Krug)*

In Table 11.2 is a concrete list of variables associated with the nanomaterials that should be reported because they influence the toxicity outcome. The more variables specified in the paper the more reliable the data.

Variables associated with the toxicity assay:

A. selection of the correct test system regarding the biological endpoints
B. different test systems for the same biological endpoint
C. controls: adapted negative controls adapted positive controls comparison to reference materials
D. testing of possible interferences of the NP with the biological test system binding of indicator molecules light absorption or fluorescence of the materials etc.
E. not considered measurement uncertainty: round robins, calibration with standards or reference material

Table 11.2 Scoring scheme variables associated with the nanomaterial

Variables	*Information available* **Make a choice:** *yes=1/no=0*	*Information available* **Make a choice:** *yes=1/no=0*
Composition core/shell	*1 or 0*	*1 or 0*
Composition coating	*1 or 0*	*1 or 0*
Purity (metal basis)	*1 or 0*	*1 or 0*
Purity (endotoxin)	*1 or 0*	*1 or 0*
Material synthesis	*1 or 0*	*1 or 0*
Primary size	*1 or 0*	*1 or 0*
Primary size distribution	*1 or 0*	*1 or 0*
Porosity	*1 or 0*	*1 or 0*
Surface area	*1 or 0*	*1 or 0*
Agglomeration/Aggregation in water	*1 or 0*	*1 or 0*
Surface charge in water	*1 or 0*	*1 or 0*
Solubility in water	*1 or 0*	*1 or 0*
Crystallinity	*1 or 0*	*1 or 0*
Shape	*1 or 0*	*1 or 0*
Sum	***Score characterisation***	***Score measurement device***

Table 11.3 Scoring scheme variables associated with the toxicity assay

Variables	*Information available* **Make a choice:** *yes=1/no=0*
Choice of correct assay regarding the toxicological endpoint	*1 or 0*
Interference of material with assay	*1 or 0*
Negative control	*1 or 0*
Positive control	*1 or 0*
Used different test systems for same biological endpoint	*1 or 0*
Sum	***Score toxicity assay***

The following table (Table 11.3) is a concrete list of variables associated with the toxicity assay that should be reported because they influence the toxicity outcome.

Variables associated with the biological system:

A. selection of the biological system
B. cell lines: selection criteria identification age and storage number of passages etc.
C. primary cells/organ systems: donor dependency donor variability culture conditions
D. culture conditions during the experiments: temperature humidity gas concentrations (O_2, CO_2) salinity etc.
E. biological parameter: cell density volume of the medium serum content of the medium compatibility of the solvent or dispersion medium

Table 11.4 Scoring scheme variables associated with the biological system

Variables	*Information available* **Make a choice:** *yes=1/no=0*
Selection of the appropriate test system	*1 or 0*
Origin of the cells	*1 or 0*
Passage of cells	*1 or 0*
Cell density	*1 or 0*
Primary cells, cell lines or cancer cells	*1 or 0*
Dispersion method of nanomaterials	*1 or 0*
Test medium	*1 or 0*
Amount of serum test medium	*1 or 0*
Agglomeration/Aggregation in exposure medium	*1 or 0*
Sum	***Score biological system***

The following scoring system (Table 11.4) was elaborated for each biological system:

11.3 Description of the Database

11.3.1 The Articles

The 68 articles used to construct the database were from 35 different journals. The most represented journals were "Nanotoxicology" (9 papers), "Toxicology in vitro" (6 papers) "Tox Letters" (4 papers), and "Tox Sciences" (4 papers). Other high ranked journals e.g. "Particle & Fibre Toxicology" had a surprisingly low success rate (2 papers). This can be due to the fact that these higher ranked journals are less tolerant to publish studies with insufficient physicochemical data on the NPs or have a broader scope.

11.3.2 The In Vitro Assays

Articles are stored in ISA-Tab-Nano describing 148 different silica NPs (or NPs with a silica shell) and their biological impact in 230 different in vitro assays. Fourty three percent of the assays investigated viability, 16 percent apoptosis/necrosis, 26 percent oxidative stress, 10 percent pro-inflammation and 6 per cent genotoxicity (Fig. 11.3).

The different types of assays per type of toxicological endpoint represented in the database are listed in Table 11.5:

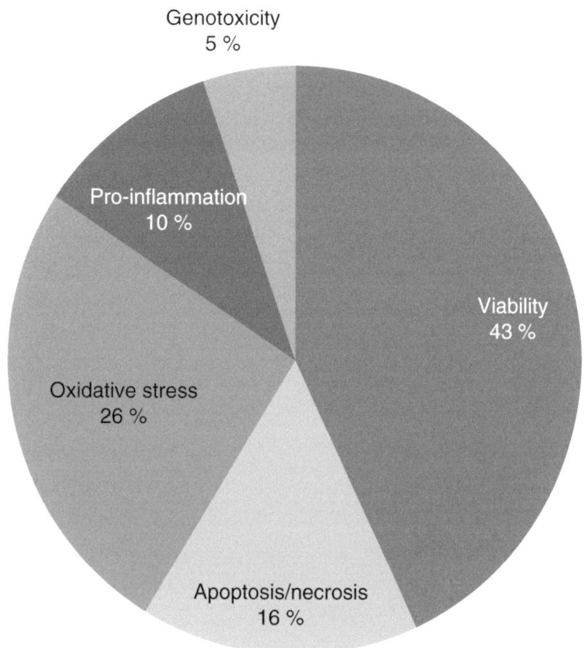

Fig. 11.3 Distribution of the 230 in vitro assays over the toxicity endpoints

11.3.3 The Particles

One hundred and fourty eight particles are stored in the database. The characteristics stored in the database are believed to be important determinants of nanotoxicity: chemical composition, primary size, primary size distribution, shape, crystallinity, solubility, surface charge, surface area, and agglomeration/aggregation. Table 11.6 gives an overview of these characteristics, how these were measured and the frequency of these measurements in the database. As a preposition for an article to be accepted for storage in the database was the characterisation of the primary size, shape, composition and crystallinity all the 148 particles have these characteristics measured (relative frequency is 1).

Few simplifications were made to make that data more homogenous and comparable:

1. diameter ranges were converted to a mean diameter by taking the arrhythmic mean of the upper and lower limit of the range
2. for NPs with a bimodal size distribution only the smallest mean diameter was taken into account for analysis
3. for ellipsoidal/cylindrical particles the average of the Feret min diameter and Feret max diameter is used as the mean diameter

Table 11.5 Different types of assays per toxicological endpoint included in the database

Cell viability	
Lactate dehydrogenase release assay Trypan blue cell staining MTT/MTS/XTT/WST-1/WST-8 reduction assay Propidium iodide uptake assay Neutral red uptake assay Annexin V-propidium iodide assay	Resazurin reduction assay ATP single parameter assay Cell proliferation assay Clonogenic assay Sulforhodamine B assay
Apoptosis/Necrosis	
qRT-PCR Western blot Annexin V- propidium iodide assay Annexin V assay	ELISA Cell morphology assay Caspase activity assay Mitotic cell cycle arrest assay
Oxidative stress	
ROS/RNS assay *Thiobarbituric acid assay* *Glutathione assay* *Hydroxyl radical assay*	SOD activity assay Western blot qRT-PCR
Pro-inflammation	
ELISA	qRT-PCR
Genotoxicity	
Micronucleus assay	Comet assay

Table 11.6 Frequency table of the physicochemical characterisation particles

Characteristic	Measurement	Freq.	Relative freq.*100
Chemical composition	Composition core	148	100
	Composition shell	148	100
	Composition coating	148	100
	Purity	24	16
	Material synthesis	97	66
Primary size	Mean diameter	148	100
	SD diameter	85	57
Shape	Shape	148	100
Crystallinity	Major crystalline fraction	148	100
	Major crystalline fraction proportion	148	100
	Minor crystalline fraction	148	100
	Minor crystalline fraction proportion	148	100
	Amorphous fraction proportion	148	100
Surface area	Mean specific surface area	75	51
	SD specific surface area	10	8
	Mean external surface area	23	16

(continued)

Table 11.6 (continued)

Characteristic	Measurement	Freq.	Relative freq.*100
Porosity	Porosity (porous/nonporous)	68	46
	Pore volume	8	5
	Pore size	21	14
Surface charge	Mean zeta-potential in water	52	35
	SD zeta-potential in water	22	15
	Mean zeta-potential in physiological saline	6	4
	SD zeta-potential in physiological saline	1	1
	Mean zeta-potential in medium	32	22
	SD zeta-potential in medium	6	4
	Mean zeta-potential in medium with serum	17	11
	SD zeta-potential in medium with serum	7	5
Agglomeration/aggregation	Mean hydrodynamic diameter in water	66	45
	SD hydrodynamic diameter in water	35	24
	Mean hydrodynamic diameter in physiological saline	8	5
	SD hydrodynamic diameter in physiological saline	4	3
	Mean hydrodynamic diameter in medium	62	42
	SD hydrodynamic diameter in medium	25	17
	Mean hydrodynamic diameter in medium with serum	37	25
	SD hydrodynamic diameter in medium with serum	21	14
Solubility	Solubility in water	1	1

11.3.3.1 Chemical Composition

All particles are made of amorphous silica or have an amorphous silica shell. The core of 15 particles is not pure silica but contains Fe3O4 (2), Fe3O4/Fe2O3 (2), dansylamide (1), rhodamine (7) or redF (3). Only eight particles have a coating: BSA (6), ethylene (1) and Al2O3 (1).

Of the 24 particles that were described in purity; 18 particles were said to have purity higher than 98 % or higher, the others were just mentioned to be pure.

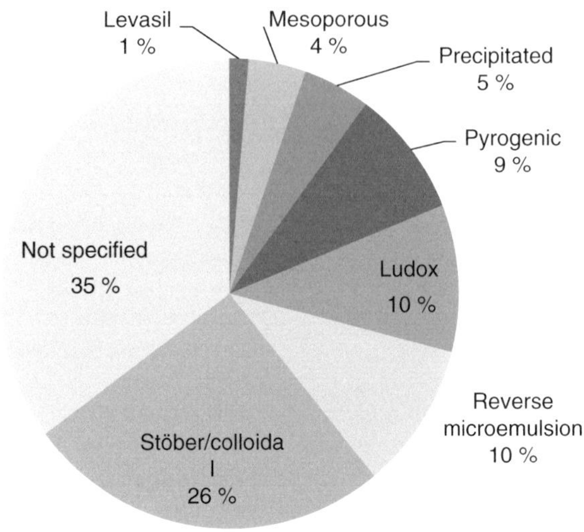

Fig. 11.4 Relative frequency of synthesis method used to make amorphous silica NPs

The material synthesis is important for the composition of nanomaterials, especially the surface chemistry of the silica particles. The density of silanol groups on the particle surface depends on the temperature. At high temperature which is for example needed to make pyrogenic silica there is a dehydration of the silanol groups on the surface area resulting in a lower density of silanol groups on the surface and a possibly different reaction with the environment [17].

For 97 of the particles the way they were synthesised was specified in the article (Fig. 11.4): 36 % of these particles were made by the Stöber process, 16 % were ludox® silica, 16 % made by the reverse microemulsion process, 14 % is pyrogenic silica, 7 % precipitated.

11.3.3.2 Primary Size

The frequency distribution of the mean diameter of the particles is depicted in Fig. 11.5. For only 57 % of the particles a standard deviation is reported.

11.3.3.3 Shape

55.4 percent of the particles are spherical, 35.1 % polyhedral, 0.7 % cylindrical, 5.4 % irregular and 3.4 ellipsoidal. The particles were described with these terms in the article or the T/S EM pictures of the particles were used to assign the particles to one of these shape classes.

For the other measured characteristics there were too many missing values, it does not seem usefull to report any descriptive statistics on these.

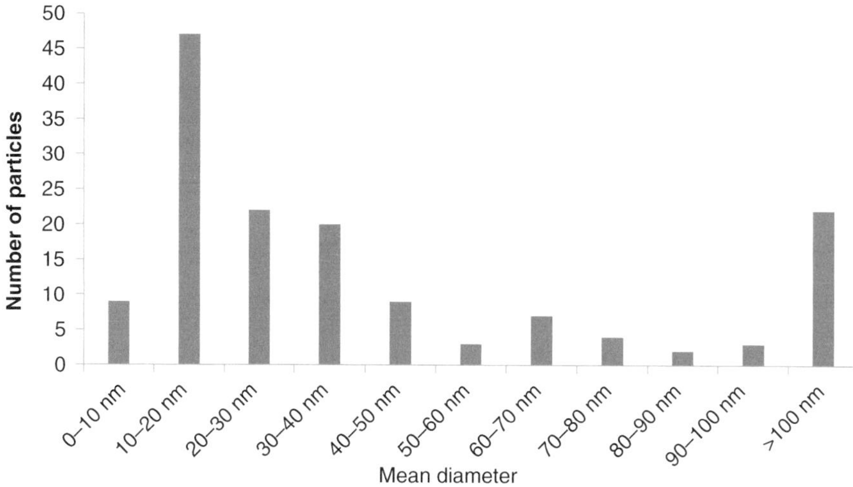

Fig. 11.5 Frequency distribution of the mean diameters of the 148 particles in the database

11.3.4 Data Evaluation

11.3.4.1 Evaluation Variables Associated with Nanomaterials

The scoring scheme developed in 11.2.3.2 to evaluate variables associated with nanomaterials was filled out for each amorphous silica NP (Table 11.2). Figure 11.6 is a frequency table of the "Score characterisation" of the 148 amorphous silica nanomaterials. Articles were allowed in the database under the condition that the composition of the core and shell, the composition of the coating, the shape, the crystallinity and the primary size of the particles were known. Therefore, no characterisation scores less than five were observed. None of the particles has a maximum characterisation score of 14.

The percentage of characteristics for which the measurement device was specified is calculated for each particle ("Score measurement device"/"Score characterisation"*100 - Fig. 11.7).

The average of this calculation for all the NPs was only 45 % implying that in more than 50 % of the reported characteristics the method used was not clearly specified.

11.3.4.2 Evaluation Variables Associated with the Toxicity Assay and Test System

The two scoring schemes developed in to evaluate variables associated with the toxicity assay and test system were filled out for the MTT reduction assays (30 assays). Figure 11.8 are the frequency tables of the "Score toxicity assay" and the "Score test system" of the 30 MTT reduction assays included in the database.

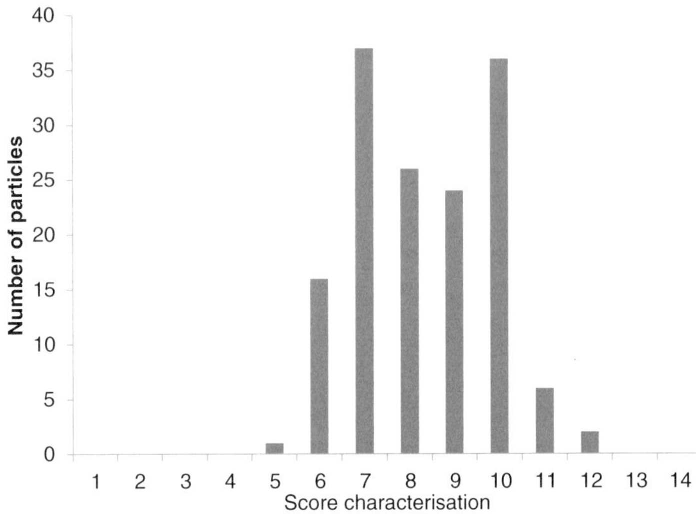

Fig. 11.6 Frequency distribution of the "Characterisation score" of the 148 particles in the database

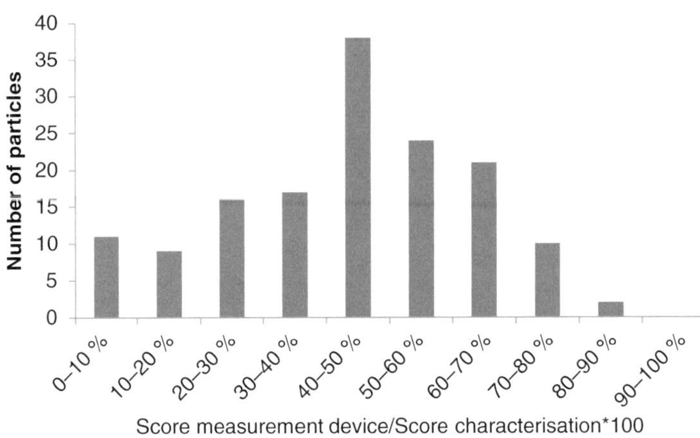

Fig. 11.7 Frequency distribution of the percentage of characteristics for which the measurement device was specified

11.4 Discussion

This case study describes the construction of an ISA-Tab-Nano formatted database, collecting data of in vitro toxicological studies published in peer reviewed journals. In total after triaging the data of 68 articles were included (starting from more than 600 hits). The characteristics (physicochemical and biological) of 148 different amorphous silica NPs are described in the database.

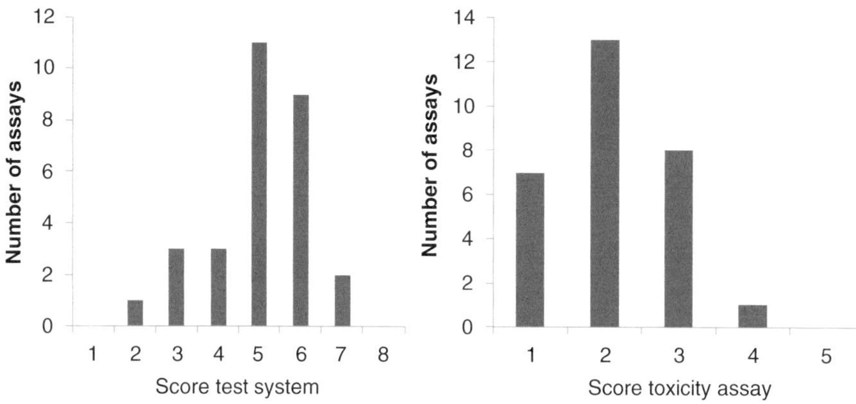

Fig. 11.8 Frequency distribution of the Score test system and Score toxicity assay for the MTT reduction assays in the database

Several issues and according needs came up during construction of the database:

11.4.1 Purpose Made Materials vs Commercial (Standard/ Benchmark) Materials

Many different particles were identified (although only amorphous silica NPs were included) and not much overlap in the use of nanomaterials was found. This is due to the fact that several silica NPs were "in house made" (or purpose made) and therefore only used in one experiment/study. On one hand this increases the scope of the database but on the other also increases the amount of data gaps and reduces the amount of overlap in the database hindering to create models generating data with high confidence. The use of a benchmark material would significantly increase the number of overlapping data points and make comparison between experiments, researchers, and particles possible.

11.4.2 Quality Evaluation of the Existing Data

In search of the critical properties that determine the toxicity of NPs, numerous investigations have been undertaken. Despite these efforts our knowledge on the possible hazardous effects of nanotechnology and its applications lags far behind the progress of nanotechnology. This is, mainly in the early years of nanotoxicology, caused by toxicity testing without appropriate material characterization and the lack of standardized dispersion and experimental protocols. Researchers underestimated the complexity of nanomaterials and experiments were set-up in similar ways as with chemicals. Inadequate characterisation made it impossible to link nanomaterials' characteristics to their biological effect. Likewise, the question whether the existing (biological) assays for chemicals were appropriate for nanomaterials was not sufficiently addressed. Lately a lot of attention went to the importance of physicochemical characterisation of nanomaterials. Papers were published on minimal information criteria [2, 18]. Still little attention went to the validation of the (biological) assays for testing nanomaterials. Now, most researchers are aware of the importance of a detailed characterisation of both the nanomaterials and the toxicity assays, and the validation of the methods used to characterise them. The scientific community has started to fill the data and knowledge gaps.

This awareness has also initiated the search for good and workable quality criteria to assess the reliability of the already published toxicological data. The criteria used in our database are not integrated into the final database in ISA-Tab-Nano since these are not solely objective data but are the result of unavoidable subjective expert judgment. Researcher using the database can easily implement his/her own quality judgment to the database before applying it for modelling purposes.

11.4.3 Standardized Formats

Finally, it can be recommended that all researchers should deliver the data presented in a scientific paper as an annex in ISA-Tab-Nano format. Now we have curated manually 68 papers, which is a tedious job, prone on reporting errors. The process of curation of data from a scientific paper includes extracting data from text (in which often not all measured data are given), figures (the small scale does often not allow precise extraction of numeral data, and smoothing of curves does not allow correct extraction), and tables (providing in general the most detailed data). Not only the different sections (text vs tables and figures) deliver different quality of data, also the fact that most experiments, although repeated several times (n-value of repeated measures) are reported as one mean value (± SD); this data is still valuable in a database but it would be better to collect the data per experimental run. Therefore, if researchers are encouraged to deliver – together with a published paper/report – a small ISA-TAB Nano compatible database, this would deliver more useful details into the database and would reduce significantly the number of 'human' errors.

The current database will be made available to the other researchers within other modelling projects and with research institutes such as JRC in Italy. The database will be used in the MOD-TOX-ENP project, where algorithm(s) will be developed to identify those physicochemical properties determining the hazardous effects of NPs.

This study describes the construction of an ISA-Tab-Nano formatted database, collecting data of in vitro toxicological studies published in peer reviewed journals. The study revealed some issues and future needs:

- Lately a lot of attention went to the importance of physicochemical characterisation of nanomaterials, to make the link between materials' characteristics and biological effect. This should not result in a loss of quality of the biological characterization, including interference verification, of the nanomaterial.
- Due to a general lack of good reference/benchmark data in publications it remains difficulty in comparing and combining data of different peer reviewed papers into a databases.
- Standardised database formats will improve the exchange and integration of data – here we used ISA-Tab-Nano format.
- It can be recommended that all researchers should deliver, as an annex to any publication, the data presented in ISA-Tab-Nano format (or another standard format).

References

1. caNanoLab [Internet]. Cited 6 Aug 2015. Available from: https://cananolab.nci.nih.gov/caNanoLab/#/
2. Mills KC, Murry D, Guzan KA, Ostraat ML (2014) Nanomaterial registry: database that captures the minimal information about nanomaterial physico-chemical characteristics. J Nanopart Res 16(2):1–9
3. NIOSH topic: nanotechnology : nanoparticle information library | CDC/NIOSH [Internet]. Cited 6 Aug 2015. Available from: http://www.nanoparticlelibrary.net/
4. NBI Knowlegebase [Internet]. Cited 6 Aug 2015. Available from: http://nbi.oregonstate.edu/analysis.php
5. Napierska D, Thomassen LC, Lison D, Martens JA, Hoet PH (2010) The nanosilica hazard: another variable entity. Part Fibre Toxicol 7(1):39
6. PubMed relevance sort [Internet]. Cited 1 Sep 2015. Available from: http://www.nlm.nih.gov/pubs/techbull/so13/so13_pm_relevance.html
7. Thomas DG, Gaheen S, Harper SL, Fritts M, Klaessig F, Hahn-Dantona E et al (2013) ISA-TAB-Nano: a specification for sharing nanomaterial research data in spreadsheet-based format. BMC Biotechnol 13(1):2
8. Klimisch H-J, Andreae M, Tillmann U (1997) A systematic approach for evaluating the quality of experimental toxicological and ecotoxicological data. Regul Toxicol Pharmacol 25(1):1–5
9. ToxRTool – Toxicological data Reliability Assessment Tool — EURL ECVAM [Internet]. Cited 31 Jul 2015. Available from: https://eurl-ecvam.jrc.ec.europa.eu/about-ecvam/archive-publications/toxrtool
10. Schneider K, Schwarz M, Burkholder I, Kopp-Schneider A, Edler L, Kinsner-Ovaskainen A et al (2009) "ToxRTool", a new tool to assess the reliability of toxicological data. Toxicol Lett 189(2):138–144
11. Oberdörster G, Maynard A, Donaldson K, Castranova V, Fitzpatrick J, Ausman K et al (2005) Principles for characterizing the potential human health effects from exposure to nanomaterials: elements of a screening strategy. Part Fibre Toxicol 2(1):8
12. Publications in the series on the safety of manufactured nanomaterials – OECD [Internet]. Cited 2015 Aug 7. Available from: http://www.oecd.org/science/nanosafety/publicationsintheseriesonthesafetyofmanufacturednanomaterials.htm
13. Monteiro-Riviere NA, Inman AO, Zhang LW (2009) Limitations and relative utility of screening assays to assess engineered nanoparticle toxicity in a human cell line. Toxicol Appl Pharmacol 234(2):222–235

14. Wörle-Knirsch JM, Pulskamp K, Krug HF (2006) Oops they did it again! Carbon nanotubes hoax scientists in viability assays. Nano Lett 6(6):1261–1268
15. Lubinski L, Urbaszek P, Gajewicz A, Cronin MTD, Enoch SJ, Madden JC et al (2013) Evaluation criteria for the quality of published experimental data on nanomaterials and their usefulness for QSAR modelling. SAR QSAR Environ Res 24(12):995–1008
16. Krug HF (2014) Nanosafety research—are we on the right track? Angew Chem Int Ed 53(46):12304–12319
17. Vansant EF, Voort PVD, Vrancken KC (1995) Characterization and chemical modification of the silica surface. Elsevier, Amsterdam/New York, p. 573
18. Luyts K, Napierska D, Nemery B, Hoet PHM (2013) How physico-chemical characteristics of nanoparticles cause their toxicity: complex and unresolved interrelations. Environ Sci Process Impacts 15(1):23–38

Index

L. Tran et al. (eds.), *Modelling the Toxicity of Nanoparticles*, Advances in Experimental Medicine and Biology, Vol. 947, DOI 10.1007/978-3-319-47754-1

O

MIX
Papier aus verantwortungsvollen Quellen
Paper from responsible sources
FSC® C105338

If you have any concerns about our products,
you can contact us on
ProductSafety@springernature.com

In case Publisher is established outside the EU,
the EU authorized representative is:
Springer Nature Customer Service Center GmbH
Europaplatz 3, 69115 Heidelberg, Germany

Printed by Libri Plureos GmbH
in Hamburg, Germany